ADVANCES IN CEMENT-BASED MATERIALS

T0264893

PROCEEDINGS OF THE INTERNATIONAL CONFERENCE ON ADVANCED CONCRETE MATERIALS, STELLENBOSCH, SOUTH AFRICA, 17–19 NOVEMBER 2009

Advances in Cement-Based Materials

Editors

G.P.A.G. van Zijl & W.P. Boshoff
Stellenbosch University, South Africa

CRC Press
Taylor & Francis Group
Boca Raton London New York

CRC Press is an imprint of the
Taylor & Francis Group, an **informa** business

CRC Press
Taylor & Francis Group
6000 Broken Sound Parkway NW, Suite 300
Boca Raton, FL 33487-2742

First issued in paperback 2017

CRC Press/Balkema is an imprint of the Taylor & Francis Group, an informa business

Typeset by Vikatan Publishing Solutions (P) Ltd., Chennai, India

Published by: CRC Press/Balkema
P.O. Box 447, 2300 AK Leiden, The Netherlands
e-mail: Pub.NL@taylorandfrancis.com
www.crcpress.com – www.taylorandfrancis.co.uk – www.balkema.nl

ISBN-13: 978-0-415-87637-7 (hbk)
ISBN-13: 978-1-138-11404-3 (pbk)

**Visit the Taylor & Francis Web site at
http://www.taylorandfrancis.com**

**and the CRC Press Web site at
http://www.crcpress.com**

Advances in Cement-Based Materials – van Zijl & Boshoff (eds)
© 2010 Taylor & Francis Group, London, ISBN 978-0-415-87637-7

Table of contents

VI

Advances in Cement-Based Materials – van Zijl & Boshoff (eds)
© 2010 Taylor & Francis Group, London, ISBN 978-0-415-87637-7

Foreword

The International Conference on Advanced Concrete Materials (ACM2009) brings together international experts and practitioners in concrete materials from the broad spectrum in the Civil Engineering industry including academics, consultants, contractors, scientists and suppliers. It provides a forum for the exchange of the latest research results, design and application procedures, methods for structural analysis and case studies on advanced cement-based materials. While the world rallies in the current economic recession, infrastructure development holds the key to future economic growth. Non-durable, inefficient use of construction materials, and no or inappropriate consideration of economical and ecological effects over the full life cycle of infrastructure will burden societies and governments for generations. Relatively recent advances in construction materials such as high performance concrete, high performance fibre reinforced concrete and self compacting concrete, hold key potential. Their thorough understanding, characterisation of short and long term behaviour and optimisation are objectives of the ACM2009, held in Stellenbosch, South Africa during 17–19 November, 2009.

The conference main themes are *High Performance Concrete, High Performance Fibre-Reinforced Concrete* and *Self Compacting Concrete*. Within these themes, the following topics are addressed:

- high and ultra-high strength concrete
- self compacting / self levelling concrete
- self compacting / self levelling fibre-reinforced concrete
- super-absorbent polymers
- fibre-reinforced concrete
- hybrid fibre concrete
- strain-hardening cement composites
- ultra-high strength fibre-reinforced concrete

The ACM2009 has been organised jointly by Stellenbosch University and the Concrete Society of Southern Africa (CSSA), and supported by the Cement and Concrete Institute (C&CI). The organising committee of representatives of these three organisations is praised for its professional and enthusiastic service.

Special acknowledgements are due to the following organisations:

- Pretoria Portland Cement
- BASF Construction Chemicals
- Mapei

Authors from 13 countries representing 5 continents have prepared the 38 papers bound in these proceedings. All the papers were peer-reviewed by members of the International Scientific Committee. Their efforts to ensure quality papers relevant to the conference themes are gratefully acknowledged.

Lastly, the authors should be thanked for their efforts in delivering papers of such high standard.

Gideon P.A.G. van Zijl &
W.P. (Billy) Boshoff
Editors

Francois Bain
President of the Concrete Society
of Southern Africa
August 25, 2009

Advances in Cement-Based Materials – van Zijl & Boshoff (eds)
© 2010 Taylor & Francis Group, London, ISBN 978-0-415-87637-7

ACM2009 committees

Conference Chair

Prof. Gideon van Zijl, *Stellenbosch University*

Organising Committee

This conference is organised by the Concrete Society of Southern Africa (CSSA) and Stellenbosch University (SU) and supported by the Cement and Concrete Institute (C&CI), represented by the members:

Dr. Billy Boshoff, *Chair person, SU*
Mrs. Natasja Pols, *Co-chair, CSSA*
Mr. Francois Bain, *President of CSSA*
Ms. Jeanine Kilian, *CSSA*
Mrs. Amanda de Wet, *SU*
Ms. Natalie Scheepers, *SU*
Mr. John Sheath, *C&CI*
Mrs. Hanlie Turner, *C&CI*

Scientific Committee

M.G. Alexander, *South Africa*
Y. Ballim, *South Africa*
H.W. Bennenk, *The Netherlands*
H. Beushausen, *South Africa*
J.E. Bolander, *USA*
W.P. Boshoff, *South Africa*
E. Denarié, *Switzerland*
G. de Schutter, *Belgium*
G. Fischer, *Denmark*
R. Gettu, *India*
P. Kabele, *Czech Republic*
B.L. Karihaloo, *UK*
E.P. Kearsley, *South Africa*
V.C. Li, *USA*
V. Mechtcherine, *Germany*
H.-W. Reinhardt, *Germany*
K. Rokugo, *Japan*
P. Rossi, *France*
S.P. Shah, *USA*
V. Slowik, *Germany*
L.J. Sluys, *The Netherlands*
L. Taerwe, *Belgium*
R.D. Toledo Filho, *Brazil*
J.G.M. Van Mier, *Switzerland*
G.P.A.G. van Zijl, *South Africa*
F.H. Wittmann, *Germany*

Advances in Cement-Based Materials – van Zijl & Boshoff (eds)
© 2010 Taylor & Francis Group, London, ISBN 978-0-415-87637-7

Contributing institutions

The International Conference on Concrete Materials 2009

Organised by

Stellenbosch University

Concrete Society of Southern Africa

Supported by

Cement and Concrete Institute

Advanced materials technology & processing

Advances in Cement-Based Materials – van Zijl & Boshoff (eds)
© 2010 Taylor & Francis Group, London, ISBN 978-0-415-87637-7

Controlling properties of concrete through nanotechnology

S.P. Shah

Center for Advanced Cement-Based Materials, Northwestern University, Evanston, IL, USA

ABSTRACT: This article is a summation of recent work at the Center for Advanced Cement-Based Materials (ACBM) at Northwestern University. ACBM's areas of focus currently include self-consolidating concrete (SCC) and nano-modification of paste matrices with carbon nanotubes. Concerning SCC, three projects discussed here include reducing formwork pressure through use of nanoclays, quality control of fiber-reinforced SCC using an AC-IS method, and development of an improved slipform paving concrete. All of these topics share innovative processing techniques to ensure superior concretes that can further today's growing need for reliable high performance concretes.

1 INTRODUCTION

The future of today's high strength and high performance concrete relies on both new processing techniques as well as new types of admixtures specifically geared towards the nanostructure of cement paste. This article is a summation of recent work at the Center for Advanced Cement-Based Materials (ACBM) at Northwestern University. ACBM's areas of focus currently include self-consolidating concrete (SCC) and nano-modification of paste matrices with carbon nanotubes. SCC has seen a growing share in the concrete industry, and at ACBM, research has focused on formwork pressure, fiber-reinforcement and modification for slipform applications. Both of these topics share innovative processing techniques to ensure superior concretes that can meet today's growing need for reliable high performance concretes.

2 REDUCED FORMWORK PRESSURE OF SCC

A key advantage of SCC is accelerated casting and placing since vibration is not required for consolidation. However, faster casting rates may lead to higher lateral pressure on formwork; this is a major concern for cast-in-place applications, especially when casting tall elements, and has raised questions about the adequacy of using current formwork design practices for SCC. Since the development of formwork pressure is not fully understood in SCC, construction codes in the USA require design of formwork to withstand full hydrostatic pressure due to the fluidity of the concrete. However, it has been demonstrated that the formwork pressure of

SCC can be less than hydrostatic (Fedroff & Frosch 2004) due to the rebuilding of a three-dimensional structure when the concrete is left at rest (Sun et al. 2007). The mechanisms behind this stiffening phenomenon are of particular interest to users of SCC. Ideally, SCC should be flowable enough to self-consolidate, then immediately stiffen to gain green strength (or strength right after casting) once at rest. This will prevent formwork pressures from reaching hydrostatic pressures during the casting process and allow a more efficient design of formwork. Underestimating the pressure may cause deformed structural elements or even formwork collapse, while overestimating the pressure leads to unnecessary costs due to over-built formwork.

Formwork pressure and structural rebuilding of SCC is highly influenced by the mixture proportioning of the paste matrix. With proper design, it may be possible to achieve significant reductions in lateral pressure development. In order to test formwork pressure for different mixes, a pressure device was developed that subjects a sample of plastic concrete to a vertical load using a universal testing machine. Lateral pressure and pore water-pressure transducers are used to determine the total lateral pressure and pore-water pressure, respectively, generated by the vertical loading. This pressurized cylinder setup, referred to as the piston method, consists of a cylinder 300 mm in height that enables it to measure lateral pressure variations over time in the laboratory for concrete subjected to various vertical loads that correspond to different casting heights, and is shown in Figure 1.

Using this device, two SCC mixtures were tested; both were proportioned using a fine-to-coarse aggregate ratio of 0.47, w/b ratio of 0.35,

Figure 1. Formwork pressure device.

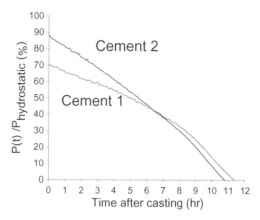

Figure 2. Evolution of pressure decay, where P(t) is the formwork pressure of the concrete at a specific time; $P_{hydrostatic}$ is constant and corresponds to the total vertical pressure applied at the end of casting (approximately 240 kPa). The casting height was 10 m (33 ft), and the casting rate was 7 m/hr (23 ft/hr).

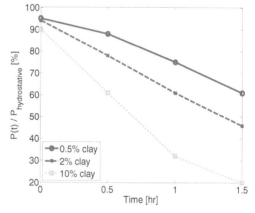

Figure 3. Influence of a metakaolin clay on formwork pressure.

and paste volume of 36%. A constant VMA dosage of 400 ml/100 kg of binder was used in all of the mixtures, and the superplasticizer content was adjusted to an initial slump flow of 670 ± 12.5 mm. The formwork pressure results are shown in Figure 2 (Ferron et al. 2007).

The main difference between the mixtures was the composition of the paste matrix, specifically the type of cement used in the concrete. Although both cements were ASTM Type 1 cements, the alkali and C_3A content of the mixtures was significantly different (Cement 1 had a lower alkali/C_3A content). As seen in Figure 2, a 30% reduction in formwork pressure (with respect to the hydrostatic pressure) was produced by altering the paste matrix. In this case, Cement 1, with a lower alkali/C_3A content, required less superplasticizer in order to maintain the same slump flow diameter as Cement 2 (Ferron 2008).

ACBM is currently investigating how the addition of small amounts of processed clays can increase the reduction of formwork pressure. Previously, clay particles have been shown to enhance green strength and allow a stiffer structure in concrete to develop (Curcio & DeAngelis 1998). Figure 3 shows the fraction of vertical pressure of SCC transferred as lateral pressure for 0.5, 2.0, and 10.0% replacement of cement with a metakaolin clay.

3 FRESH STATE MICROSTRUCTURE OF SCC

To understand why changes in alkali/C_3A or clay amounts affect the structural rebuilding, research has focused on the microstructure, and more recently, the nanostructure, of fresh state concrete. Characterizing the structure of cement suspensions is difficult due to the polydisperse, high solids concentration and hydration characteristics. Thus, there is a lack of knowledge about the fresh state structure of cementitious materials, which is especially important during the processing of concrete. Furthermore, the initial fresh state microstructure

Table 1. Mix compositions for FBRM pastes.

Mix	w/b	SP/b [%]	VMA dosage
P1	0.4	–	–
P1-SP	0.4	0.8	–
P1-VMA_H	0.4	–	high
P1-SP-VMA_H	0.4	0.8	high
P1-SP-VMA_M	0.4	0.8	medium

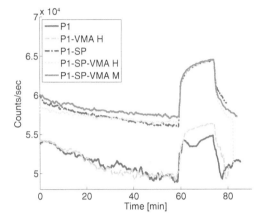

Figure 4. Influence of chemical admixtures on floc evolution.

affects the final microstructure, thereby influencing mechanical properties of hydrated paste or concrete (Struble 1991). Rheology of concrete is related to the degree of flocculation/coagulation of the paste matrix, which in turn is a function of the interparticle forces. Thus, perhaps the most representative parameter for studying the flocculation process (and indirectly the interparticle forces) is to monitor the change in size of the particle flocs.

A novel experimental method using a focused beam reflectance measurement (FBRM) probe was recently used by researchers at ACBM to examine the floc size evolution of concentrated cement paste suspensions subjected to shear (Ferron 2008). This is one of the first experiments for in-situ investigations of the microstructural response of concentrated cement paste suspensions subjected to shear-induced stresses. Results of compositions shown in Table 1 are presented in Figure 4.

The FBRM floc size measurements were conducted while subjecting the sample to a 40 rpm mixing intensity followed by a 400 rpm mixing intensity. Generally, higher count numbers indicate mixtures with fresh state microstructures that are less agglomerated.

As seen in Figure 4, no substantial changes in the number of chords counted were seen when VMA was added (compare P1 and P1-VMA_H), which is an indication that additional flocculation of the cement particles was not due to the incorporation of this VMA. Rather, the superplasticizer was shown to be the dominating factor affecting the chord length measurements—a significant increase in the number of counts occurred when superplasticizer was used (compare P1 with P1-SP). This shows that the superplasticizer molecules directly interact with the cement particles such that the cement particles are deflocculated, which increases the number of particles in the system. When both VMA and superplasticizer are used in a paste, the flocculation behavior is more similar to that of a paste with just superplasticizer. This behavior is seen regardless of the VMA dosage (compare P1-SP, P1-SP-VMA_M, and P1-SP-VMA_H). It can be concluded that the VMA did not interact with the cement particles, or if it did, it did not have any influence on the flocculation properties. Thus, it is likely that the increase in cohesiveness when this particular VMA is used is garnered from the polymers binding to the water phase.

Developing a quantitative relationship between paste, mortar, and concrete rheology is perhaps the most fundamental issue concerning concrete rheology. The relationship among paste, mortar, and concrete rheology is complex, but the ability to link these three behaviors is beneficial because this would allow for the prediction of concrete rheology solely from the characterization of the paste or mortar phase.

4 FIBER-REINFORCED SCC

Currently, the concrete industry is interested in the possible use of steel fibers as a partial or even total replacement of secondary reinforcement in concrete. The negative effects of fibers on concrete workability as well as improper placement and compaction may cause poor fiber distribution (Ferrara & Meda 2006). Regions with reduced amounts of fibers act as flaws, triggering early failure and activating unforeseen mechanisms. This leads to compromised structural performance, e.g in terms of deflection stiffness, crack opening toughness, and load-bearing capacity. The advantage of adding steel fibers to SCC lies in the self compactability of SCC as well as the rheological stability of SCC in the fresh state (Ferrara et al. 2008a). It has been shown that with an adequate mix design (Ferrara et al. 2007) fibers can be oriented along the flow direction. By suitably tailoring the casting process to the foreseen application, fiber orientation can be designed to match the anticipated stress pattern (direction of the principal tensile stresses) within the structural element during service (Ferrara et al. 2008b, Stahili et al. 2008). The possibility of modeling the casting of fresh concrete, e.g. through Computational Fluid Dynamics (CFD) (Roussel et al. 2007), can help predict the direction of flow lines along which fibers may orient and optimize the whole process. Monitoring fiber dispersion related issues through suitable non destructive methods, such as the Alternating Current Impedance Spectroscopy (Ozyurt et al. 2006) would also be crucial for reliable quality control.

Thorough investigation on these subjects has been performed jointly by Politecnico di Milano and ACBM. Four $1 \text{ m} \times 0.5 \text{ m} \times 0.1 \text{ m}$ slabs were cast with a self-consolidating steel fiber reinforced concrete (SCSFRC) containing 50 kg/m^3 of steel fibers. The fibers were 60 mm long with a 0.8 mm diameter. The slump flow diameter of each slab was 600 mm. Each slab was cut into beams and the fibers were counted on the beam side faces to determine fiber orientation factor, $\alpha = n_{fibers} V_{fiber}/A_{fiber}$, where n_{fiber} is the specific number of fibers on the examined surface, V_{fiber} is fiber volume fraction (0.67% in this case), and A_{fiber} area of the fiber cross section.

The casting flow process, as shown in the upper part of Figure 5, was simulated by a CFD code

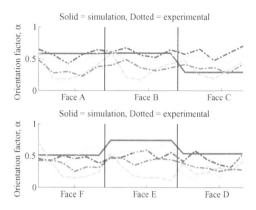

Figure 6. Comparison between experimental orientation factors in the 4 test slabs (A-F 1 to 4) and numerical ones, computed from shear rate vectors from CFD modeling.

(Polyflow 3D), shown in the lower part of Figure 5. For the sake of simplicity, a 2-D homogeneous single fluid simulation was performed by suitably calibrating the input parameters (Bingham fluid $\tau_0 = 100 \text{ Pa}$, $\mu = 100 \text{ Pas}$) for the fresh state behavior of the SCSFRC composite. The orientation of the fluid concrete flow lines (shear rate vectors) as computed with respect to relevant surfaces within the slab, has been compared with the fiber orientation factor. Considering the 2D features of the simulation, numerically computed orientation factors always provide an upper bound estimate of the experimental ones.

The results (Figure 6) are encouraging and stand as an interesting step in paving the way towards a more widespread use of simulation of fresh concrete flow, including at the industrial scale.

5 IMPROVED SLIPFORM PAVING WITH SCC

The slip-form paving process is used extensively to create highways and other pavements worldwide. This process combines concrete placing, casting, consolidation, and finishing into one efficient process. The stiff concrete used in slipform paving machines requires extensive internal vibration, a process that can lead to durability issues. It has been shown that over-consolidation caused by the internal vibrators contribute to the formation of premature cracks. Typically, pavements are designed to last 25–30 years, however in several instances in the United States, slip-cast pavements have shown significant cracks at three years. ACBM, in conjunction with the Center for Portland Cement Concrete Pavement at Iowa

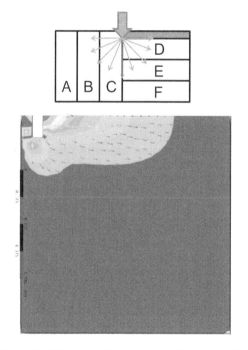

Figure 5. Schematic of the casting of the test-slab and results of CFD modeling (shear rate vectors highlighted).

State University, has developed several promising mixture proportions of a low compaction energy concrete which is tentatively called slip-form self-consolidating concrete, SF-SCC (Pekmezci et al. 2007).

Fundamental research on particle packing and flocculation mechanisms provided insight on how to eliminate internal vibration and durability issues associated with longitudinal cracking along the vibration trail. The development of SF-SCC required changing the microstructure by combining concepts from particle packing, admixture technology, and rheology. Specifically, the addition of different materials such as nanoclays and fly ash to the composition made it possible to maintain a balance between flowability during compaction and stability after compaction (Tregger et al. 2009). For this research, a model minipaver that simulates the slipform paving process without the application of internal or external vibration was developed. Concrete slabs of mixes modified with fly ash or fly ash and clay showed much better shape stability and surface smoothness than the slab with a standard slipform concrete mix as shown in Figure 7.

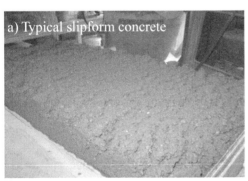

Figure 7. (a) Minipaver slab with typical slipform concrete. Rough surfaces indicate poor consolidation. (b) Minipaver slab with SF-SCC mix. Smooth surface indicates proper consolidation while straight edges indicate adequate green strength.

It was also demonstrated that very small amounts of clays (0.3% by volume of concrete) resulted in large increases in green strength (as high as 30%) while maintaining fluidity (Tregger et al. 2009). This was also shown to be the case for mixes used in formwork pressure tests containing clays. Future work at ACBM is focused on understanding the reasons for the change in green strength due to clays using rheology as well as FBRM methods.

6 NANO-MODIFICATION WITH CARBON NANOTUBES

Cement-based materials are typically characterized as quasi-brittle materials that exhibit low tensile strength. Typical reinforcement of cementitious materials exists at the millimeter scale and/or at the micro scale using macrofibers and microfibers, respectively. However, cement matrices still exhibit flaws at the nanoscale. The development of new nanosized fibers, such as carbon nanotubes (CNTs), has opened a new field for nanosized reinforcement within concrete. The remarkable mechanical properties of CNTs suggest that they are ideal candidates for high performance cementitious composites. The major drawback however, associated with the incorporation of CNTs in cement based materials is poor dispersion (Groert 2007). To achieve good reinforcement in a composite, it is critical to have uniform dispersion of CNTs within the matrix (Xie et al. 2005). Few attempts have been made to add CNTs in cementitious matrices at an amount ranging from 0.5 to 2.0% by weight of cement. Previous studies have focused on the dispersion of CNTs in liquids by pre-treatment of the nanotube's surface via chemical modification (e.g. Cwirzen et al. 2008). Preliminary research has shown that small amounts of CNTs can be effectively dispersed in a cementitious matrix (Konsta-Gdoutos et al. 2008).

At ACBM, the effectiveness of the dispersing method was investigated through nanoimaging of the fracture surfaces of samples reinforced with 0.08% CNTS by weight of cement. Results from SEM images of cement paste samples reinforced with CNTs that were added to cement as received (without dispersion) and CNTs that were dispersed following the method described elsewhere (Konsta-Gdoutos 2008) are presented in Fig. 8. As expected, in the samples where no dispersing technique was used [Figure 8 (a)] CNTs appear poorly dispersed, forming large agglomerates and bundles. On the other hand, in the samples where dispersion was achieved by applying ultrasonic energy and using a surfactant [Figure 8 (b)] only individual CNTs were identified on the fracture surface. The results indicate that the application of ultrasonic energy

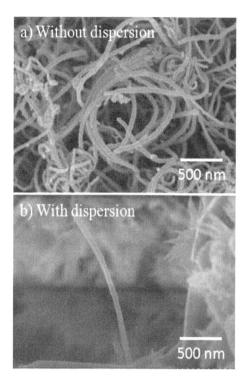

Figure 8. SEM images of cement paste reinforced with CNTs dispersed with (b) and without (a) the application of ultrasonic energy and the use of surfactant.

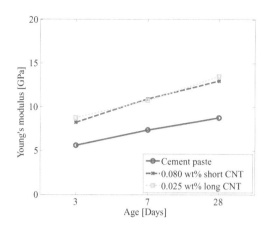

Figure 9. Young's modulus results from fracture mechanics tests of CNTs nanocomposites which exhibit the best mechanical performance among the different mixes tested.

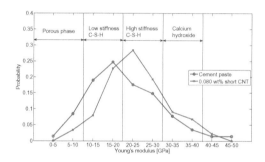

Figure 10. Probability plots of the Young's modulus of 28 day cement pastes with and without 0.08 wt% short CNTs for a w/c of 0.5.

and the use of surfactant can be employed to effectively disperse CNTs in a cementitious matrix.

To evaluate the reinforcing effect of CNTs, fracture mechanics tests were performed using MWCNTs with aspect ratios of 700 and 1600 for short and long CNTs, respectively. Additionally, to investigate the effect of CNTs concentration, cement paste samples reinforced with lower and higher amounts of CNTs (0.048 wt% and 0.08 wt%, respectively) were tested. The Young's modulus results from the fracture mechanics tests are shown in Figure 9. In all cases, the samples reinforced with CNTs exhibit much higher Young's modulus than plain cement paste. More specifically, it is observed that the specimens reinforced with either short CNTs at an amount of 0.08 wt% or long CNTs at an amount of 0.048 wt% provide the same level of mechanical performance. Generally, it can be concluded that the optimum amount of CNTs depends on the aspect ratio of CNTs. When CNTs with low aspect ratio are used a higher amount of close to 0.08 wt% by weight of cement is needed to achieve effective reinforcement. However, when CNTs with high aspect ratio are used, a smaller amount of CNTs of close to

0.048 wt% is required to achieve the same level of mechanical performance.

Comparing the 28 days Young's modulus of the nanocomposites with that of the plain cement paste, a 50% increase is observed. Based on the parallel model [16] the predicted Young's modulus of cement paste nanocomposites reinforced with either 0.048 wt% or 0.08 wt% CNTs at the age of 28 days (~9.1 GPa) is much lower than the experimental values obtained (~13 GPa). In addition, nanoindentation was performed on samples with and without CNTs. Figure 10 shows the probability plot of the 28 days Young's modulus of plain cement paste and cement paste reinforced with 0.08 wt% short CNTs.

The probability plots are in good agreement with results from the literature (Constantindes & Ulm 2008, Mondal et al. 2008). Young's modulus values less than 50 GPa represent four different phases of cement paste corresponding to the porous phase,

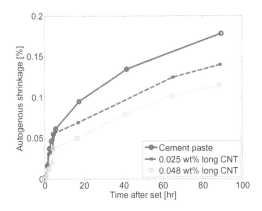

Figure 11. Improvement of autogenous shrinkage for cement pastes containing 0.025 and 0.048 wt% of CNTs for w/c of 0.3.

low stiffness C-S-H, high stiffness C-S-H and calcium hydroxide phase, while values greater than 50 GPa are attributed to unhydrated particles (Constantinides & Ulm 2008, Mondal et al. 2008). The different phases have been found to exhibit properties that are considered as inherent material properties and are independent of the mix proportions. As expected, the peak of the probability plots of plain cement paste with w/c = 0.5 falls in the area of the low stiffness C-S-H, which is the dominant phase of cement nanostructure. On the other hand, the peaks of the probability plots of the nanocomposites are in the area of 20 to 25 GPa which corresponds to the high stiffness C-S-H, suggesting that the addition of CNTs results in a stronger material with increased amount of high stiffness C-S-H. Moreover, it is observed that the probability of Young's modulus below 10 GPa is significantly reduced for the samples with CNTs for both water-to-cement ratios. The nanoindentation results provide an indirect method of estimating the volume fraction of the capillary pores, indicating that the CNTs reduce the amount of fine pores by filling the area between the C-S-H gel (Shah et al. 2009), which may be one reason why current research has shown other improved properties with CNTs including autogenous shrinkage. Autogenous shrinkage measurements are shown in Figure 10, comparing plain cement paste with paste containing long CNTs at 0.025% and 0.048% by weight, all with a w/c of 0.3.

7 CONCLUSIONS

To meet the increasing need for high-performance, durable construction materials, ACBM has taken a back-to-basics approach to improve understanding of the properties of cement-based materials at a small scale and to develop new materials. Innovative ways to improve SCC in terms of formwork pressure, fiber dispersion and green strength have been developed through concepts of particle packing and flocculation, admixture technology, and rheology. Incorporation of carbon nanotubes has seen very promising results, while the nanoscale characterization of cement paste samples showed that the mechanical properties of the C-S-H gel—the glue in concrete—vary in a wide range, requiring complex modeling. Yet understanding cement-based materials at this scale can provide new ways to improve the high-strength and high performance concretes of today.

ACKNOWLEDGEMENTS

The author would like to acknowledge the researchers that contributed to the work presented in this paper: Mark Beacraft, Professor Maria Konsta-Gdoutos, Professor Liberato Ferrara, Professor Raissa Ferron, Zoi Metaxa and Nathan Tregger.

REFERENCES

Curcio, F. & DeAngelis, B. 1998. Dilatant behavior of superplasticized cement pastes containing metakaolin. *Cement and Concrete Research* 28(5): 629–634.
Constantinides, G. & Ulm, F. 2007. The nanogranular nature of C-S-H. *Journal of the Mechanics and Physics of Solids* 55(1): 64–90.
Cwirzen, A., Habermehl-Chirzen, K. & Penttala, V. 2008. Surface decoration of carbon nanotubes and mechanical properties of cement/carbon nanotube composites. *Advances in Cement Research* 20(2): 65–73.
Fedroff, D. & Frosch, R. 2004. Formwork for self-consolidating concrete. *Concrete International* 26(10): 32–37.
Ferrara, L. & Meda, A. 2006. Relationships between fibre distribution, workability and the mechanical properties of SFRC applied to precast roof elements. *Materials and Structures* 39(4): 411–420.
Ferrara, L., Park, Y.D. & Shah, S.P. 2007. A method for mix-design of fiber reinforced self compacting concrete. *Cement and Concrete Research* 37(6): 957–971.
Ferrara, L., Park, Y.D. & Shah, S.P. 2008. Correlation among fresh state behaviour, fiber dispersion and toughness properties of SFRCs. *ASCE Journal of Materials in Civil Engineering* 20(7): 493–501.
Ferrara, L., di Prisco, M., & Khurana, R.S. 2008. Tailoring optimum performance for the structural use of self consolidating SFRC. In Proc. BEFIB08, 7th RILEM International Symposium on Fiber Reinforced Concrete: Design and Applications, Chennai, India: 739–750.
Ferron, R. 2008. Formwork Pressure of Self-Consolidating Concrete: Influence of Flocculation Mechanisms, Structural Rebuilding, Thixotropy and

Rheology. Department of Civil and Environmental Engineering. Evanstion, Ph.D. dissertation, Northwestern University.

Ferron, R. Gregori, A., Sun, Z. & Shah, S.P. 2007. Rheological method to evaluate structural buildup in self-consolidating concrete cement pastes. *ACI Materials Journal* 104(3): 242–250.

Groert, N. 2007. Carbon nanotubes becoming clean. *Mateials Today* 10(1–2): 28–35.

Konsta-Gdoutos, M.S., Metaxa, Z.S. & Shah, S.P. 2008. Nanoimaging of highly dispersed carbon nanotube reinforced cement based materials. In Proc: BEFIB08, 7th RILEM International Symposium on Fiber Reinforced Concrete: Design and Applications, Chennai, India: 125–131.

Mondal, P., Shah, S.P. & Marks, L.D. 2008. Nanoscale characterization of cementitious materials. *ACI Materials Journal* 105(2): 174–179.

Ozyurt, N., Woo, L.Y., Mason, T.O. & Shah, S.P. 2006. Monitoring fiber dispersion in fiber reinforced cementitious materials: comparison of AC-Impedance Spectroscopy and Image Analysis. *ACI Materials Journal* 103(5): 340–347.

Pekmezci, B.Y., Voigt, T., Kejin, W. & Shah, S.P. 2007. Low compaction energy concrete for improved slipform casting of concrete pavements. *ACI Materials Journal* 104(3): 251–258.

Roussel, N., Geiker, M., Dufour, F., Thrane, L. & Szabo, P. 2007. Computational modeling of concrete flow: General overview. *Cement and Concrete Research* 37(9): 1298–1307.

Shah, S., Konsta-Gdoutos, M., Metaxa, Z. & Mondal, P. 2009. Nanoscale modification of cementitious materials. In Bittnar, Z., Bartos, P., Nemecek, J. Smilauer, V., Zeman, J. (eds.), *Nanotechnology in Construction 3; Proceedings of NICOM3, Prague, May 31-June 2, 2009,* Springer Berlin Heidelberg.

Stähli, P. Custer, R. & van Mier, J.G.M. 2008. On flow properties, fibre distribution, fibre orientation and flexural behaviour of FRC. *Materials and Structures* 41(1): 189–196.

Struble, L.J. 1991. The Rheology of Fresh Cement Paste. In S. Mindess (ed.), *Advances in cementitious materials; Ceramic Transactions* 16: 7–29.

Sun, Z., Gregori, A., Ferron, R. & S.P. Shah. 2007. Developing falling-ball viscometer for highly flowable cementitious materials. *ACI Materials Journal* 104(2): 180–186.

Tregger, N., Knai, H. & Shah, S.P. 2009. Flocculation behavior of cement pastes containing clays and fly ash. *ACI Special Publication* 259(10).

Xie, X.L., Mai, Y.W. & Zhou, X.P. 2005. Dispersion and alignment of carbon nanotubes in polymer matrix: A review. *Mater. Sci. Eng. Rep.* 49: 89–112.

Advances in Cement-Based Materials – van Zijl & Boshoff (eds)
© *2010 Taylor & Francis Group, London, ISBN 978-0-415-87637-7*

Reducing the cracking potential of ultra-high performance concrete by using Super Absorbent Polymers (SAP)

L. Dudziak & V. Mechtcherine
Institute of Construction Materials, TU Dresden, Germany

ABSTRACT: Ultra-High Performance Concrete (UHPC) has a number of superior properties adding up to this new material's considerable potential in numerous practical applications. It is, however, subject to very high autogenous shrinkage at an early age and resultantly to a pronounced tendency to crack. In the research project reported here, the effectiveness of the internal curing of UHPC to mitigate these autogenous deformations has been investigated. Super Absorbent Polymers (SAP) were used first to absorb water in the fresh mix and subsequently to release it into the pore system during hydration as the relative humidity decreases. The results obtained show that the addition of dry SAP and an amount of extra water during mixing leads to a pronounced reduction in deformation due to autogenous shrinkage. Moreover, the paper reports the effects of the addition of SAP on the rheological behaviour of fresh UHPC, its mechanical properties in the hardened state, and changes in its microstructure.

1 INTRODUCTION

Ultra-High-Performance Concrete (UHPC) has superior mechanical properties and very high durability due to its dense microstructure. Hence, it is suitable for use in harsh environments and operating conditions. However, this new material is prone to early-age cracking, resulting primary from high autogenous shrinkage, which may significantly decrease material resistance to the ingress of corrosive media. Therefore, autogenous shrinkage must be mitigated in order to assure the increased durability of UHPC.

Conventional methods in the curing of concrete cannot contribute substantially to mitigating autogenous shrinkage of compositions with a low water-cement ratio even if intensive wetting is applied. Since the microstructure of UHPC is very dense even at early ages, it does not allow the sufficiently rapid transport of curing water into the interior of the concrete members.

With that in mind, it has been proposed to use so-called internal curing by adding materials with high water storing capacity to the mixture. These materials, finely distributed over the concrete volume, can store water until the hydration process leads to a shortage of free water in the pore system (often described as self-desiccation). By giving up the water to the surrounding matrix, such tiny reservoirs cure the concrete internally.

Super Absorbent Polymers (SAP) appear to be most appropriate for use as a water-regulating additive. These polymers consist of cross-linked chains which have dissociated ionic functional groups facilitating the absorption of large amounts of water. Following the pioneering work by Jensen & Hansen (2002), they and other researchers tested various types of SAP and demonstrated that certain types show a very pronounced ability to mitigate autogenous shrinkage of high-strength mortar and concrete (Lura et al. 2006, Mechtcherine et al. 2006) and UHPC (Mechtcherine et al. 2009). The advantage of SAPs is that they can be in principle engineered for the special purposes of internal curing by designing the necessary size and shape of the particles, water absorption capacity and other properties. Only small amounts of SAP plus some additional water for internal curing are added directly to the fresh mix.

This paper presents recent findings from an ongoing research project studying the use of SAP for internal curing of UHPC. It focuses on the results of the internal curing on autogenous shrinkage as well as on the effect of SAP addition on the development of stresses due to restrained autogenous deformations of UHPC. Furthermore, the influences of the internal curing on the mechanical properties and the material microstructure are reported.

2 MIX PROPORTIONS AND PROPERTIES OF FRESH CONCRETE

The mixtures examined were composed based on the recipe for one fine-grain UHPC and one containing coarser aggregates, both developed under the aegis of the Priority Program 1182 "Sustainable Building with Ultra High Performance Concrete (UHPC)" (Fehling et al. 2005). Besides different graining of aggregates in both compositions, the mixtures differed in their water-cement ratios, one alternative using two types of silica fume as well as superplasticizers. In contrast to the original recipe, steel fibres were not used in the experiments with fine-grain concrete, cf. Table 1.

In total, seven mix compositions were experimentally evaluated in this study: two reference concrete mixes with no internal curing (fine-grain reference concrete F-R and coarse-grain, fibre-reinforced concrete Cf-R) and five UHPC mixes with internal curing (fine-grain UHPC: F-S.4, F-S.3.04, F-S.3.05, and F-S.4.07, cf. Table 1; coarse-grain UHPC: Cf-S.3.04, cf. Table 2). With regard to internal curing, the variable parameters were the amount of internal curing agent (0.3% and 0.4% SAP, related to the mass of cement) and the amount of additional water (0.04 to 0.07 in reference to the mass of cement). An exception is the mix identified as F-S.4, to which purposefully only SAP material was added but no extra water for internal curing was introduced; see Table 1.

As the SAP suspension polymerised, spherical particles of average size 150 μm were selected. Due to the small portion of internal curing material used, the volume of the SAP was neglected in calculations.

Contrarily, the volume of the water pores formed by the absorption of extra water by SAP was considered in the mix design under the assumption that water pores have the density of free water. The amount of water for internal curing was estimated with the intention of compensating the loss of workability due to the absorption of mixing water by SAP. The workability was measured by means of slump-flow testing according to DAfStb standards for self-compacting concrete. The same slump-flow values were aimed at for mixtures with SAP as measured for the finely and coarse-grained reference mixes.

The second part of Table 1 gives the properties of fresh fine-grain concrete obtained. A great number of slump-flow tests have been performed in the framework of the project. Due to a high sensitivity of the reference material a noticeable scattering of the results was observed for the selfsame material tested on different dates. This explains partly the relatively high standard deviation values. The slump-flow values of the reference concrete F-R scattered between 750 mm and 830 mm. The average values of the UHPC with internal curing (addition of SAP and extra water) fall into this range. Only in the exceptional case of the addition of pure SAP with no extra water (mixture F-S.4), a much lower slump-flow average value of 555 mm was recorded. This can be traced back to the high absorption of mix water by SAP. Furthermore, the UHPC mixes with internal curing showed a slight increase in air pore content in comparison to the reference mix F-R, while the density slightly

Table 1. Compositions and properties of fresh fine-grain UHPC mixes with and without SAP. Standard deviations are given in parentheses.

Components		F-R	F-S.4	F-S.3.04	F-S.3.05	F-S.4.07
Cement	kg/m³	853.4	853.4	824.5	817.6	804.2
SF	kg/m³	138.5	138.5	133.8	132.7	130.5
Water	kg/m³	170.3	170.3	164.5	163.1	160.5
SAP	% m.c.	–	0.4	0.3	0.3	0.4
W_{IC}*	kg/m³	–	–	33.0	40.9	56.3
W/C*		–	0.22	0.22	0.26	0.27	0.29
W/(C + SF)*		–	0.19	0.19	0.23	0.23	0.25
Quartz flour I	kg/m³	212.3	212.3	205.1	203.4	200.1
Quartz sand	kg/m³	1000	1000	966.3	958.2	942.4
SP	kg/m³	30.2	30.2	29.1	28.9	28.4
Slump flow	mm	780	555	750	830	830
		(55)	(–)	(38)	(–)	(–)
Air content	%	4.6	5.4	5.0	5.2	4.7
		(0.7)	(0.7)	(0.5)	(–)	(1.0)
Density	kg/m³	2371	2330	2322	2319	2274
		(21)	(18)	(17)	(24)	(16)

*Total value, including IC water and water from superplasticizer (65%).

Table 2. Compositions and properties of fresh, coarse-grain UHPC mixes with and without SAP. Standard deviations are given in parentheses.

Components		Cf-R	Cf-S.3.04
Cement	kg/m³	650	633.4
SF	kg/m³	177	172.5
Water	kg/m³	158	154
SAP	% m.c.	–	0.3
W_{IC}	kg/m³	–	25.3
W/C*		–	0.27	0.31
W/(C + SF)*		–	0.21	0.25
Quartz flour I	kg/m³	325	316.7
Quartz flour II	kg/m³	131	127.7
Quartz sand	kg/m³	354	345
Basalt	kg/m³	597	581.8
Steel fibres	kg/m³	192	187.1
SP	kg/m³	30.4	29.6
Slump flow	mm	680 (3)	695 (8)
Air content	%	2.1 (0.2)	2.3 (0.4)
Density	kg/m³	2593 (16)	2549 (20)

*Total value, including IC water and water from superplasticizer (65%).

decreased due to the addition of SAP and internal curing (IC) water.

Similar tendencies in respect of the effect of internal curing on the air content and density were observed also in the experiments on UHPC with coarse aggregates (cf. Table 2). The average slump flow value for UHPC with internal curing (mix Cf-S.3.04) was 695 mm, which was slightly higher than the corresponding value of 680 mm for the reference mix Cf-R.

3 EARLY AGE AUTOGENEOUS DEFORMATIONS

The measurement of autogenous shrinkage on fine-grain UHPC was performed using the method developed by Jensen and Hansen in 1995, where the special design of the measuring device (dilatometer) and usage of polyethylene corrugated, tube-shaped moulds enable continuous monitoring of the concrete's deformations starting immediately after filling and encapsulating the tubes, cf. Figure 1a. A similar setup, rather with tubes with a larger diameter, was used for investigating the shrinkage behaviour of coarse-grain UHPC, cf. Figure 1b.

Since deformations when the concrete is still fluid have no significant influence on stress development, the evaluation of the deformations for this paper was performed starting with the final set taken at so-called time "t_0". This time was

determined using the needle penetration test (Vicat test according to DIN EN 480-2) and ultrasonic measurements.

At least three shrinkage tests were performed for each UHPC composition. Figure 2 shows the average curves for development of autogenous deformation over time obtained during the first 28 days following the final set of fine-grain UHPC mixes. All the mixtures containing SAP and extra water showed pronounced reductions in shrinkage deformations. These reductions were particularly dramatic at a very early age, indeed in the first twelve hours after the final set. The addition of 0.3% SAP by mass of cement plus extra water (mixes F-S.3.04 and F-S.3.05) resulted in a decrease in autogenous shrinkage from approximately 1100 μm/m (F-R) to approximately 300–350 μm/m in the first day of measurement. This effect is particularly remarkable when one considers the relatively low tensile strength of concrete at this age and, consequently, its low material resistance to cracking.

Furthermore, Figure 2 demonstrates that an increase in the SAP addition from 0.3 to 0.4% by mass of cement in combination with an increase in the addition of extra water (mix F-S.4.07) not only further reduced the magnitude of autogenous shrinkage deformations but also caused changes in their development over time. After reaching the age of one or two days, UHPC with 0.4% SAP, showed only a very minor increase in autogenous deformations while 0.3% SAP already followed in general the course of the corresponding curves for UHPC without SAP addition, i.e. they showed approximately the same increase in autogenous shrinkage over time.

Figure 3 shows individual curves obtained from the measurements on UHPC mixes with coarse aggregates for the first seven days after the final set. The absolute values for autogenous shrinkage in the reference mixture Cf-R are approximately

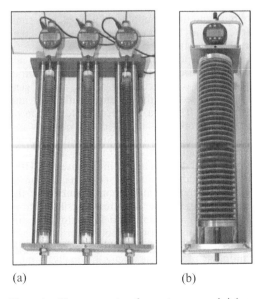

(a)   (b)

Figure 1. Test apparatus for autogenous shrinkage measurements on: (a) fine-grain UHPC, (b) UHPC with coarse aggregates.

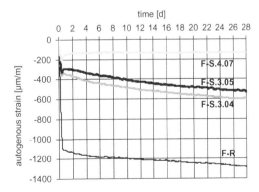

Figure 2. Autogenous shrinkage of fine grained UHPC measured beginning after the final set.

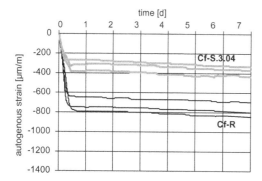

Figure 3. Autogenous shrinkage of coarse grained UHPC measured starting after the final set.

35% lower in comparison to the fine-grain reference mixture F-R. This is likely due to the lower binder content of the coarser mix as well as the presence of stiff coarse aggregates and steel fibre, which counteract the shrinkage deformation of the matrix to some extent. The addition of SAP and extra water (mix Cf-S.3.04) leads to a reduction of autogenous shrinkage deformations to approximately the same level as measured in the fine-grain mixture with the same percentage of SAP and IC water addition (mix F-S.3.04).

4 MECHANICAL PROPERTIES

To provide information on the potential effect of internal curing on the mechanical behaviour of hardened concrete, a series of compression tests on cubes with side lengths of 100 mm was performed. Additionally the halves of small beams were tested in compression (referred in Table 3 "prisms", cross-section 40 mm × 40 mm). These specimens were obtained from the foregoing three-point bend tests on beams with the dimensions 160 mm × 40 mm × 40 mm, while the span was 100 mm. Furthermore, Young's modulus was measured on prisms with the same dimensions as the small beams used in bend tests. All specimens were sealed in plastic foil immediately after demoulding at an age of one day and subsequently stored in a standard laboratory climate.

Table 3 gives the average values and the standard deviations obtained for compressive strength. A noticeable relative decrease in compressive strength for concretes containing SAP was measured at early ages. However, this difference diminished with increasing age of the UHPC. The measurements at the age of 28 days exhibited only a minor decrease in strength in the cases when the addition of SAP and extra water was relatively

Table 3. Compressive strength of investigated concretes standard deviations are given in parentheses.

| Mixture | Compressive strength [MPa] | | | | | |
| | Prisms | | | Cubes | | |
	3d	7d	28d	1d	7d	28d
F-R	106	122	141	–	127	160
	(5)	(7)	(14)	–	(18)	(14)
F-S.4	104	112	134	–	118	156
	(5)	(5)	(10)	–	(10)	(14)
F-S.3.04	95	111	137	–	121	155
	(6)	(6)	(7)	–	(–)	(8)
F-S.3.05	88	108	142	–	125	147
	(2)	(4)	(7)	–	(–)	(6)
F-S.4.07	87	94	120	–	95	128
	(3)	(4)	(8)	–	(13)	(13)
Cf-R	142	–	225	64	117*	188
	(6)	(–)	(8)	(2)	(3)	(5)
Cf-S0.3.04	124	–	207	42	109*	181
	(3)	(–)	(7)	(6)	(7)	(6)

* Tested at the age of 3 days.

small (0.3% SAP). This holds true for both fine-grain and coarse-grain UHPC. A higher dosage of SAP (0.4%) and curing water (mix F.4.07) led however to a considerable decrease in compressive strength. The reduction of the strength for the mix F-S.4 (SAP only, no extra water) was again much less pronounced.

Table 4 gives the measured values of the flexural strength and the E-modulus. In general, very similar tendencies can be observed here as with compressive strength.

The interpretation of these results is not straightforward. On the one hand, a reduction in the strength of the concrete matrix can be generally expected for the mixtures with SAP and extra water. This is a result of the formation of entraining pores, initially filled with curing water and subsequently dried out. Such voids affect the strength negatively. Also the total amount of capillary and gel pores is higher in the case of UHPC with SAP addition. On the other hand, due to the reduction of the autogenous shrinkage of the cement paste, the internal stresses resulting from the hindrance of shrinkage deformations by stiff aggregates must be considerably lower in the case of specimens with SAP. This is positive from the standpoint of concrete strength. Finally, it must be said that the entire pore system of the cement paste has an effect on the strength of concrete. Similar effects should be considered when discussing the influence of the internal curing on the Young's modulus.

Table 4. Flexural strength and Young's modulus for the concretes investigated. Standard deviations are given in parentheses.

Mixture	Flexural strength [MPa]			Young's modulus [GPa]	
	3d	7d	28d	7d	28d
F-R	15.7	16.5	18.5	47.1	48.8
	(1.5)	(1.5)	(2.9)	(1.3)	(1.2)
F-S.4	13.4	14.0	13.0	45.9	49.3
	(0.8)	(1.5)	(2.8)	(–)	(0.9)
F-S.3.04	12.5	14.8	15.5	–	46.8
	(1.4)	(1.4)	(2.1)	(–)	(1.2)
F-S.3.05	12.8	12.1	13.5	–	–
	(0.7)	(0.2)	(0.2)	(–)	(–)
F-S.4.07	12.2	13.9	13.8	38.7	42.5
	(0.4)	(0.8)	(2.0)	(2.5)	(1.9)
Cf-R	28.6	–	35.3	49.1	54.7
	(3.1)	(–)	(2.6)	(–)	(1.0)
Cf-S0.3.04	26.8	–	34.8	47.4	53.4
	(2.4)	(–)	(2.9)	(0.4)	(0.6)

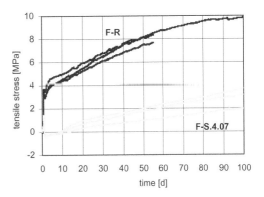

Figure 4. Development of tensile stresses due to restraint in sealed concrete specimens made of mixes F-R and F-S.4.07 during the first 100 days after casting.

5 RESTRAINED SHRINKAGE MEASUREMENTS

Instrumented ring tests were performed to assess the magnitude of the tensile stresses developed due to restrained autogenous deformations of UHPC, both with and without SAP addition, hence, to estimate quantitatively the tendency of these concretes to crack. A concrete annulus was cast around a steel ring of dimensions standardized in ASTM C1581-04. Negative volumetric changes in the concrete induce pressure at the steel ring, resulting in its contraction. Simultaneously tensile stresses in the concrete annulus arise since the steel ring restrains the free deformation of the concrete. The strains were measured continuously by four strain gauges glued to the inner side of the steel ring, providing information on the magnitude of these deformations. These values were used for calculating tensile stresses in concrete according to the equations proposed in Hossain & Weiss 2004.

Since the stresses due to restrained autogenous deformation should be investigated at the place of origin, an outer steel ring was used in addition to the inner ring. It served as a part of the "mould" when producing the concrete annulus and subsequently as protection against desiccation at the circumference of concrete annulus. The top of the concrete specimen was sealed using two layers of self-gluing aluminium foil. The advantage of using the outer steel ring is that it also enables measuring the expansive behaviour of concrete.

Figure 4 depicts initial results obtained from ring tests on two types of fine grained UHPC—the reference fine-grain concrete F-R and the mixture with internal curing F-S.4.07—for continuous measurements up to a concrete age of 100 days. According to these results, considerable tensile stresses of approximately 3 to 4 MPa developed in specimens made of the reference UHPC at an age of 24 hours, while the stresses in the concrete with internal curing (F-S.4.07) rise hardly above 1 MPa. Thus, the positive effect of internal curing could be observed already at very early ages.

In subsequent days of measurement, the mitigation of autogenous shrinkage using internal curing exhibited a continued trend toward dramatic reduction in stresses caused by restraint: at a concrete age of 100 days the stresses induced in the specimen made of the mix F-S.4.07 containing SAP were approximately 3 MPa and the corresponding stresses in the reference concrete F-R more than 3 times this value at the same age. This demonstrates clearly a considerable reduction of the cracking potential of UHPC due to internal curing.

6 PORE STRUCTURE

An investigation into the porosity of UHPC was performed for a better understanding of the mechanisms leading to the results presented in the foregoing sections. The effect of the addition of SAP and curing water on changes in the UHPC microstructure was studied using the Mercury Intrusion Technique. Additionally, the microstructure in the surroundings of SAP particles was investigated by REM/ESEM.

Two fully automated Mercury Porosimeters, one for the pressure range from 0.01 to 300 kPa and another for the higher range up to 400 MPa, were used to investigate porosity and pore size distribution. Using this equipment pore diameters ranging from 3.4 nm to 130 μm could be measured.

The investigations were performed at concrete ages of 1 day, 3 days, 7 days, 28 and 90 days; however, in this paper only the results obtained for the concrete age of 3 day and 28 days, respectively, are presented and discussed. The samples were taken from sealed prisms. When the test time approached, the prisms were shattered and afterwards gently dried at 40°C for 24 hours to stop further hydration.

The results of first evaluation showed as expected, a decrease in porosity over time for all UHPC compositions (cf. Mechtcherine et al., 2009). However, at all ages an increase in total porosity was detected for UHPC produced using SAP and extra water. In the following a qualitative comparison of the cumulative pore volume curves as well as of the pore size distribution is presented for selected parameters in order to provide a more comprehensive view on the pore systems in the material with and without SAP.

Figures 5 and 6 show cumulative curves for concretes F-R, F-S.4, F-S.3.04 and F-S.4.07 at the ages of 3 days and 28 days, respectively. The individual curves were obtained form the measurement on different batches. Only pore diameters less than 10 μm are considered here. With pore diameter logarithmically scaled, the cumulative pore volume increases nearly linearly with decreasing pore diameter. The slope of the curve is very flat, indicating a rather low volume of capillary macropores (diameter d > 0.1 μm). This holds true down to the pore diameter of approximately 30–40 nm at a concrete age of three days and down to approximately

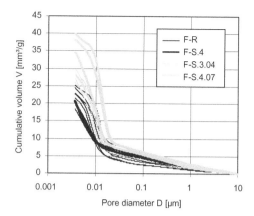

Figure 6. Cumulative pore volume curves for selected UHPC mixtures at a concrete age of 28 days.

15–25 nm at a concrete age of 28 days, while the curves for UHPC with SAP addition (F-S.4, F-S.3.04, and F-S.4.07) are slightly above the curves for the reference material F-R. For smaller pore diameters the rise in all curves is much more pronounced. At a concrete age of 3 days the mixtures F-R and F-S.4 (both without extra water) attain nearly equal cumulative pore volumes, which are, however, less than the corresponding values obtained for the mixtures F-S.3.04 and F-S.4.07 (both with addition of SAP and extra water), cf. Figure 5. The measurements on the 28-day-old specimens also show a distinct rise in pore volume with fine pore diameters. The mixtures F-S.3.04 and F-S.4.07 (both produced with SAP and the addition of extra water) provided again the most pronounced rise in the cumulative pore volume, followed by the reference mixture F-R and finally by the mixture F-S.4 (SAP addition, but no IC water), cf. Figure 6.

Figures 7 and 8 illustrate the pore size distributions for 3-day-old and 28-day-old UHPC mixes, respectively. This usual mode of presentation, i.e. covering all measured pore diameters and using a logarithmically scaled axis for the abscissa, does not enable a clear distinction between the curves for the individual parameters under investigation since the curves are too close to each other. Furthermore, since the peak values in the region of small pore diameters clearly dominate the diagram, no difference can be recognised between the values obtained for larger pore diameters (all these values are very close to the abscissa).

For the sake of better readability, only the values for the pore diameter smaller then 40 nm are presented in the graphs in Figures 9 and 10. To make the distinction even clearer, a linear scale was used for the abscissa instead of a logarithmic one.

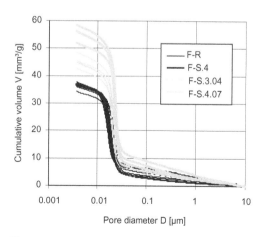

Figure 5. Cumulative pore volume curves for selected UHPC mixtures at a concrete age of 3 days.

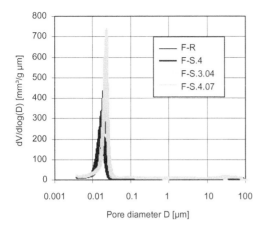

Figure 7. MIP pore size distribution for selected UHPC mixtures at a concrete age of 3 days, complete graph.

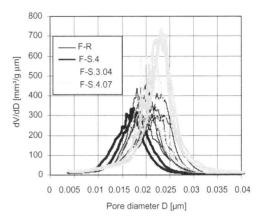

Figure 9. MIP pore size distribution for selected UHPC mixtures at a concrete age of 3 days, small pore diameters only.

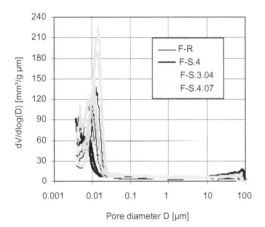

Figure 8. MIP pore size distribution for selected UHPC mixtures at a concrete age of 28 days, complete graph.

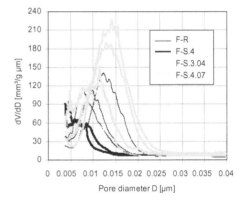

Figure 10. MIP pore size distribution for selected UHPC mixtures at a concrete age of 28 days, small pore diameters only.

The evaluation of the most frequent pore diameter (maximum of the pore size distribution curve) showed, as expected, that for all mixtures investigated this representative diameter decreases with increasing concrete age, and the corresponding height of the peaks becomes lower. The effect of internal curing on this material characteristic was much less pronounced. The most frequent pore diameters are approximately equal for the reference mixture F-R and the mixture with moderate internal curing F-S.3.04. The corresponding values are about 20 nm for the 3-day-old concrete and about 10 nm for 28-day-old concrete. However, for 3-day-old concrete, the curves for the mixture F-S.3.04 (with SAP and IC water) show considerably higher maximum values as the curves obtained for the reference mixture F-R.

The mixture F-S.4, to which SAP only was added (no IC water), yields at both concrete ages presented the lowest maxima. Furthermore, the most frequent pore diameter is clearly shifted towards the lower pore diameter values, which indicates that it possesses a finer pore system. In contrast to this, the UHPC F-S.4.07 with its relatively high addition of SAP and IC water has both the largest most frequent pore diameter and the highest maximum values (see Figures 9 and 10).

Of interest as well is the comparison of differential pore volume at the lowest measured pore diameters of 3.4 nm, which might give some indication of the possible portion of the gel pores in the UHPC matrix at an age of 28 days and so provide some additional information on the effect of internal curing on the pore system. All mixtures containing SAP show on average higher values than the reference mixture F-R. This can be interpreted as a sign of a densification of the matrix due to the internal curing.

To compare the porosity values obtained for larger pore diameters, the diagrams presented in previous figures were rescaled as shown in Figures 11 and 12. The graphs show that all mixtures with SAP possess higher volumes of larger pores (D > 40 nm) in comparison to the reference mixture. One possible explanation for this phenomenon is the presence of different artefacts caused by the SAP addition. Figure 13 shows as an example a void produced by an SAP particle. The particle expanded as it absorbed water from the fresh mix, but subsequently it seems to have shrunk while at this stage giving its water (now

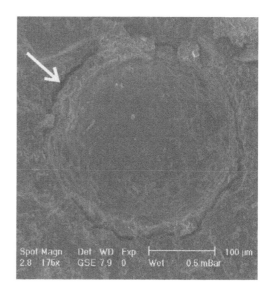

Figure 13. ESEM image of an SAP void and surrounding concrete matrix (F-S.3.04, concrete age 28 days); the arrow indicates the gap between the mineralised "mantle" of the particle and the matrix.

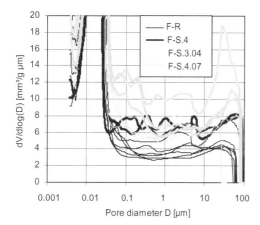

Figure 11. MIP results for selected UHPC mixtures at a concrete age of 3 days, rescaled to show pore size distribution for larger pores.

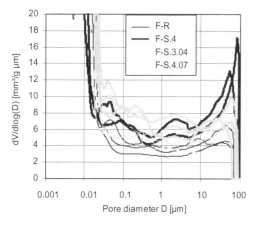

Figure 12. MIP results for selected UHPC mixtures at a concrete age of 28 days, rescaled to show pore size distribution for larger pores.

as internal curing water!) back to the hardening, self-desiccating matrix. It is likely that due to the volume reduction of the particle its mineralised "mantle" delaminates from the matrix at some places, building some kind of a split or gap. Such gaps can be misinterpreted as pores during MIP measurements.

Another particularity is a noticeable increase of the pore volume values for the pore diameters between 10 and 100 µm in the cases when SAP was used. This can be likely traced back to the fact that some quantity of small SAP particles fell in this size region.

Further investigations on the effect of internal curing on UHPC microstructure are being performed in an ongoing research project.

7 CONCLUSIONS

The following conclusions can be drawn from the results of this investigation with regard to the effect of the internal curing using SAP and additional water on behaviour of UHPC:

- The addition of SAP has no negative effect on the workability of UHPC if a sufficient amount of extra water is used.
- Measurement of the rheological behaviour of fresh concrete serves as a reliable indicator of additional water demand due to the presence of SAP.

- Internal curing using SAP and an extra amount of water reduces the autogenous shrinkage of UHPC dramatically. This effect becomes even more pronounced with increasing amounts of SAP and extra water.
- There is only a slight decrease in the compressive strength and Young's modulus of UHPC on using SAP if an appropriate amount of extra water is added for the purpose of internal curing.
- Internal curing seems to have relatively little effect on the pore size distribution in UHPC, while the total pore volume tends to increase with increased amount of extra water for internal curing. There are some indications that internal curing leads to an increase in gel pores' volume.

Investigations into the effect of internal curing on the microstructure of high-performance concretes are as yet incomplete, and further research is necessary using additional experimental techniques.

ACKNOWLEDGEMENTS

The financial support of the German Research Foundation in the framework of the Priority Program 1182 "Sustainable Building with Ultra-High-Performance Concrete (UHPC)" is gratefully acknowledged. Furthermore, the authors thank Ms Simone Hempel for performing the porosimetry measurements.

REFERENCES

Fehling, E. et al. 2005. *Entwicklung, Dauerfestigkeit und Berechnung Ultra-Hochfester Betone (UHPC)*. DFG-Forschungsbericht (Kennwort: FE 497/1-1).

Hossain, A.B. & Weiss, W.J. 2004. Assessing residual stress development and stress relaxation in restrained concrete ring specimens. *Cem. Conc. Comp* (26) 531–540.

Jensen, O.M & Hansen, P.F. 2002. Water-entrained cement-based materials—II. Experimental observations. *Cement and Concrete Research* (32) 973–978.

Jensen, O.M. & Hansen, P.F. 1995. A dilatometer for measuring autogenous deformation in hardening Portland cement paste. *Materials and Structures* (28) 406–409.

Lura, P. et al. 2006. Autogenous strain of cement pastes with superabsorbent polymers. *Volume Changes of Hardening Concrete: Testing and Mitigation. RILEM Proceedings PRO 52*, O.M. Jensen et al. (eds.), RILEM Publications S.A.R.L., 57–66.

Mechtcherine, V. et al. 2006. Internal curing by Super Absorbent Polymers—Effects on material properties of self-compacting fibre-reinforced high performance concrete. *RILEM Proceedings PRO 52*, O.M. Jensen et al. (eds.), RILEM Publications S.A.R.L., 87–96.

Mechtcherine, V., Dudziak, L. & Hempel, S. 2009. Mitigating early age shrinkage of concrete by using Super Absorbent Polymers (SAP). In: *CONCREEP-8*, T. Tanabe et al. (eds.), Taylor & Francis Group, London, 847–853.

Advances in Cement-Based Materials – van Zijl & Boshoff (eds)
© 2010 Taylor & Francis Group, London, ISBN 978-0-415-87637-7

Mechanical effects of rice husk ash in ultra-high performance concretes: A matrix study

M.D. Stults, R. Ranade & V.C. Li
Advanced Civil Engineering Materials Research Laboratory, University of Michigan, Ann Arbor, MI, USA

T.S. Rushing
US Army Engineer Research and Development Center, Vicksburg, MS, USA

ABSTRACT: Rice husk ash (RHA) has been observed to greatly improve both the mechanical and durability performance of conventional concrete mixtures, including increases in compressive strength while improving resistance to chloride ion transport and decreased permeability. This study explores the effects of using RHA as a mineral admixture in an ultra-high-performance concrete (UHPC) matrix, for which very high compressive strengths have already been achieved through dense particle packing and selection of materials. RHA is a promising material because of its widespread availability, its effectiveness as a pozzolanic material, and its potential for creating more environmentally sustainable building materials. Results of this study indicate that through the inclusion of RHA, the compressive strength of various mixtures at various water-binder ratios was improved and that RHA should be considered further for its use as a pozzolan in UHPC.

1 INTRODUCTION

Over the past several years there have been two threads of research within concrete material development, all providing beneficial information in their own right. The first of these developments has been that of ultra-high-performance concretes (UHPC). Designed with the intent of providing an alternative to steel construction and to concrete intensive applications, these UHPCs have been able to attain very high compressive and flexural strengths. These mixtures are typically characterized by very small, densely packed particles, intense cement requirements, and special curing regimes. Through creating a higher performing concrete, less concrete can be used overall, and by creating elements in a precast facility, on-site construction times can be reduced.

The second trend in research has been the inclusion of supplementary cementitious materials (SCM) into blended cements and general mixtures. SCMs provide a way to reduce the cement demand while maintaining a specified performance. Many of the materials considered for SCMs are industrial wastestream products. While these materials have very appealing benefits, they are often expensive, especially outside of industrialized nations. Additionally, many of these products require further processing before they can be used in cementing applications, which may reduce their effectiveness as a non

energy-intensive alternative to cement. In addition to reducing the cement demand of a particular mixture, these SCMs are able to give the same durability benefits of using densified (DSP) mixtures, namely reduced porosity and the associated effects.

2 RESEARCH SIGNIFICANCE

The research undertaken by this study seeks to create a bridge between the two areas of research discussed above, integrating an agricultural waste stream product into an already high-performing concrete. In doing so, it is expected that more research will be undertaken to expand the current knowledge of the use of such agricultural SCMs in creating UHPCs. This paper seeks to pose the question as well as provide some initial findings.

3 EXPERIMENTAL PROCEDURES

3.1 Specimen preparation

For this study a mixture proposed by Neeley & Walley (1995) and modified by O'Neil (2008) was further adjusted; this mixture can be found in Table 1. It should be noted that this study focuses only on the composite matrix, thus fibers were omitted from the mixture design. Each matrix prepared for this study

Table 1. Mixture proportions (ratio by mass) from O'Neil (2008).

Cement	Silica fume	Silica sand	Silica flour	Water	HRWRA
1	0.389	0.967	0.277	0.22	0.016

was mixed in a 10 liter capacity Hobart mixer. The dry constituents were added in order of increasing density, to avoid loss of fines upon mixer start up, as suggested by O'Neil (2008). The dry materials were mixed for 3 minutes to ensure uniform distribution prior to the addition of water. After 3 minutes, half of the water was added to the mixture, then all of the HRWRA, then the remainder of the water was used to rinse out the HRWRA container, to ensure that all of the measured material was added to the mixture. Depending upon the amount of water and RHA, each mixture had a different time-to-paste time, but most mixtures required 5-10 minutes of continuous mixing before a plastic paste was achieved. The amount of HRWRA added varied from mixture to mixture, as each mixture exhibited different levels of demand, but the water contributed to mixture with an increased amount of HRWRA is considered negligible, and is therefore not included in the mix water.

It should be noted, however, that some of the mixtures with 20% RHA did not achieve a plastic state prior to the completion of mixing. With the addition of more HRWRA this would have been possible, but it was observed that above a certain limit of HRWRA addition, the setting of the resulting mixture was significantly retarded, to the extent that the use of such a mixture simply would not be practical.

3.2 Evaluation of compressive strength

Within this study, the two primary tests that were completed were those for compressive strength and fracture toughness. Compression testing was carried out on a Forney F-50F-F96 force control machine. These tests were conducted on no less than three 2-inch cubes in accordance with ASTM C 109, with the exception that the loading rate of 35 ± 7 psi per second was used. Due to rate effects, the compressive strengths reported herein can be taken as conservative.

3.3 Evaluation of fracture toughness

Currently, there are no existing standard methods for measuring the stress intensity factor for concrete, so ASTM E 399, *Linear-Elastic Plane-strain Fracture Toughness K_{IC} of Metallic Materials*, is adopted for this purpose. The use of the method

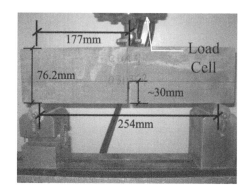

Figure 1. Experimental test set-up for measuring matrix fracture toughness.

has been previously explored and confirmed for use with cementitious matrices by Li et al. (1995). Figure 1 illustrates the testing arrangement used in this study. For each matrix evaluated in this way, at least four 3-inch by 1.5-inch by 12-inch specimens were prepared with a midspan notch cut to approximately 30 mm and loaded at a rate corresponding to 0.002 mm/s on an MTS 810 using load transducer in a closed loop cycle. The actual notch length was measured and used for calculating the fracture toughness.

The fracture toughness is calibrated for use with the geometry employed and calculated using the following equation:

$$K_Q = \frac{P_Q S}{B W^{3/2}} \cdot f\left(\frac{a}{W}\right) \qquad (1)$$

where P_Q = peak load, S = span, B = specimen thickness, W = specimen depth, and $f(a/W)$ is the geometric calibration factor, which ranged between 1.91 and 2.18 depending upon the actual crack length of each specimen.

4 EFFECTS OF RHA IN UHPC MATRIX

The study contained herein was primarily focused on the basic mechanical effects of including rice husk ash into a modified UHPC mixture design. The first part of the study sought to find an optimized value of rice husk ash addition, i.e., the mixture giving the highest compressive strength, keeping in mind workability concerns. The effect on fracture toughness was also considered, as it is an influential behavior in the tensile performance of fiber reinforced cement-based composites. The main part of this study will examine various levels of RHA addition at different water/binder (w/b) ratios. Once the critical w/b ratio and addition levels were established, mixtures considering the replacement of cement with RHA were also made and tested. After an optimized value of RHA was determined, the effects of various curing conditions were examined to see what effects, if any, the inclusion of RHA had in these curing conditions.

4.1 Optimization of RHA content and w/b ratio

The mixture proportions (Table 1) from O'Neil (2008) were adopted as the reference mixture (with w/b = 0.20, rather than 0.16 as given), with adjustments using RHA as described below. The physical and chemical properties of cementitious materials used in this study are given in Table 2.

The mixtures were cured in the "7-4-2" scheme, as established by O'Neil, i.e., the specimens were cured after set in room temperature water until seven days age, then in 90°C water for four days, followed by 90°C air for two days. The specimens were then tested at 14 days age, assuming that the elevated temperatures had allowed them to reach sufficient maturity.

Several researchers have reported that there is a significant increase in the water demand of a mixture for a given workability when RHA is used. In consideration of the increased water demand,

w/b ratios of 0.20, 0.25, and 0.30 were used, although these are typically considered much too high for typical UHPC mixtures. This increase in water demand was attributed to RHA's very high specific surface area and carbon content (Zhang & Malhotra, 1996). The typical chemistry of RHA produced by the power plant used for this study can be found among the physical and chemical properties of the other cementitious materials used in this study in Table 2. The aggregates used in this investigation were silica flour (40 µm average size) and silica sand (269 µm average size). The superplasticizer used was a low-viscosity, polycarboxylate-based liquid high range water reducing admixture.

In his seminal work, P.K. Mehta (1978) suggested that rice husk ash does not need to be ground prior to its incorporation into cementing, though it was recommended to avoid its aggressive water absorption, such as was found by the authors in the present work. The majority of researchers to date have employed various methods of grinding RHA prior to using it, but this is typically done with the motivation that smaller particles are better for packing, in keeping with the densified small particle (DSP) philosophy.

There has been no clear consensus on an optimal level of grinding, but it has been observed that even with a significant amount of grinding, the unground specific surface area of a fluidized bed burned rice husk ash is higher than that which has been finely ground (Meryman, 2007). As can be see in the SEM micrograph in Figure 2, unground RHA has what can be considered a naturally optimized surface area resulting from the honeycomb-like internal structure (right side of micrograph) and nodule external structure (lower left side of micrograph). This microstructure enables the RHA to play an important role in the hydration reaction

Table 2. Typical physical and chemical properties of cementitious (binder) materials used.

	Type H cement (C)	Silica fume (SF)	Rice husk ash (RHA)
Specific gravity	3.15	2.2	2.15
Specific surface area (m²/kg)	–	15,000–30,000	43,000
Mean particle size (µm)	30	0.4	120
Chemistry (% of mass)			
SiO₂	20.83	>85	92
CaO	62.91	0.426	2.1
Al₂O₃	2.8	0.208	1.1
Fe₂O₃	4.4	0.055	0.4
MgO	4.5	0.235	0.2
Na₂O	–	0.129	0.14
SO₃	2.9	–	–
TiO₂	–	–	–
K₂O	–	0.652	–
LOI	1.15	<0.7	4
Phases based on Bogue calculations			
C₂S	11.18		
C₃S	64.34		
C₃A	<0.01		
C₄AF	13.5		
CaSO₄	4.95		

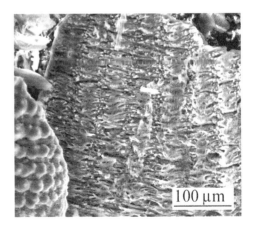

Figure 2. Micrograph of rice husk ash particles.

both by increasing its chemical reactivity and by providing host sites for the nucleation of hydration products. For this study, then, it was decided that, despite the traditional DSP theory that typically governs UHPC mixture design, a larger, unground RHA particle would be used.

The mixture proportions used for the first part of this study are similar to that given in Table 1 and are presented in Table 3. The exceptions are additions of RHA in percentages of the original binder materials (cement + silica fume), the water varied as the amount of new binder materials (cement + silica fume + RHA) changes to maintain the desired water-binder ratio, and the amount of superplasticizer used is adjusted as needed to attain a semi-flowable paste consistency.

4.2 Results of optimization of RHA and w/b ratio

As can be expected within any cementitious composite, an increase in the relative water content of a particular mixture will decrease its compressive strength; this study shows no exception. For a given w/b ratio, however, it was observed that the compressive strength of each mixture tested was generally improved by the addition of RHA (Table 3 and Figure 3), with an optimal amount dependent on the w/b ratio. From these data, it is clear that the inclusion of RHA is beneficial to these mixtures, providing a basis for its merits as a pozzolan.

As can be seen in Figure 3, there is a trend among the various mixtures indicating that as the relative amount of water in the mixture increases, the optimal (giving the highest compressive strength) level of RHA addition decreases. This is shown that while at the w/b ratio of 0.20, the compressive strength of the mixtures increased in proportion to the amount of RHA added. Conversely, at the w/b ratio of 0.30 the compressive strength of the mixtures decreased in proportion to the amount of RHA added. The behavior seen of the mixtures at the w/b ratio of 0.25 can then be thought of as a transitional region, where neither the high nor low amount of RHA addition gives as much benefit to compressive strength as does the median level of addition.

This behavior could be due to absorbent nature of the ash. Having a high amount of semi-internal surface area, RHA has been observed to be a particularly effective absorbent, and it is even marketed as such by Agrilectric Inc., the supplier of the material used in this study. The significance of this behavior is that when the water is initially added to the dry mixture, there is opportunity for the RHA to absorb an amount of water out of the mixture, preventing it from immediately participating in the hydration of the cement particles.

Once hydration begins to start in earnest, however, the hydroxide ions released into the hydration

Table 3. Proportions and results of mixtures used in study.

Mixture name	Water-binder ratio kg/kg	RHA % (C+SF)	f'c MPa	K_{IC} MPa√m	Dynamic Young's modulus GPa
Optimization study					
RHA0-20	0.20	0	129*	1.1	49
RHA5-20	0.20	5	114	1.02	46
RHA10-20	0.20	10	137*	1.01	46
RHA20-20	0.20	20	155	0.88	43
RHA0-25	0.25	0	101	0.88	41
RHA5-25	0.25	5	83	1.01	39
RHA10-25	0.25	10	145	0.88	38
RHA20-25	0.25	20	112	0.75	40
RHA0-30	0.30	0	86	0.85	37
RHA5-30	0.30	5	109	0.70	34
RHA10-30	0.30	10	94	0.82	35
RHA20-30	0.30	20	87	0.60	26
Cement replacement study					
RHA-C10-20	0.20	10	116	1.08	46
RHA-C20-20	0.20	20	112	1.07	38
RHA-C30-20	0.20	30	–	–	–
Curing study					
RHA0-20 14H₂O	0.20	0	90	1.06	53
RHA0-20 742	0.20	0	129*	1.11	49
RHA0-20 742+	0.20	0	137	1.28	45
RHA10-20 14H₂O	0.20	10	109	1.05	47
RHA10-20 742	0.20	10	137*	1.01	46
RHA10-20 742+	0.20	10	168	1.32	39
Unmodified averages					
RHA0-20			116	1.16	
RHA0-20 742			141	1.04	
RHA10-20			140	0.87	
RHA10-20 742			133	1.15	

* These values represent the average of the tests conducted in both the optimization study as well as the curing study. This was done to improve the statistical strength of the results by expanding the number of samples.

solution begin to attach to the silicate ions on the surface of the pozzolans (Taylor, 1997). For silica fume, this results in nothing more than decomposition. However, in RHA, due to that water that was initially absorbed into the semi-internal honeycomb, the decomposition of the silicate structure

24

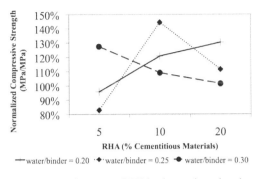

Figure 3. Performance of RHA mixtures in various levels of addition and at various water-binder ratios (normalized to relative control mixture).

of the particle may lead to the release of additional water into the matrix.

The benefit of this delayed hydration is that by this time in the hydration process, C-S-H is likely to have already formed on the surface of the cement particles, inhibiting the diffusion of this water that would allow further hydration of the cement. This water, instead may allow increased transport of the dissolved portlandite to the pozzolans where further conversion to C-S-H can occur.

This effect, however, becomes less potent as the relative amount of water provided in the mixture increases. As the mix water increases, there is an increase in abundance of "free" water available for cement hydration, and results in the formation of capillary pores. The result of which may be that the extra water may now enable the initial consumption of more silica fume, which could have otherwise been relegated to a more physical rather than pozzolanic task.

So the RHA may now act more like a weak site within the matrix, being large and relatively weak in structure, potentially having fewer products formed within it to reinforce it, and less silica fume to pack around it. In this situation, the strength gained from the additional pozzolan could be offset by the weakness of this large particle.

The above discussion may explain why as the water-binder ratio increases it is seen that the optimum level of RHA addition decreases in kind. Initially, when there is less abundance of water in the matrix (at a w/b ratio of 0.20), high amounts of RHA are more effective. But at a relatively higher amount of water within the matrix relatively lower amounts of RHA are beneficial.

A possible explanation for why the RHA5-series mixtures do not show any improvement in strength until the highest water-binder ratio level test, could be that in such a small quantity relative to the rest of the mixture, the absorption effect is not so significant as it is with greater

amounts of RHA. This would result in the matrix strength being dominated by the effects of small particle packing, leaving the RHA, though strong, as flaw-like sites being such a large and weak material.

In the above reasoning, there is the implication that there is an optimal balance for each water-binder ratio, occurring at each one for various reasons. This study, then, proposes that further research ought to be conducted to understand what the strengthening mechanisms behind the inclusion of RHA are, and how to best utilize those mechanisms for the specific aim of each application in which it is applied.

While the above discussion is highly speculative, as is any discussion on the subject of cementitious hydration kinetics, there are some observations made while using an SEM that may suggest the occurrence of such a process. Small specimens of size 0.5 cm × 0.5 cm × 0.5 cm were prepared using a water-cooled precision diamond saw and studied under an SEM in low vacuum mode. The observation is that in a mixture containing no RHA, there was a prevalence of small, well-defined crystals throughout the surface of the sample. Using the energy-dispersive x-ray spectroscopy (EDAX EDS) feature of the SEM to determine the approximate atomicity of these crystals, it was found that the ratio of atoms seems to suggest that these crystals are calcite; the micrograph and accompanying spectrum can be seen in Figures 4 & 5.

It is quite unlikely that these crystals were formed during the initial hydration, but in the process of cutting and air-drying the samples in preparation for use in the SEM, it is possible that reaction of carbonic acid (from atmospheric CO_2 dissolved in the cutting water) with CH could have enabled this. In samples containing RHA, there were very little, if any, of these crystals observed (Figure 6). This may indicate the efficacy of including RHA for consuming CH, and supports the theory presented above in explanation of this improved performance.

4.3 Replacing cement with rice husk ash

This study also sought to consider the potentials of not only using RHA to add to the matrix as a pozzolan, but also to examine the possibility of using it instead of cement, thereby reducing the overall demand. As was confirmed by the optimization results, the most prominent results of the RHA inclusion were at the lowest water-binder ratio tested, 0.20. Consequentially, this ratio will be used to evaluate the use of replacing cement with RHA. Within this part of the study, three potential mixtures were evaluated, replacing cement at 10%, 20%, and 30%, their specific mixture proportions can be found in Table 4.

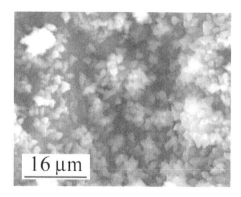

Figure 4.   Micrograph of potential calcite growth.

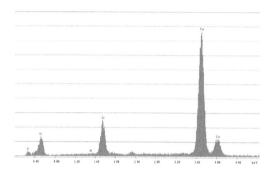

Figure 5.   EDS spectrum of potential calcite growth on specimen surface.

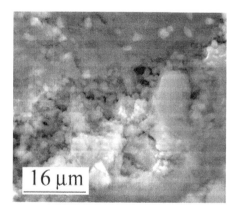

Figure 6.   Micrograph of RHA20-20 showing little potential calcite growth.

In the process of mixing it was found that the mixture containing 30% RHA replacing cement could not reach a castable state without incurring a massive amount of superplasticizer, which would have delayed the setting time enough to preclude

Table 4.   Cement replacement mixture proportions.

| | Mass ratio by mixture | | |
Constituent	RHA-C10-20	RHA-C20-20	RHA-C30-20
Cement	0.9	0.8	0.7
Silica Fume	0.389	0.389	0.389
Rice Husk Ash	0.1	0.2	0.3
Silica Flour	0.277	0.277	0.277
Silica Sand	0.967	0.967	0.967
Water	0.2778	0.2778	0.2778
HRWRA	0.055	0.084	>0.1

it from the curing scheme used in this study, and probably most other applications. As a result, this mixture was abandoned for the present study. Each of the other mixtures was cured in the 7-4-2 manner described above, and compression and fracture toughness tests were carried out as previously described.

4.4   *Effects of replacing cement with rice husk ash*

The results of our study indicate that cement can be successfully replaced by both 10% and 20% RHA (Table 3). Furthermore, it is interesting to note that despite the loss of cement, and the decrease in fracture toughness, the compressive strength of the resulting matrix was maintained. The maintenance of compressive strength is in keeping with the results above, where at a water-binder ratio of 0.20 significant increases in compressive strength were seen at both 10% and 20% addition of RHA. With the decrease in strength that comes with the removal of cement, it is seen that this is adequately compensated by the addition of RHA.

Previous researchers have found similar results for mixtures of higher water-binder ratios. Two separate studies have both found that while using ground RHA, increases in compressive were found when any cement was replaced by RHA, but both agree that the best performance was gained when the replacement level was 20% (Bui, et al. 2005, Ganesan, et al. 2008). Though the mechanisms of strengthening are assumed to be different in nature, it is interesting to see that at lower water-binder ratios the 20% level of replacement still gives good performance. The implications of this observation are that there may yet be the possibility of reducing the cement demand in UHPC design that normally requires extensive amounts of the same due to the absence of coarse aggregates and to ensure high strength.

4.5 The effect of various curing regimes on a selected RHA mixture

Building upon the results of the optimization study, it was decided that the mixture containing 10% RHA by addition at a water-binder ratio of 0.20 had the most promise for use, and therefore it was selected for this portion of the study. Although the mixture at the same relative water level containing 20% RHA performed better, the mixture with 10% was much easier to work with and cast, and it was estimated that that difference would become increasingly important when a mixture containing fibers is considered.

This portion of the study included the RHA10-20 mixture as well as its relative control, RHA0-20, to examine the effects of three different curing conditions. In addition to the 7-4-2 scheme described above, the specimens were cured for 14 days submerged in water; a 7-4-2⁺ scheme was also adopted. The latter consists of the same seven days in water and four days in 90°C water, but the final two days were adjusted to include two days in 200°C air. The increase in temperature was performed to increase the possibility of the formation of the higher ordered C-S-H phases. These mixtures were prepared with the same proportions given above, scaled for the larger mixed used; the matrices were mixed in a 2 cu. ft capacity pan mixer, with 3 cubes and 4 beams cast for each curing scheme.

4.6 Results of curing study

In completing the aforementioned curing schemes, it was observed that RHA10-20 outperformed the control for both strict water curing and the 7-4-2⁺ curing schemes, but exhibited a lower strength after the 7-4-2 regime. This can be accounted for in part by the fact that some of the RHA10-20 cubes adhered together during curing and when separated some minor damage was incurred. As a result, the averages of both mixtures' compressive strength from this portion of the study and the optimization study will be used; the initial values are reported as well, the data are presented in Table 3 above.

It is no surprise that RHA10-20 outperformed the control mixture in a water curing scheme, given the highly reactive nature of RHA resulting from its high surface area and high content of amorphous silica (Mehta & Pitt, 1976). In this, then, there is the possibility of using RHA for mixtures that must be cast-in-place where specialized curing regimes are not possible, but high strengths are still required.

Again, it is interesting to note that the mixture with RHA seems to be more receptive of the higher temperature curing conditions found in the 7-4-2⁺ curing scheme. One possible explanation for this is the increased amount of silica within in the mixture for the possible formation of higher ordered C-S-H phases.

4.7 Effects of RHA on fracture toughness

When designing materials for specific uses, it is not always enough to improve the compressive strength of the bulk material, but there are occasions on which it is important to adjust the compressive and tensile performances independent of each other. General material design theory has accepted the conclusion that an increase in compressive strength leads to an increase in fracture toughness. However, the effect of RHA on the matrices in which it has been included proves itself to be a unique material in that the fracture toughness of the matrix, generally speaking, decreases in proportion to the amount of RHA used while the compressive strength increases; these results can be found in Table 3. Another researcher (Giaccio, 2007) found similar behavior in conventional high-strength concretes incorporating RHA, ranging in water-binder ratios from 0.28 to 0.56, though the reductions in fracture energy reported therein were not as significant as those found in this study.

The decreases in fracture toughness are most consistent at the lowest water-binder ratio, 0.20, which can be attributed to this base matrix being the most densely packed. With the inclusion of these larger and weaker structured particles, though the compressive strength may increase because of hydration products forming in and around this weak carbon structure, when put under tension this "composite" is easily pulled apart. Due to the densification of the matrix around it, the bulk material fracture toughness is quite high, as indicated by the control, yet the toughness of the individual RHA could be relatively smaller. This contrast between the toughness of each material could lead to stress concentration at these weaker sites, leading to the significant reductions in the overall matrix fracture toughness observed here. It is also possible that the shape of the RHA may have a tendency to induce flaw-like defects more sensitive to tension than compression, giving similar behavior.

4.8 Effects of RHA on dynamic Young's modulus

From the notched beams broken for finding the fracture toughness, 38 mm × 38 mm × 127 mm prisms were cut and the dynamic Young's Modulus was calculated using the longitudinal resonance frequency, in accordance with ASTM C 215, the results of which are given in table 3. It will be seen that there are no appreciable drops in the stiffness of the material with the addition of RHA, the greatest being a 20% reduction for RHA20-30. This shows, then, that though the density of the matrix is decreased by the inclusion of RHA, the stiffness of the material is maintained.

When considering the stiffness of the matrices in which cement is replaced by RHA, it will be noticed that there are some significant decreases. This can be expected since there is less cement and therefore less unhydrated cement in the matrix, which can result in lower modulus, as unhydrated cement has been attributed as a stiffening factor in low water-binder ratio mixtures (O'Neil, 2008). If RHA is to be used as a cement replacement, we recommended that no more than 10% be replaced for high performance mixtures, as the Young's modulus is an important factor for structural design.

5 SUMMARY AND CONCLUSIONS

In the this study of the inclusion of RHA into potential UHPC matrices, we observed the following:

1. RHA can be used in an unground form with beneficial results.
2. RHA is an effective pozzolan, even at lower water-binder ratios as the addition of RHA into matrices maintained or increased compressive strength.
3. The addition of RHA did not lead to significant reductions in matrix stiffness, but did lead to reductions in the material's fracture toughness.
4. RHA can be used as a replacement for cement even at low water-binder ratios.
5. RHA may allow the independent tailoring of matrix compressive and tensile performances.
6. The increased water demand of RHA may be limited by the amount added as well as the desired water-binder ratio of the matrix.

From the above results it can be seen that though unsupported by the prevailing theories in ultra high performance material design, rice husk ash can play an important role in designing more sustainable materials for the future. Even if functioning simply as a material included to increase the yield of a mixture, RHA is effective at maintaining, if not increasing compressive strength. Additionally, by decreasing the density of the matrix while maintaining both stiffness and compressive strength at lower water-binder ratios, the inclusion of RHA could mean reduced dead-weight in elements made therefrom, potentially reducing transport weight in precast elements while maintaining the desired performance. That rice husk ash mixtures are also more responsive to higher heat treatment could also be of benefit to precast applications.

Aside from mechanical performance benefits, it should be noted that, at the present time, RHA is an agricultural waste-stream product that is available in many countries where other pozzolans may not be as common or inexpensive. In instances such as Agrilectric Power Partners, the provider of the RHA used in this study, the hulls from the milled rice have already been used to generate power and the resulting ash that can be used in concreting applications is then a tertiary use, maximizing the benefit of what would otherwise be waste.

ACKNOWLEDGEMENTS

The authors would like to express their gratitude to the US Army Engineer Research and Development Center for funding this work, Dr. Wenping Li of Agrilectric R&D for providing the rice husk ash used in this study, Jim Fete of Beckman Coulter for his assistance in the particle size characterization of the materials used for this study, and Daoxuan "Sven" Yan for his invaluable assistance in preparing the mixturees. The authors also wish to acknowledge NSF grant #DMR-0320740 for the purchase of the Quanta 3D ESEM & Edax EDS instruments used in this study.

REFERENCES

Bui, D.D., Hu, J. & Stroeven, P. 2005. Particle Size Effect on the Strength of Rice Husk Ash Blended Gap-Graded Portland Cement Concrete, *Cement & Concrete Composites*, Vol. 27, pp. 357–366.

Ganesan, K., Rajagopal, K. & Thangavel, K. 2008. Rice Husk Ash Blended Cement: Assessment of Optimal Level of Replacement for Strength and Permeability Properties of Concrete, *Construction and Building Materials*, Vol. 22, No. 8, pp. 1675–1679.

Giaccio, G., Rodriguez de Sensale, G. & Zerbino, R. 2007. Failure Mechanism of Normal and High-Strength Concrete with Rice-Husk Ash, *Cement & Concrete Composites*, Vol. 29, pp. 566–574.

Li, V.C., Mishra, D.K. & Wu, H.C. 1995. Matrix Design for Pseudo Strain-Hardening Fiber Reinforced Cementitious Composites, *RILEM Materials and Structures*, Vol. 28, No. 183, pp. 586–595.

Mehta, P.K. & Pitt, N. 1976. Energy and Industrial Materials Made from Crop Residues, *Resource Recovery and Conservation*, Vol. 2, pp. 23–28.

Mehta, P.K. 1978. Siliceous Ashes and Hydraulic Cements Produced Therefrom, US Patent 4,105,459.

Meryman, H. 2007. Concrete for a Warming World, M.S.E. Thesis, Department of Civil and Environmental Engineering, University of California, Berkeley.

Neeley, B.D. & Walley, D.M. 1995. Very High-Strength Concrete, *The Military Engineer*, Vol. 87, No. 572.

O'Neil, E.F. 2008. On Engineering the Microstructure of High-Performance Concretes to Improve Strength Rheology, Toughness, and Frangibility, Ph.D. Thesis, Department of Civil Engineering, Northwestern University, Evanston, Illinois.

Taylor, H.F.W. 1997. *Cement Chemistry*, 2nd Ed., London: Thomas Telford Publishing.

Zhang, M. & Mahotra, V.M. 1996. High Performance Concrete Incorporating Rice Husk Ash as a Supplementary Cementing Material, *ACI Materials Journal*, Vol. 93, No. 6, pp. 629–636.

Smart dynamic concrete: New approach for the ready-mixed industry

Rashid Jaffer, Roberta Magarotto & Joana Roncero
BASF Construction Chemicals, South Africa

ABSTRACT: The use of Self-Compacting Concrete (SCC) as an "everyday" concrete in the Ready-Mixed Concrete Industry is still limited and focused on certain specific applications. The introduction of Smart Dynamic Construction, an innovative technology able to provide the benefits of the traditionally vibrated concrete along with those of SCC including ease of production with enough robustness for everyday use in a cost-effective way, can contribute to expand the use of SCC in the Ready Mixed Industry.

In the present paper, some practical experiences performed in Ready-Mixed Concrete plants are shown. The results shown that the production of Smart Dynamic Concrete using fines lower than 380 kg/m^3 is feasible by means of an innovative Viscosity Modifying Admixture called RheoMATRIX©.

1 INTRODUCTION

The use of Self-Compacting Concrete (SCC) in the Precast Industry has increased significantly from its introduction in the market in the mid '90 s'. It is estimated that, in Europe, more than 50% of the concrete used for structural Precast elements is SCC. Nevertheless, the picture is quite different in the Ready Mixed Concrete Industry where less than 1% of the concrete produced is SCC.

In spite that the guidelines published (1–3) define a wide range of SCC both considering its composition, as well as its fresh properties and application, the reality of the market indicates that SCCs are mainly characterized by high content of fines (particles with diameter lower than 0,125 mm), usually in the range of 450 to 600 kg/m^3, along with low volume of coarse aggregates. This high content of fines, which provides high volume of cement paste (i.e., between 35–40%), is needed in order to provide the appropriate cohesion able of preventing bleeding and/or segregation.

Although the composition of these SCCs varies significantly from one country to another (4, 5), in general, the content of fines is usually higher than 500 kg/m^3 and it is characterized by high plastic viscosity being, mainly, suitable for heavily reinforced structures. Nowadays, the use of SCC with high content of fines represents a technology extensively implemented specially in the Precast industry. Along these lines, these concretes provide high mechanical and durability performance, as well as, the required fresh properties.

Nevertheless, the statistics of ERMCO (6) reveal that the mechanical requirements of the "everyday" concrete are significantly lower that those provided by "standard" SCC. Along these lines, SCC is usually supplied with excess in the mechanical and durability requirements in order to fulfill the fresh requirements implying higher cost of SCC than conventional vibrated concrete. This is one of the most important reasons of the limited use of SCC in Ready Mixed Industry although not the only one.

Among these reasons, one of the most important is the technical difficulties for preparing, with existing materials, robust SCC with low content of fines (cement + filler), even using Viscosity Modifying Admixtures (VMA) currently available in the market. On the other hand, the characteristics of locally-available materials (i.e., cement, aggregates and filler) for SCC production are, not always, the most appropriate and, consequently, a lot of laboratory work is needed in order to adjust the mix design with the aim of obtaining stable concrete. Also, variations in the moisture content of the aggregates and, especially, of the filler significantly influence the stability of the SCC, implying several adjustments in the mix design.

Additionally, the use of extra fines (filler) implies, sometimes, new silos along with further control of the concrete in the plant. Therefore, this, along with the high contents of cement usually used, lead to an increase in production costs.

Consequently, the use of SCC as an "every-day concrete" is, nowadays, difficult especially in the case of the Ready Mixed Concrete Industry. Along these lines, the development of the Smart Dynamic Construction (SDC) system, by means of an innovative VMAs that allows designing concrete with self-compacting properties without incorporating extra fines (filler) and, consequently, having a total content of fines (cement + filler) in the range of 350–380 kg/m³, providing the required stability in the concrete. The use of this new family of VMAs called RheoMATRIX, along with superplasticisers especially designed for SCC, leads to an improvement in the cost-effectiveness of the concrete that can contribute to expand the use of SCC in the Ready Mixed Concrete Industry.

In the present study, some practical experiences with SDC on Ready-Mixed plants are presented. Results show that the preparation of concretes with low fines content but maintaining self-compacting properties can be feasible.

2 EXPERIENCES IN READY-MIXED CONCRETE PLANTS

A new VMA able to provide enough stability (bleeding and segregation) in concrete but without sacrificing the fluidity of the concrete as usually occurs with conventional VMAs was developed in the facilities of BASF Construction Chemicals. After the laboratory phase, several series of field tests in Ready Mixed plants was performed at various global sites. In the present paper, details and results of four of these experiences performed in Spain and South Africa are presented. Note that the aim of these tests is different in each case. Further field tests in other countries are also shown in (7, 8).

2.1 Decrease of cement content to obtain conventional mechanical performance

The objective of these industrial tests consists of minimizing the cement content from a reference SCC with the aim of obtaining conventional strength in the range of 25–35 MPa. Tests in two different plants (A and B) were performed.

In plant A, the reference composition of the SCC consists of 412 kg/m³ of cement without filler using crushed limestone aggregates. The composition of the reference C used in the plant is shown in Table 1 (Reference-A). The objective consists on decreasing the cement content in order to obtain concrete with self-compacting properties in the range of 25–30 MPa (SDC).

The use of RheoMATRIX has permitted decreasing the cement content up to 310 kg/m³ while maintaining the stability of the concrete without bleeding and/or segregation. The composition and properties of this concrete are shown in Table 1 (SDC-A). Note, also, that the use of the new VMA has avoided the presence of fly ash in the surface of the concrete contributing to

Table 1. Composition in kg/m³ and properties of the reference SCC and SDC with low fines content in plant A.

Composition in kg/m³	Reference-A	SDC-A
0–2 mm sand	330	350
0–4 mm sand	760	840
4–12 mm gravel	700	700
CEM II B/M 42.5R	412	310
Water	175	190
Superplasticizer (Glenium type)	7,6	5,5
VMA (RheoMATRIX)	–	0,5
Properties		
Slump flow	63 cm	57 cm
Appearance	Slight bleeding. Presence of fly ash on the surface	Good without bleeding or segregation
Compressive strength at: 7 days	39,9 MPa	25,6 MPa

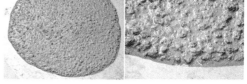

a) Reference-A

b) SDC-A

Figure 1. Appearance of the SCC of reference (a) and SDC with low fines content (b) prepared in plant A.

maintain them integrated in the bulk concrete. It is important to note also the appearance of SDC-A concrete, being less sticky than SCC with high fines content. The appearance of the concretes described in Table 1 is shown in Figure 1.

Similar experience was performed in plant B where, as in the previous case, the aim was to decrease the cement content in order to obtain a concrete with self-compacting properties but with compressive strengths in the range of 25–30 MPa. The composition of the reference SCC, with 400 kg/m³ of cement, is shown in Table 2.

Table 2. Composition in kg/m³ and properties of the reference SCC and SDC with low fines content in plant B.

Composition in kg/m³	Reference-B	SDC-B
0–2 mm sand	558	558
0–4 mm sand	596	646
4–10 mm gravel	604	604
CEM II B/M (V-S-LL) 42.5R	400	350
Water	260	255
Superplasticizer (Glenium type)	7,0	7,0
AMV (RheoMATRIX)	–	1,0
Properties		
Slump flow	77 cm	73 cm
Appearance	Good without bleeding or segregation	Good without bleeding or segregation
Compressive strength at: 7 days	38,0 MPa	22.0 MPa
28 days	47,1 MPa	32.2 MPa

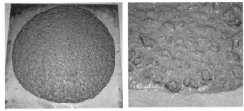

a) Reference-C

b) SDC-C

Figure 2. Appearance of the SCC of reference (a) and SDC with low fines content (b) prepared in plant B.

Table 3. Composition in kg/m³ and properties of the reference SCC and SDC with low fines content in plant C.

Composition, in kg/m³	Reference-C	SDC-C
0–6 mm natural siliceous sand	1015	1150
6–12 mm crushed limestone gravel	680	620
Limestone filler	170	–
CEM II A/V 42.5R	330	400
Water	196	225
Superplasticizer (Glenium type)	7	8,5
AMV (RheoMATRIX)		0,40
Properties		
Slump flow	63 cm	62 cm
Compressive strength at: 1 day	15,5 MPa	13,6 MPa
7 days	29,0 MPa	29,2 MPa
28 days	33,8 MPa	33,2 MPa

Table 4. Composition in kg/m³ and properties of the reference SCC and SDC with low fines content in plant D.

Composition in kg/m³	Reference D	SDC-D
0–2 mm natural sand	275	235
0–5 mm dolomite crushed sand	835	995
4–13 mm dolomite crushed stone	710	710
Cem I 42,5	318.5	220.5
PFa	136.5	94.5
Water	195	205
Superplasticiser (Glenium type)	7.73	5.6
VMA (Rheomatrix)	–	1.2
Properties		
Slump Flow	710 mm	690 mm
Viscosity (V Funnel)	10 sec	14 sec
Compressive strength at: 7 days	29.3 MPa	19.4 MPa
28 days	42.8 MPa	28.5 MPa

2.2 Replacement of filler in SCC maintaining mechanical requirements

In plant C, the objective of this series of tests consists of eliminating the filler in the composition of the reference SCC providing an alternative composition with similar strength development. This will avoid the requirement of one silo for filler that would be used for a second cement type and, therefore, this would improve the logistics of the plant. It is important to note that, in this case, the sand available is of natural siliceous origin with fines content (<0.125 mm) of about 1%. The composition of the reference concrete is shown in Table 3 and, as can be seen, 330 kg/m³ of cement and 170 kg/m³ of limestone filler are being used.

The use of the new VMA has permitted a new composition of concrete with self-compacting properties but using 100 kg less of fines; in this case 400 kg/m³ of cement, when compared to the reference concrete. The appearance of the concretes is shown in Figure 2. As can be seen, both concretes show similar appearance.

Reference Mix-D: Cast in place

SDC-D: Cast in place

SDC-D: Finishing by rodding surface

Figure 3.   Placement and finishing of SCC.

2.3 *Decrease in total cementitious content and decrease in fine filler ratio to obtain conventional mechanical performance*

The primary objective of these trials was to mini-mise the total cementitious content to obtain the design requirement of 25–30 MPa. The laboratory trials highlighted an improved paste quality using the VMA, RheoMATRIX. Further refinement of the mix design concentrated on the reduction of the fine filler proportion of the total sand content.

The plant trials were done through a conventional dry batch plant, with mixing been done in the ready-mix truck. To ensure adequate mixing, and concrete homogeinity the concrete was batched in three, two cubic metre increments. The composition of the reference used in the plant I shown in Table 4 (Reference D). The objective to decrease the total cemenititious content, and reduce the fine filler proportion with self-compacting properties in the range 25–30 MPa is highlighted in SDC-D.

The use of Rheomatrix improved the paste quality adequately to reduce the high cementitious content required for workability and mix cohesion. The reduction of total cementitious content by 140 kg/m³, significantly improved the mix economics. In the conventional approach the crushed aggregate required additional fines to improve the cohesion, and reduce bleeding of the concrete. Applying the VMA RheoMATRIX allowed reducing the blend ratio of the fine filler sand by 6% without any significant decrease in the concrete workability, and stability.

The overall mix economics was reduced by 19%.

3   CONCLUSIONS

Self Compacting Concrete has been limited by high material costs particularly cement, to specialised applications. Conventional applications in the of 25–30 MPa concrete range, has not been cost effective due to the high fines content required to produce self compacting concrete. The use of a new generation Viscosity Modifying Admixture, called RheoMATRIX increases the scope of applying the Self Compacting Concrete for conventional concrete applications, with all the benefits associated with this new innovation. Now self compacting concrete can be produced with a lower requirement for total fines i.e., less than 380 kg/m³.

This concrete, Called Smart Dynamic Concrete, can contribute to significantly expand the use of Self-Compacting Concrete in the Ready-Mixed Concrete Industry as an "every day concrete".

REFERENCES

(1) Japan Society of Civil Engineers, 'Recommendation for Self-Compacting Concrete', Tokyo, Japan, August 1999.
(2) EFNARC (European Federation for Specialist Construction Chemicals and Concrete Systems), 'The European Guidelines for Self-Compacting Concrete. Specification, Production and Use', May 2005.

(3) ACI Committee 237, 'Self-Consolidating Concrete', technical committee document 237R-07, 2007, 30 pp.

(4) Wallevik, O.H., 'Rheology—A Scientific Approach to Develop Self-Compacting Concrete', 3rd International RILEM Symposium on Self-Compacting Concrete, Reykjavik, Iceland, August 2003, p. 10.

(5) Zerbino, R., Agulló, L., Barragán, B., Garcia, T. and Gettu, R. 'Caracterización reológica de Hormigones Autocompactables', published by Dept. of Construction Engineering, Technical University of Catalunya, 2006.

(6) ERMCO, 'European Ready-Mixed Concrete Industry Statistics. Year 2006', 2007.

(7) Khurana, R.S., Magarotto, R. and Moro, S., 'Smart Dynamic Concrete: an innovative approach for the construction industry', to be presented at 7th International Congress on Concrete: Construction's Sustainable Option, Dundee, Scotland, July 2008.

(8) Corradi, M., Khurana, R.S. and Magarotto, R., 'Low-fines Self-Compacting Concrete', 5th Intnl. RILEM Symposium on SCC, Ghent (Belgium), 2007.

Cracking properties of steel bar-reinforced expansive SHCC beams with chemical prestress

Hiroo Takada, Yuji Takahashi, Yuichiro Yamada, Yukio Asano & Keitetsu Rokugo
Department of Civil Engineering, Gifu University, Gifu, Japan

ABSTRACT: An expansive additive was added to strain-hardening fiber-reinforced cement composites (expansive SHCC) to achieve a higher cracking strength when used in combination with reinforcing bars. Flexural loading tests on steel bar-reinforced SHCC beams were conducted after selecting the expansive additive contents based on length change tests on the material and uniaxial tension tests on steel bar-reinforced dumbbell-shaped specimens. Within the range of this study, as the expansive additive content increased from 0% to 10%, the cracking load significantly increased, the number of cracks during loading decreased, and the deflection at the time of tensile reinforcement yielding decreased.

1 INTRODUCTION

Strain-hardening fiber-reinforced cement composites (SHCC) are ductile materials that form densely distributed fine cracks, showing quasi-strain-hardening properties under uniaxial tensile stress. SHCC characterized by their self-control of crack width have been applied to various structures to utilize their high tensile resistance and deformability, as well as to inhibit penetration of deteriorative substances into concrete (Kanda et al. 2006). SHCC include ECC (engineered cementitious composites), which were developed by Li (2002) of Michigan University et al. based on micromechanics and fracture mechanics as the design principles. When ECC is combined with reinforcing bars, it is reported to form numerous distributed cracks and continues to deform in an integrated manner with the bars even after their yielding (Li 2002).

In this study, SHCC containing an expansive additive (EX) were defined as expansive SHCC and used to apply chemical prestress to steel bar-reinforced SHCC beams to achieve a higher cracking strength. This is intended for reinforced concrete structures that are required to retain high watertightness for a long time, such as water tanks. The cracking behavior of such beams was investigated by flexural loading tests.

2 OUTLINE OF EXPERIMENTS

2.1 *Type and proportioning of SHCC*

A volume ratio of 1.5% was selected as a standard for high-density polyethylene fibers for SHCC. The EX was added at different ratios to the mass

Table 1. Materials for SHCC.

Material	Properties and specifications
High-density Polyethylene fibers (PE)	Diameter: 0.012 mm, Length: 12 mm Density: 0.98 g/cm³, Strength: 2.6 GPa Elastic modulus: 88 GPa
Cement (C)	JIS R5210 High-early-strength Portlandcement, Density: 3.13 g/cm³
Expansive additive (EX)	JIS A6202 Ettringite and calcium hydrate combined formation type Density: 3.05 g/cm³
Silica Sand (S)	Maximum diameter: 0.2 mm, Density: 2.60 g/cm³
Admixture (Ad)	JIS A6204 High-range water reducing admixture, Polycarboxylate type
Methylcellulose (MC)	Nonionic water-soluble type

of the binder in place of cement so that the binder content would become constant. Table 1 gives the materials for the SHCC. Table 2 contains the standard mixture proportions.

2.2 *Length change of expansive SHCC*

Length change tests were conducted on SHCC prisms of size 40 by 40 by 160 mm containing an EX with a ratio of 0, 2, 6, and 10% in accordance with JIS A1129-3 (Dial gauge method). The length change was measured with respect to the

Table 2. Standard mixture proportions.

| W/B (%) | Fiber volume (%) | Unit content (kg/m³) | | | | | |
		W*	Binder C + EX	S	Ad	MC	PE
30	1.5	379	1264	395	37.9	0.9	14.6

W* is water with dissolved admixture.

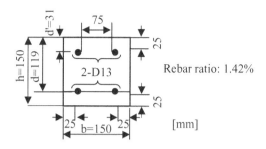

Rebar ratio: 1.42%

Figure 1. Cross-section of beam.

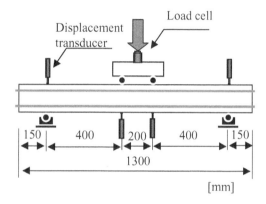

Figure 2. Test setup.

as-demolded datum length. After being water-cured at 20°C until an age of 7 days, specimens were stored in an atmosphere with a humidity of 58% equilibrated using a saturated solution of sodium bromide in a curing room at 20°C in accordance with Appendix 1 of JIS A6202. Similar length change tests were also conducted on the matrix mortar of SHCC specimens with the same EX content but excluding fibers to examine the effect of fibers on the length change. Cylindrical specimens were simultaneously fabricated to confirm the effect of the EX on the compressive strength.

2.3 Uniaxial tension tests on dumbbell-shaped specimens

Based on the information on the EX content and length change obtained in section 2.2, uniaxial tension tests were conducted on dumbbell-shaped SHCC specimens with an EX content of 0, 6, 8, and 10%. The test areas of dumbbell specimens measured 30 by 30 mm in cross-sectional area and 80 mm in length as recommended by JSCE (2008). The SHCC was placed in molds while being vibrated with a table vibrator specified in JIS R5201. Specimens were demolded at an age of 1 day, cured in water at 20°C until an age of 7 days, and then air-cured in the laboratory until the testing at an age of 13 days. Uniaxial tension tests were also conducted on dumbbell-shaped specimens with an EX content of 0 and 8%, each having a D6 reinforcing bar (SD295A) in the longitudinal direction near the center, in order to examine their cracking properties with expansion being restrained by reinforcing bars.

2.4 Flexural loading test on steel bar-reinforced SHCC beams

2.4.1 Fabrication of specimens and measurement of chemical prestrain

Steel bar-reinforced SHCC beams were fabricated with an EX content of 0, 6, 8, and 10%. The cross-section of the beams is shown in Figure 1. Longitudinal bars were provided near the top and bottom sides to prevent beam warping due to expansion, while excluding transverse bars. The SHCC was placed in inclined molds, being allowed to flow continuously from one end toward the other and consolidated with a form vibrator. After being demolded at an age of 1 day, specimens were wet cloth-cured in a curing room at 20°C until an age of 7 days.

Two strain gauges with a test length of 3 mm were glued face-to-face to a point at a rebar of each specimen and covered with waterproofing tape to measure the strain due to the EX. Thermocouples were also provided at the same positions to measure the temperature of the members near the strain gauges.

2.4.2 Flexural loading tests

Two beam specimens were fabricated for each set of conditions to test one at an age of 12 days to examine the changes in the chemical prestress and mechanical properties over time. Four-point loading was applied to beam specimens as shown in Figure 2, with the loading and support span lengths being 200 and 1,000 mm, respectively. The load, the displacement at the supports and loading points, and the strain on the compression edge were measured using load cells, displacement transducers, and strain gauges, respectively. In order to pinpoint the load at the bending crack onset, four strain gauges with a test length of 60 mm were applied to the tension edge, with ends overlapping, so that the entire uniform bending moment zone could be covered.

In the loading tests, loads were increased in increments up to the crack onset, up to 30 kN, and up to 40 kN. Specimens were then unloaded and reloaded up to 50 kN, up to 60 kN, and to yielding. After being re-unloaded, specimens were loaded again to a deflection of 10 mm and then to failure. The positions and widths of cracks on the tension edge within the loading span were observed and measured at each stage by using a microscope.

3 TEST RESULTS

3.1 Length change tests of expansive SHCC

Figure 3 shows the results of length change tests. "SH" and "MM" in the legend denote the SHCC and matrix mortar, respectively, and the attached figures denote the EX content. All specimens tended to shrink in an atmosphere with an equilibrium relative humidity of 58% after water curing. As for the relationship between the EX content and the length change, the effect of the EX was scarcely found with an EX content of 2%, whereas the shrinkage was mitigated by 300×10^{-6} with an EX content of 6%. With an EX content of 10%, an expansion ratio of around 800×10^{-6} was retained even after shrinkage. As to the effect of fibers, the length changes of the SHCC were smaller than those of the matrix mortar, suggesting that fibers in the SHCC restrain both the shrinkage and EX-induced expansion. The shrinkage of the SHCC

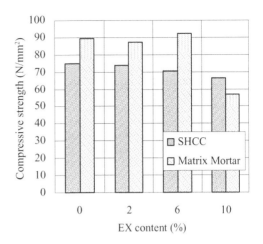

Figure 4. Relationship between EX content and compressive strength.

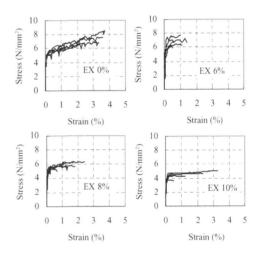

Figure 5. Uniaxial tensile stress-strain curves.

with no EX was equivalent to that of the matrix mortar containing 6% EX.

Figure 4 shows the relationship between the EX content and the compressive strength. With an EX content of 10%, the compressive strength of the matrix mortar significantly decreases due to excessive expansion, but the reductions in the compressive strength of the SHCC are moderate due to the restraint of the expansion as shown in Figure 3.

3.2 Uniaxial tension tests on dumbbell-shaped specimens

Figure 5 shows the results of uniaxial tension tests on dumbbell-shaped SHCC specimens with an EX content of 0, 6, 8, and 10%. The results

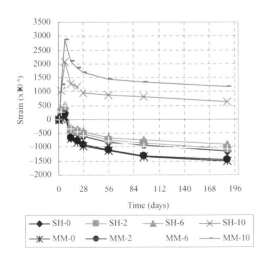

Figure 3. Results of length change tests.

of uniaxial tension tests on steel bar-reinforced dumbbell-shaped SHCC specimens are shown in Figure 6. No appreciable effect of adding an EX was observed in free-expanded SHCC specimens without reinforcing bars. In specimens with an EX content of 10%, multiple fine cracks concentrated near the first crack. As shown in Figures 3, 4, and 5, this is presumably because an EX content of 10%, which caused large free expansion, reduced the strength, resulting in small differences between the cracking strength and tensile strength as reported in past studies, thereby inhibiting the cracks to disperse.

When reinforced with steel bars, however, specimens with an EX content of 8% showed higher cracking strengths than specimens with no EX as given in Table 3, presumably because the EX-induced expansion of the SHCC was restrained by reinforcing bars, generating chemical prestress.

3.3 *Flexural loading tests on steel bar-reinforced SHCC beams*

3.3.1 *Measurement of steel strain*
Figure 7 shows the time-related changes as the average of the readings of two strain gauges glued to compression bars in each beam specimen.

Normal SHCC beams with no EX showed a steel compressive strain of 356×10^{-6} due to autogenous and drying shrinkage, whereas the steel compressive strain of expansive SHCC beams with an EX content of 6% was only 5×10^{-6}, being smaller by about 350×10^{-6}. With a higher EX content of 8% and 10%, steel tensile strain of 178×10^{-6} and 347×10^{-6}, respectively, were measured, which correspond to chemical prestresses in the SHCC of 0.8 N/mm^2 and 1.6 N/mm^2, respectively.

For fabricating beam specimens, a batch size of 70 liters of the SHCC was mixed in a pan-type mixer with a capacity of 100 liters. Since the amount of the binder per batch was 88.48 kg, 79.63 kg of cement and 8.85 kg of the EX were mixed for the case of an EX content of 10% to fabricate two beams, for each using approximately 30 liters of the mixture. It was expected that the heat due to the large amount of binder would accelerate the chemical reaction, leading to a length change behavior different from that of small specimens previously tested. However, the relationship between the EX content and the expansion of SHCC beams turned out to be of a similar tendency as the results obtained from 40 by 40 by

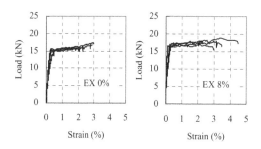

Figure 6. Uniaxial tensile load-strain curves of steel bar reinforced dumbbell-shaped SHCC specimens.

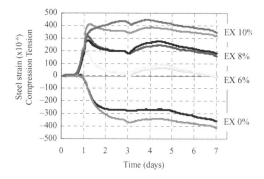

Figure 7. Readings of strain gauges glued to steel bars in each beam specimen.

Table 3. Results of uniaxial tension tests on steel bar re-inforced dumbbell-shaped SHCC specimens.

Specimen no.	1	2	3	4	5	Mean	Standard deviation
EX content 0%							
Cracking load (kN)	3.7	2.4	3.1	3.8	3.9	3.4	0.56
Yield load (kN)	15.1	15.6	15.5	14.9	15.1	15.2	0.27
Maximum load (kN)	16.8	17.4	16.0	15.8	16.0	16.4	0.61
Ultimate strain (%)	2.9	3.0	2.1	2.0	2.3	2.4	0.41
EX content 8%							
Cracking load (kN)	5.1	5.4	4.7	4.6	4.3	4.8	0.43
Yield load (kN)	15.4	14.6	14.8	15.0	15.6	15.1	0.41
Maximum load (kN)	16.2	15.3	16.1	15.3	16.9	16.0	0.68
Ultimate strain (%)	2.3	2.1	3.1	0.6	3.4	2.3	0.98

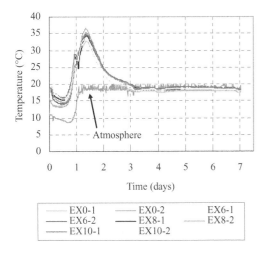

Figure 8. Temperature of the steel bars in SHCC and in the atmosphere.

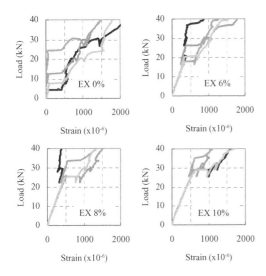

Figure 9. Relationship between load and measurements of four strain gauges glued to tension surface.

Table 4. Loads and deflections of beams at loading points.

EX content (%)	Initial crack load (kN)	Yielding Load (kN)	Yielding Deflection (mm)	Maximum load (kN)
0	4.31	69.4	4.1	86.2
6	18.1	66.1	3.7	77.2
8	25.6	65.1	3.5	75.8
10	28.7	64.3	3.2	75.1

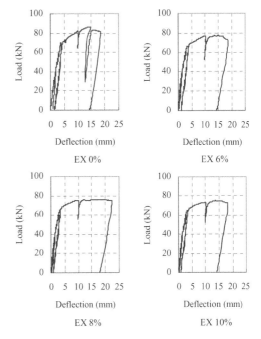

Figure 10. Load-deflection curves.

160 mm specimens described in section 3.1. Also, the temperature of the members measured near the compression bar under a cover depth of 25 mm showed similar histories regardless of the EX content, reaching the peak of around 35°C at an age of about 1.3 days as shown in Figure 8.

3.3.2 Flexural loading tests
Figure 9 shows the relationship between the load and the measurements of four strain gauges glued to the tension surface within the loading span of each specimen. When loading a specimen while monitoring the relationship between the load and the four strain gauge measurements, the four lines elastically rose with the same slope, but one of the paths instantaneously shifted sideways at a certain load. The loading was stopped to microscopically observe the test zone of the strain gauge showing an abrupt increase in the measurement, and a crack was recognized in the zone. It was also found that, when a crack opens in an adjacent strain gauge zone, the gauge shows a reduction in the strain reading. Thus, the load and the location of crack onset were easily pinpointed by this test.

Also, the four strain gauges simultaneously showed large strains at the time of beam yielding,

39

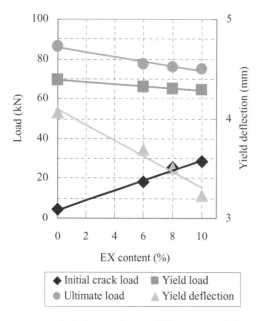

Figure 11. Relationship between EX content and characteristic values of the bending tests.

Legend:
◆ Initial crack load   ■ Yield load
● Ultimate load   ▲ Yield deflection

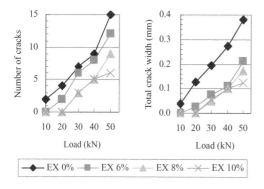

Figure 12. Number of cracks and total crack width related to load.

Legend: ◆ EX 0% ■ EX 6% ▲ EX 8% ✕ EX 10%

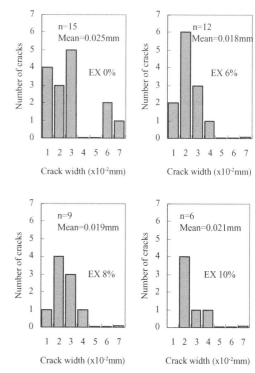

Figure 13. Histograms of crack widths.

allowing the monitoring of the instant when the beam shifted to the plastic deformation mode. The strain gauges can follow the deformation of the member, presumably because the SHCC is capable of deforming beyond the elastic deformation of the reinforcement as shown in Figure 5, inhibiting crack localization at the yield point of the reinforcement, while keeping the openings small due to the multiple fine cracks.

Table 4 gives the loads applied to the beams at the loading points and the deflection of the beams pinpointed by the above-mentioned method.

Figure 10 shows the relationship between the load and the deflection during flexural loading. Figure 11 shows the relationships between the EX content and characteristic values of the bending tests. Figure 12 shows the number of cracks and the total crack width related to the load. As the EX content increased, the yielding load of the reinforcement slightly decreased, and the maximum load (corresponding to the initiation point of SHCC crushing) also decreased. These can be attributed to the fact that the chemical prestress slightly shifted the neutral axis toward the tension side, thereby reducing the distance between the resultant compressive force and resultant tensile force in the cross-section (arm length). As the EX content increased, the cracking load significantly increased due to the effect of shrinkage compensation or chemical prestress. Also, an increase of the EX content reduced the number of cracks, thereby substantially retaining the rigidity of the beams. The deflection at the point of tensile reinforcement yielding was thus definitely reduced. Figure 13 shows the histograms of the widths of cracks that occurred in the loading span of steel bar-reinforced SHCC beams under a load of 50 kN.

Though no significant difference is observed between average crack widths with different EX contents, an addition of an EX reduced crack widths greater than 0.04 mm, narrowing the crack width distribution, while reducing the number of cracks as the EX content increases. Residual cracks in the expansive SHCC after the load test were characterized by areas having densely concentrated fine cracks as shown in Photograph 1. It is inferred that the restrained expansion densified the matrix, restraining the slipping of the fibers. This presumably reduced the number of cracks until yielding, but the denseness of cracks subsequently increased in the range that underwent large plastic deformation after reinforcement yielding.

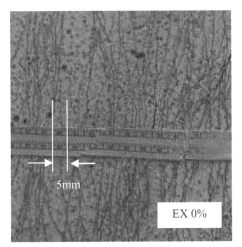

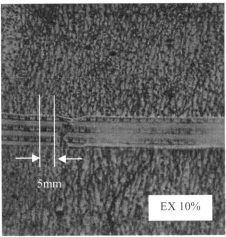

Photograph 1. Residual cracks after loading test.

4 CONCLUSIONS

Findings obtained in this study are summarized as follows:

1. The length change of SHCC due to expansion induced by an expansive additive and shrinkage during drying were restrained by fibers. An expansive additive content of 8% or more was necessary for inducing significant expansion.
2. Uniaxial tension tests on dumbbell-shaped specimens reinforced with a D6 bar revealed that the SHCC retained its integrity with the bar and that the deformation continued after yielding of the bar. When an expansive additive was used, the cracking load increased due to the restraint of the expansion by the bars.
3. The generation of chemical prestress was confirmed in a steel bar-reinforced expansive SHCC containing an expansive additive at a ratio of 8% and 10%.
4. Monitoring using strain gauges attached to the tension edges of steel bar-reinforced SHCC members was effective in pinpointing crack onset and reinforcement yielding.
5. The cracking load of steel bar-reinforced expansive SHCC beams significantly increased as the expansive additive content increased, due to shrinkage compensation or chemical prestressing.
6. The average and maximum crack widths in a steel bar-reinforced expansive SHCC until reinforcement yielding were 0.02 mm and 0.04 mm, respectively. The number of cracks decreased as the expansive additive content increased.

REFERENCES

JSCE. 2008. Recommendations for Design and Construction of High Performance Fiber Reinforced Cement Composites with Multiple Fine Cracks (HPFRCC). Concrete Library 127 Japan Society of Civil Engineers.

Kanda, T., Sakata, N., Kunieda, M. & Rokugo, K. 2006. State of the Art of High Performance Fiber Reinforced Cement Composite Research and Structual Application, Concrete Journal, Vol. 44, No. 3, 3–10.

Li, V.C. 2002. "Reflections on the Research and Development of Engineered Cementitious Composites (ECC)," Proceedings of the JCI International Workshop on Ductile Fiber Reinforced Cementitious Composites (DFRCC), Gifu, Japan, Japan Concrete Institute, 1–22.

Flexural creep behavior of SHCC beams with or without cracks

Yuichiro Yamada, Tomohiro Hata, Takuya Ohata & Keitetsu Rokugo
Department of Civil Engineering, Gifu University, Gifu, Japan

Tadashi Inaguma
JR Tokai Consultant, Co. Ltd, Nagoya, Japan

Seung-Chan Lim
Deros-Japan, Co. Ltd, Kanazawa, Japan

ABSTRACT: Strain-hardening fiber-reinforced cement composites (SHCC) are materials that form multiple fine cracks by the cross-linking effect of fibers under tensile stress, showing quasi-strain-hardening behavior. SHCC beams were fabricated using high strength polyethylene fibers and subjected to long-term flexural loading tests under different target loading strain levels to investigate the flexural creep behavior and cracking properties of SHCC. The tensile creep coefficient of SHCC in the elastic region before crack onset was approximately 1.2. The creep strain did not level off by the end of a loading period of 1,600 h with post-cracking strain levels. When flexural loading began with a load of a level greater than the crack onset strain, the tensile creep strain progressed with increases in both the number and width of cracks.

1 INTRODUCTION

Strain-hardening fiber-reinforced cement composites (SHCC) are materials that form multiple fine cracks by the cross-linking effect of fibers under tensile stress, showing quasi-strain-hardening behavior. In contrast to conventional cementitious composite materials, SHCC are capable of bearing tensile forces, allowing design incorporating the tensile performance. Recommendations for design and construction of SHCC have been published by JSCE (2007) both in a book and on web.

When SHCC are applied in expectation of bearing the tensile forces, they may be in tension or may be even forming multiple fine cracks when the permanent load acts on them. As tensile stress after crack onset is maintained by the cross-linking stress of fibers, the creep behavior of SHCC under permanent loads is considered to be closely related to the properties of fibers contained. The creep behavior of SHCC should thus be investigated under not only compressive but also tensile stress. It is desirable to examine the tensile creep under uniaxial tensile stress, but this is not easy in consideration of the necessity for ensuring the accuracy and effects of secondary bending in uniaxial tension testing. The tensile creep of SHCC has been investigated by Boshoff & van Zijl (2007a,b and

2008a,b). Since structures are primarily subjected to bending stress in actual use, it is considered effective to evaluate the creep behavior under bending stress close to that acting on actual structures.

SHCC beams were fabricated using high strength polyethylene fibers and subjected to long-term flexural loading tests under different target loading strain levels to experimentally investigate the flexural creep behavior and cracking properties.

2 OUTLINE OF EXPERIMENT

In this study, long-term flexural loading tests were conducted to investigate the flexural creep behavior of SHCC. Prior to these long-term tests, static bending tests and uniaxial tension tests were also conducted to grasp the static mechanical behavior and strength properties, respectively.

2.1 *Materials and mixture proportions*

Tables 1 and 2 outline the materials and mixture proportions, respectively. High-strength polyethylene fibers with a tensile strength of 2,600 N/mm^2 were used at a volumetric ratio of 1.5% as organic fibers. Fly ash was used in place of part of cement to reduce the cement content so as to suppress the hydration heat. An expansive additive and

Table 1.	Materials for SHCC.
Material	Properties and specification
High-density	Diameter: 0.012 mm, Length: 12 mm
Polyethylene fibers	Strength: 2.6 GPa, Elastic modulus: 88 GPa
Cement	JIS R 5210 High-early-strength Portlandcement
Fly ash	JIS A 6201 TypeB
Expansive additive	Ettringite type
Silica sand	Maximum diameter: 0.2 mm
Admixture	JIS A 6204 High-range water reducing admixture polycarboxylate type
Shrinkage-reducing	Alkylene oxide alcohol agent

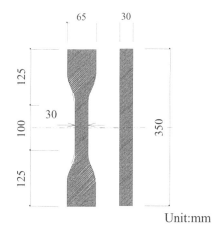

Figure 1. Dumbbell-shaped specimen.

Table 2. Mixture proportions of SHCC.

W/B (%)	Fiber volume (%)	Replacement ratio of FA (%)	Binder C+FA+EX (kg/m³)
30	1.5	30	1264

shrinkage-reducing admixture were also used to inhibit drying and autogenous shrinkages. The compressive strength and Young's modulus at an age of 6 months were 90.3 N/mm² and 20.8 kN/mm², respectively.

2.2 Uniaxial tension test

Figure 1 shows the dumbbell-shaped specimen. Both ends of dumbbell-shaped specimens were directly fixed to jigs attached to a displacement controlled hydraulic servo testing machine and subjected to uniaxial tension tests. The average strain was determined by measuring the displacement in the 100 mm gauge length zone and dividing it by the datum length. The number of specimens was six.

2.3 Static bending tests

The state of testing is shown in Photo 1. Beam specimens measuring 100 by 150 by 1,200 mm were subjected to four-point loading on a support span of 1,000 mm and with a loading span of 200 mm. The displacement was measured using π-shaped gauges with a gauge length of 100 mm fixed to brass nuts glued at midspan on both the tension and compression sides of each beam. The measured displacements were divided by the

Photo 1. Static bending test.

datum length to obtain the strains, with which the curvature ϕ was calculated by Eq. (1),

$$\phi = \frac{\varepsilon'_c + \varepsilon_t}{L + 2H} \qquad (1)$$

where ε'_c is the measured compressive strain, ε_t is the measured tensile strain, L is the specimen depth, and H is the nut depth. The strains on the compression and tension edges of each specimen were also calculated in consideration of the nut depth.

Cracks that occurred during loading were also photographed using a digital camera to examine the crack distribution by image analysis.

2.4 Long-term flexural loading test

Table 3 gives the test conditions. The stressed condition of SHCC can be divided into three regions: the elastic region up to the first crack onset;

Table 3. Test conditions.

Series	Specimen	Strain level	Load (kN)	Initial strain (μ)
Level-1	Level-1–1	2/3 of first	4.0	118
	Level-1–2	crack onset		134
Level-2	Level-2–1	Yield strain of	12.5	1897
	Level-2–2	reinforcement		3147
Level-3	Level-3–1	1/5 of softening	19.4	10495
	Level-3–2	onset strain		14411

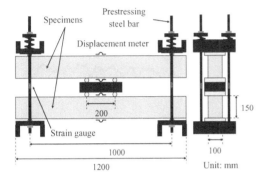

Figure 2. Flexural creep test.

strain-hardening region after cracking; and strain-softening region after the ultimate strain. As it is essential to combine SHCC with reinforcing bars in actual work to prevent brittle failure, the upper limit of the serviceability of SHCC is therefore considered to be around 2,000 μ, the yield strain of reinforcement, in the strain-hardening region.

In view of this, three strain levels were selected: Level 1: about 2/3 of the first crack onset strain; Level 2: a strain level of the yield strain of reinforcement; and Level 3: a strain level of about 1/5 of the softening onset strain (1.1%). Two specimens each were subjected to each strain level at an age of around 5 months. Each specimen was demolded two days after placing and then moist-cured for about a month in a curing room. In order to minimize the effect of drying shrinkage on the strain that would increase under sustained stress, specimens were subjected to air curing for around 4 months. Tests were conducted in a room with an air temperature of 20 ± 2°C to minimize the effect of temperature changes.

Figure 2 shows the outline of loading testing. The specimen size and measuring method were the same as in the static bending testing. Two specimens were combined in parallel, with the load being applied to both by fastening four prestressing steel

bars. The load under which the strain on the tension edges of specimens reached the target loading strain level was maintained. The applied load was controlled by the strain reading of strain gauges attached to the prestressing steel bars based on the load-strain relationship obtained beforehand by tension testing on steel bars.

The number, distribution and width of cracks were measured using a digital camera and microscope, at appropriate intervals.

3 RESULTS AND DISCUSSION

3.1 Tensile properties of materials

Figure 3 shows the stress-strain relationship obtained by uniaxial tension tests. All specimens showed quasi-strain hardening after cracking, with multiple fine cracks being dispersed. In specimens showing low ultimate strains, at which softening began, crack localization occurred outside of the gauge length.

3.2 Static bending test results

Figure 4 shows the load-strain relationship on the tension edge obtained from the static bending tests along with the number of cracks, which was counted on the tension edge in the midspan 80 mm zone. Photo 2 shows the crack distributions. The number of cracks increased in proportion to the strain on the tension edge.

3.3 Long-term flexural loading test results

Prior to the beginning of testing, cracking was visually checked. As a result, cracks presumably

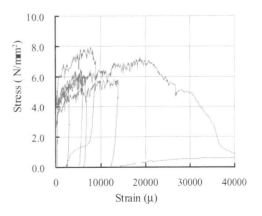

Figure 3. Stress-strain relationship obtained by uniaxial tension tests.

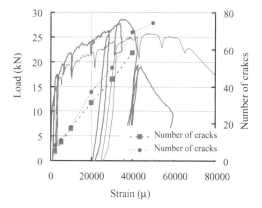

Figure 4. Load-strain relationship on tension edge obtained from static bending tests along with number of cracks.

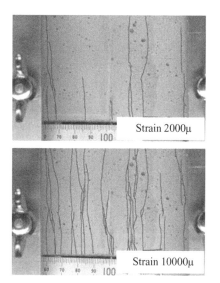

Photo 2. Crack distributions.

due to shrinkage were observed on the surfaces of all specimens. Therefore, the strain obtained from these tests includes the increases and reductions in the strain due to drying shrinkage. Nevertheless, loads were applied to these specimens when the progress of drying shrinkage was leveling off after a drying period of about 4 months. The strain to be applied on the tension edge was not less than 2,000 μ, which is substantially greater than the auto genous and drying shrinkages. It is therefore inferred that the effect of these shrinkages were relatively small when com pared with the strain that grew later under sustained stress.

3.3.1 *Flexural creep behavior before onset*

Figure 5 shows the changes in the strain over time under Level 1 loading. The compressive and tensile strains increase from the beginning of loading over time, with the tensile strain leveling off at around 1,008 h from the beginning of loading. Whereas the compressive strain of specimen Level-1-2 levels off, that of specimen Level-1-1 slowly continues, though with small increments. After loading for 1,600 h, the absence of flexural cracks in either of the specimens was visually confirmed. Accordingly, the tensile creep coefficient in the pre-cracking elastic region was investigated. The creep coefficient was determined by Eq. (2) assuming the

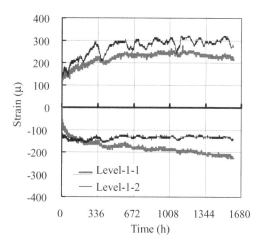

Figure 5. Changes in strain over time under Level 1 loading.

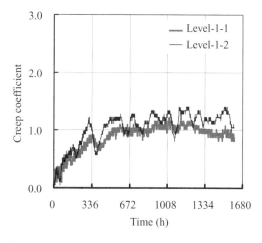

Figure 6. Time-related changes in creep coefficient.

Davis-Granville law, which is generally applied to the compressive creep of normal concrete.

$$\psi = \frac{\varepsilon_{all} - \varepsilon_{ini}}{\varepsilon_{ini}} \qquad (2)$$

where ψ is the creep coefficient; ε_{all} is a measured strain at a specific time; and ε_{ini} is the initial strain. Figure 6 shows the time-related changes in the creep coefficient. This figure suggests that the tensile creep coefficient is approximately 1.2.

3.3.2 Flexural creep behavior before crack onset

Figures 7 and 8 show the changes in the strain on the tension and compression edges, respectively, under Level 2 and Level 3 loads. Though the strains on the tension and compression edges of two specimens each were measured, large deformation with multiple cracks occurred in specimen Level-3-2 after a loading period of 280 h, causing the π-shaped gauge on the tension edge to drop off. Measurement on this edge was therefore impossible thereafter. The strain of all specimens rapidly increased in the initial stage until around 100 h, but the increases became slower thereafter, continuing at nearly constant rates. The strain tended to be larger under a larger loading strain level. Abrupt changes in the strain presumably indicate rapid increases in the width or number of cracks.

The tensile and compressive creep coefficients were then calculated using Eq. (2). In this paper, increases in the strains on the compression and tension edges under a sustained load with respect to the initial strains at the beginning of loading are referred to as creep strains, and their ratios to the initial strains are referred to as creep coefficients. Figure 9 shows the time-related changes in the creep coefficients. The creep coefficients of

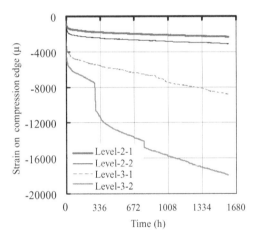

Figure 7.   Strain on tension edge.

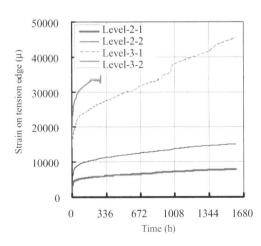

Figure 8.   Strain on compression edge.

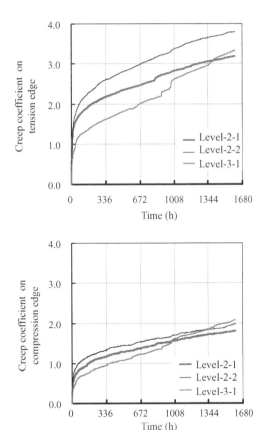

Figure 9.   Time-related changes in creep coefficient.

the three specimens were nearly the same after a loading period of 1,600 h in the loading strain level ranges of 2,000 to 11,000 μ, which were applied to specimens in the state of multiple cracking.

3.3.3 *Cracking properties after crack onset*

Since the width and number of cracks were measured at appropriate occasions, the last was made at a time around a loading period of 1,344 h. Figure 10 shows typical changes in the crack widths over time. In all specimens, the crack widths rapidly increased during the initial stage of loading and then leveled off after certain periods at values that varied from one crack to another. After loading for 1,600 h, the crack widths were as small as 0.1 mm or less and mostly 0.2 mm or less under Level 2 and Level 3 loading, respectively, though the width of some cracks under Level 3 loading exceeded 0.4 mm, which appeared to lead to crack localization.

Figure 11 shows the time-related changes in the number of cracks. Photo 3 shows the crack distribution. The number of cracks is the number of those counted on the tension edge within the gauge length of 100 mm. Their number significantly increased during the initial stage of loading, and new cracks occurred thereafter when the strain increased.

Whereas the width of relevant cracks leveled off, their number continued to increase, presumably because crack localization occurred at some cracks.

The tensile creep properties after crack onset primarily depend on the cross-linking stress of fibers. The authors intend to investigate the creep properties of SHCC having different types of fibers for comparison. Also, uniaxial tensile creep tests and compressive creep tests should be conducted

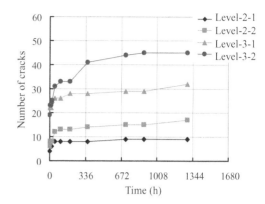

Figure 11. Time-related changes in number of cracks.

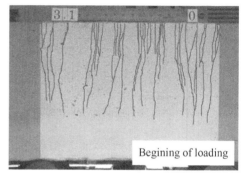

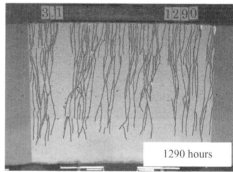

Photo 3. Crack distribution.

to elucidate the creep properties under these loads. Analysis will then be conducted based on these results to further investigate the creep behavior of SHCC.

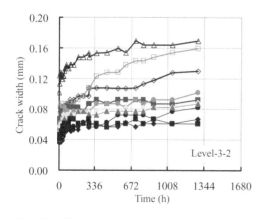

Figure 10. Typical changes in the crack widths over time.

4 CONCLUSIONS

Long-term flexural loading tests were conducted on SHCC at three loading strain levels. The results of these creep tests up to a loading period of 1,600 h are summarized as follows:

1. The tensile creep coefficient of SHCC in the elastic region before crack onset was approximately 1.2.
2. The creep strain did not level off by the end of a loading period of 1,600 h with post-cracking strain levels.
3. When flexural loading began with a load of a level greater than the crack onset strain as the initial tensile strain, the tensile creep strain progressed with increases in both the number and width of cracks.
4. The flexural crack width tended to level off over time at values of 0.1 mm or less under Level 2 loading and mostly 0.2 mm or less under Level 3 loading. Crack localization occurred under Level 3 loading, with the width of some cracks exceeding 0.4 mm. These can lead to creep failure as the ongoing tests continue.

REFERENCES

JSCE. 2007. Recommendations for Design and Construction of High Performance Fiber Reinforced Cement Composites with Multiple Fine Cracks (HPFRCC). Concrete Library 127, Japan Society of Civil Engineers. /Concrete Engineering Series 82. http://www.jsce.or.jp/committee/cocrete/e/hpfrcc_JSCE.pdf

Boshoff, W.P. and van Zijl. G.P.A.G. 2007a. Time-dependent response of ECC: Characterisation of creep and rate dependence. Cement and Concrete Research, Vol. 37, pp. 725–734.

Boshoff, W.P. and van Zijl. G.P.A.G. 2007b. Tensile creep of SHCC. RILEM PRO 53, High Performance Fiber Reinforced Cement Composites (HPFRCC5), Germany, pp. 87–95.

Boshoff, W.P. and van Zijl. G.P.A.G. 2008a. Creep of cracked strain hardening cement-based composites. Creep, Shrinkage and Durability of Concrete Structures, Japan, pp. 723–728.

Boshoff, W.P. and van Zijl. G.P.A.G. 2008b. Mechanisms of creep in fiber-reinforced Strain-Hardening Cement Coposites (SHCC). Creep, Shrinkage and Durability of Concrete Structures, Japan, pp. 753–759.

Mechanics of advanced cement-based materials

Advances in Cement-Based Materials – van Zijl & Boshoff (eds)
© *2010 Taylor & Francis Group, London, ISBN 978-0-415-87637-7*

Size-effect in high performance concrete road pavement materials

E. Denneman
CSIR Built Environment, University of Pretoria, Pretoria, South Africa

E.P. Kearsley & A.T. Visser
University of Pretoria, Pretoria, South Africa

ABSTRACT: In concrete road pavement design, the flexural strength of the material is used as a design parameter to predict fatigue life. Flexural strength however, is not a true material property as it varies with specimen size. Flexural strength also does not capture the full fracture toughness of the material, as it is an indicator of ultimate strength only. Especially for more ductile high performance fibre reinforced concrete it is appropriate to use fracture mechanics parameters to describe fracture toughness. The first objective of this paper is to quantify the size-effect for a high performance concrete, which was developed for use in an innovative pavement system. To investigate to which extent size-effect can be predicted through numerical simulation using fracture mechanics is the second objective of this paper. The results show the material to be subject to significant size effect in bending. The size-effect observed in the experiment can be satisfactorily predicted thorough numerical modelling with the embedded discontinuity method used for the study.

1 INTRODUCTION

Most current concrete road pavement design methods apply the Palmgren-Miner linear cumulative damage hypothesis (Miner, 1945), shown as Equation 1, for the prediction of high cycle fatigue failure. The hypothesis is that failure occurs when the number of stress cycles n_i applied at stress level S_i divided by the number of cycles to failure N_i at stress level S_i equals one.

$$\sum \frac{n_i}{N_i} \qquad (1)$$

Pavement design methods make use of the above concept to relate tensile stress in a full-scale pavement, calculated using elastic theory, to the experimentally determined flexural strength for the material. The flexural strength or Modulus of Rupture (MOR) is typically determined on beam specimens in four point bending (FPB). The fatigue damage prediction relation included in the latest update of the South African Concrete pavement design method (Strauss et al., 2007) is shown as Equation 2. A similar fatigue prediction equation for continuously reinforced concrete pavements is contained in the new American mechanistic empirical pavement design guide (MEPDG) (NCHRP, 2004), given in Equation 3.

$$N = a\left(\frac{\sigma}{f}\right)^b \qquad (2)$$

where N = expected life, a = damage constant determined by required serviceability of road (road category), σ = maximum horizontal tensile stress, f = flexural strength of material and b = damage calibration constant

$$\log(N_{i,j}) = C_1 \left(\frac{MR_i}{\sigma_{i,j}}\right)^{C_2} - 1 \qquad (3)$$

where $N_{i,j}$ = ultimate number of load repetitions at time i at load magnitude j, MR_i = Concrete modulus of rupture at time i, C_1 = calibration constant = 2.0, $\sigma_{i,j}$ = applied stress at time i due to load with magnitude j, and C_2 = calibration constant = 1.22.

The Palmgren-Miner linear damage hypothesis provides a convenient tool for fatigue prediction in concrete pavements. The hypothesis does however have some known limitations for use in pavement engineering:

1. It is a linear function, while fatigue distress development follows a distinctly non-linear path,
2. It does not take the sequence of loading and related residual stresses into account, and
3. The hypothesis does not take the probabilistic nature of damage into account.

Another limitation of fatigue damage laws akin to the ones shown in Equations 2 and 3 is the use of the flexural strength or MOR as a material property. Researchers have long established that flexural strength is not a true material property because its value changes with specimen size e.g. Reagel & Willis, (1931). A concise discussion of size effect in concrete is provided by Bažant & Planas (1997). Flexural strength also does not provide a sufficient indicator of the fracture toughness of the material, as it is an indicator of ultimate strength only. Roesler (2006) has shown that due to size effects and boundary conditions fatigue of full scale concrete slabs cannot be reliably predicted from tests on beam specimens. Structurally slabs have a better fatigue performance than beams, which can not be explained without the use of fracture mechanics. Fracture mechanics approaches are increasingly being applied to overcome the limitations of current pavement design methods e.g. Ioannides and Peng, (2006) and Gaedicke et al. (2009).

In South Africa, an innovative concrete pavement type known as ultra-thin continuously reinforced concrete pavement (UTCRCP) is being implemented on major highways. The material can be classified as high performance concrete with polypropylene and steel fibre reinforcement in combination with a continuous rebar mesh at the neutral axis. It is expected that existing design methodology will not yield reliable predictions of fatigue life for the material due to size-effect.

The objectives of the paper are: firstly, to experimentally determine the size effect in bending tests performed on the high performance concrete material, and secondly, to investigate to what extent the size effect can be predicted by means of numerical simulation.

The details of the laboratory experiment in terms of the properties of the material under study and the test configuration are discussed in Section 2. In Section 3 the results of the experimental study are presented, the details of the numerical model are discussed and the results of the simulation runs are compared to the experimental results. The section is ended with a discussion of the size-effect observed in the experimental and simulated data. Section 4 contains the conclusions.

2 LABORATORY EXPERIMENT

2.1 Material properties

The mix design for the high performance concrete is shown in Table 1. The maximum aggregate size used in the mix is 6.7 mm. To achieve the desired high fracture toughness for the mix, 120 kg/m³ steel fibres of 30 mm length and 0.5 mm in diameter

Table 1. Mix proportions.

Component	Type	Quantity
Cement	52.5 N	464 kg/m³
Stone	Quartzite 6.7 mm	866 kg/m³
Water	tap water	176 kg/m³
Sand	Crusher sand: Quartzite	751 kg/m³
Steel fibre	30 mm × .5 mm (l × d)	120 kg/m³
Synthetic fibre	Polypropylene	2.0 kg/m³
Admixture	undisclosed	7.4 l/m³
Silica fume		2.5 kg/m³
Fly ash		3.1 kg/m³

Table 2. Engineering properties.

Parameter	Value	Unit
Compressive strength fc	125.5	MPa
Resilient modulus E	49.7	GPa
Poisson's ratio ν	0.17	
Split tensile strength ft	7.34	MPa

were added to impede macro cracking. 2.0 kg/m³ polypropylene fibres are also included to retard micro cracking.

The engineering properties obtained at an age of 28 days for the mix are shown in Table 2. Compressive strength f_c was determined in accordance with British Standard BS 1881. The ASTM C469-02 was used to obtain the static modulus of elasticity E and Poisson's ratio v. An approximation of the tensile strength f_t was obtained using the tensile splitting test, implementing the measures to reduce size effect proposed by Rocco et al. (1999) in terms of width of the loading strip and loading rate.

2.2 Experimental plan

Three point bending (TPB) tests on notched samples were performed. The vertical displacement at the centre of the span was recorded by means of linear variable displacement transducers (LVDT) at either side of the beam. The reference frame for the displacement was mounted at half height of the beam specimen and mid-span displacement was measured relative to reference points above the supports. The crack mouth opening displacement (CMOD) was recorded using a clip-on gauge. The combination of vertical displacement and CMOD monitoring was chosen, because the CMOD clip gauge has a range of only 14 mm. In an attempt to record the full work of fracture required to completely break the ductile fibre reinforced specimen, vertical displacements of up to 20 mm were measured.

Table 3. Specimen dimensions.

Type	Length l [mm]	Width b [mm]	Height h [mm]	Span s [mm]	Notch depth a [mm]
Three Point Bending					
A	330	150	50	165	16.5
B	1000	150	150	500	50
C	1500	150	225	750	75
Four Point Bending					
B1	500	150	225	450	N/A
C1	750	150	150	675	N/A

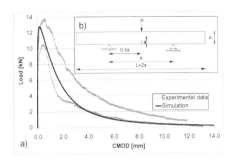

Figure 1. a) Experimental and simulated load versus CMOD for TPB tests on specimen type A, b) Sketch of TPB test setup.

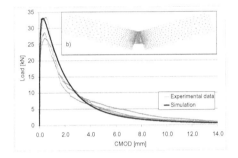

Figure 2. a) Experimental and simulated load versus CMOD for TPB tests on specimen type B, b) numerical model TPB test.

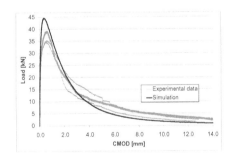

Figure 3. Experimental and simulated load versus CMOD for TPB tests on specimen type C.

$$\sigma_N = \frac{3Ps}{2bh^2} \qquad (4)$$

$$\sigma_N = \frac{Ps}{bh^2} \qquad (5)$$

To investigate specimen size-effect the TPB specimen were produced in three sizes maintaining geometry. Specimen dimensions for the TPB tests are shown in the top part of Table 3. Three specimens of each type were prepared. A sketch of the test setup can be found in Figure 1b. A beam height to span ratio of 0.3 was applied in combination with a notch depth to beam height ratio of ⅓. A beam length of twice the size of the span, compensating self weight, allows for stable tests runs up to complete fracture of the beam.

After completion of the TPB tests, the halved specimen of type B and C were subjected to four point bending (FPB) tests in the configuration shown in the lower part of Table 3 and as a sketch in Figure 4b. The FPB test is the most commonly applied in determining flexural strength (MOR). The FPB tests were performed without a preformed crack in the form of a notch. Therefore only the vertical displacement was monitored. All bending tests were run at a controlled displacement of 0.5 mm per minute.

3 EXPERIMENTAL RESULTS AND NUMERICAL SIMULATION

3.1 Laboratory results for TPB and FPB tests

The load—CMOD curves of the TPB tests on specimen types A, B and C are shown in Figures 1, 2 and 3 respectively. The load- deflection curves for the FPB tests are shown in Figures 4 and 5. The results of the numerical simulation of the tests are also shown in the figures.

Table 4 shows the average results obtained for the TPB and FPB tests. The peak load P_{max} is converted into an ultimate nominal stress σ_{Nu}. This parameter is calculated assuming an elastic stress distribution in the beam in the peak load condition and is equal to the flexural strength or MOR. σ_N (at any load) is calculated using Equation 4 for TPB tests and using Equation 5 for FPB tests.

The results for σ_{Nu}. In the table were calculated taking small deviations in the relevant dimensions for each specimen into account.

The fracture energy G_f was calculated for specimen type B. G_f is obtained from dividing the work

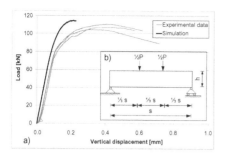

Figure 4. a) Experimental and simulated load versus vertical displacement for FPB tests on specimen type B1, b) Sketch of FPB test setup.

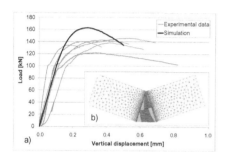

Figure 5. a) Experimental and simulated load versus vertical displacement for FPB tests on specimen type C1, b) numerical model FPB test.

Table 4. Summary of test results.

Type	N*	P_{max} [kN]	σ_{Nu} [MPa]	CV [%]	G_f [N/mm]	CV [%]
A	2	12.6	17.9	12.8		
B	3	30.5	15.2	10.8	5.15	6.6
C	3	37.4	13.1	8.3		
B1	3	107.4	14.0	4.0		
C1	6	137.3	12.0	7.1		

* Number of successfully completed tests.

of fracture W_f required to completely break the sample by the area of the beam cross-section above the notch:

$$G_f = \frac{W_f}{b(h-a)} \qquad (6)$$

W_f is defined as the area under the load – deflection curve. In weight compensated tests, the exact value of P at time is 0 is unknown. Correction for the missing part of the load-deflection tail arising

when the test is stopped short, before all fracture energy is dissipated, is also required Elices et al. (1992). The total work of fracture W_f was calculated in accordance with the methodology for weight compensated beams in Bažant and Planas (1997). Note that no G_f values are reported for the other beam types, as an objective of this paper is to investigate whether the peak load can be predicted for specimens of different sizes using the results obtained from a single size beam.

3.2 Numerical model

In the present study simulation of the TPB and FPB tests is performed using an embedded discontinuity method (EDM) based on the work by Sancho et al. (2007) and implemented into the open source finite element method framework OpenSees (OpenSees, 2008) by Wu et al. (2009). An advantage of the embedded discontinuity method over more conventional methods is that it allows cracks to propagate through elements, i.e. independent of nodal positions and element boundaries. For the post-crack behaviour of the material the EDM applies the cohesive crack approach introduced by Hillerborg et al. (1976). According to the cohesive crack approach a crack is induced when the stress in the material reaches the tensile strength f_t of the material. After the crack has formed, stresses will still be transferred over the crack, however, the amount of stress transferred reduces as the crack width increases.

Many shapes for the post crack softening function have been proposed, including linear and bilinear. For the present paper a simple exponential softening function is used:

$$\sigma = f_t e^{-aw} \qquad (7)$$

Where σ is the stress transferred over the crack, w is the crack width and a depends on the ratio of f_t and G_f, as can be shown when integrating the softening function to obtain G_f (the area under the softening curve):

$$G_f = \int_0^\infty f_t e^{-aw} dw = \frac{f_t}{a} \qquad (8)$$

$$a = \frac{f_t}{G_f} \qquad (9)$$

Due to the asymptotic behaviour of the function the stress transferred over the crack never reduces to 0. It is proposed that this is an acceptable representation of the behaviour of the fibre reinforced material under study as stress will only

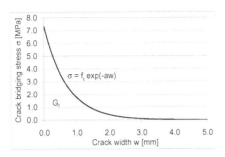

Figure 6. Softening function for material under study.

Table 5. Comparison of experimental and simulated results.

Type	Experiment σ_{Nu} [kN]	Simulation σ_{Nu} [MPa]	Error [%]
A	17.9	18.1	+1.1
B	15.2	16.5	+8.6
C	13.1	15.7	+8.3
B1	14.0	15.0	+7.1
C1	12.0	14.1	+17.5

cease to be transferred over the crack at large crack widths. The last fibres will be pulled out when the crack reaches a width of half the length of the fibre (i.e. 15 mm). Figure 6 shows the softening function determined for the material based on the value for f_t obtained from the tensile splitting tests (7.34 MPa) and the value for G_f obtained from the TPB tests for specimen type B.

The numerical simulation is performed in two-dimensional space and simple stress mode. It is possible to run the analysis using EDM elements only, but to make the calculation more efficient a narrow vertical band of triangular EDM elements was provided above the notch for the TPB model and at centre span for the FPB model. The remainder of the elements used are triangular linear elastic bulk elements. An impression of the deformed mesh for TPB and FPB at high displacement near the end of the test is provided in Figure 2b and Figure 5b respectively.

3.3 Size-effect in experiment and numerical simulation

As mentioned earlier, the results of the numerical analysis is shown graphically in Figures 1 to 5. The FPB tests, due to the absence of weight compensation, became unstable shortly after the peak-load was reached. No meaningful data was obtained beyond the point where the sudden collapse occurred. Therefore, the analysis of the FPB tests was also stopped shortly after the peak was reached.

Table 5 shows a comparison between experimentally obtained and simulated values of σ_{Nu}. As can be seen from the table and Figures 1 to 5, the model structurally over-predicts the value of the peak load. A better fit may be obtained by reducing the value for f_t. The tensile splitting test is known to yield values that are higher than the actual tensile stress (Olesen & Stang, 2006). Optimization of the fit falls beyond the scope of this paper.

For the numerical simulation of specimen type B, the stress state in the ligament above the

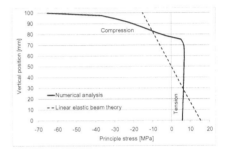

Figure 7. Simulated stress state at peak load for specimen type B.

notch at peak load is shown in Figure 7. The fracture zone has progressed ¾ of the way up the beam. The simulation gives a clear indication of the considerable difference between the proportional linear elastic stress distribution between tension and compression assumed to calculate the flexural strength or MOR for a beam and the actual stress state.

The results from Table 5 are plotted for the TPB tests in Figure 8 and for the FPB tests in Figure 9. The impact of size-effect is demonstrated by the experimental data for both TPB and FPB results. A linear trend provides the best fit to the limited experimental data available. A power function can be fitted to the results obtained from the numerical model. To show the numerical trend beyond the specimen sizes tested under the present study, the range of the modelled results was extended by simulating additional fictitious 30 mm and 360 mm high beams maintaining geometry.

The decrease in σ_{Nu} with increase in specimen size is not as steep in the numerical model as the trend displayed by the experimental results. This maybe due to the fact that the numerical model only accounts for the fracture mechanics size-effect. The fracture mechanics size-effect is the largest single source of size effect, but other sources of size effect may be at play. Bažant and Planas (1997) identify five sources of size-effect apart from the fracture

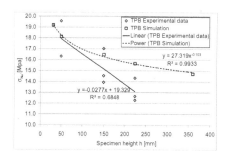

Figure 8. Comparison experimental and simulated size effect TPB tests.

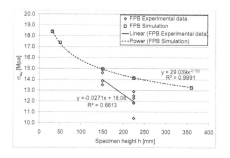

Figure 9. Comparison experimental and simulated size effect FPB tests.

mechanics size effect and each of these may have had an effect on the results.

Overall the numerical model provides a satisfactory prediction of the experimentally determined size-effect, especially taking in consideration that the fit in terms of absolute values of σ_{Nu} could further be improved by optimizing the value of f_t, as is often done.

4 CONCLUSIONS

With respect to the first objective of the paper it is concluded that the experiments show high performance fibre reinforced concrete material to be subject to significant size-effect. It is proposed that the flexural strength obtained from beam bending tests is not a suitable parameter that can be expected to be a reliable indicator of the fatigue performance of full-scale road pavements under traffic loading.

With respect to the second objective of this paper it is concluded that fracture mechanics simulation using the cohesive crack approach does provide satisfactory prediction of size-effect observed in the bending tests. The study will have to be extended to include three-dimensional concrete structures such as slabs in future to assess the adequacy of

the models to capture size-effect in full size road pavements.

The use of fracture mechanics to improve the prediction of crack propagation in concrete under static loading can be seen as a first step in an effort to replace fatigue damage models based on the Palmgren-Miner linear cumulative damage rule with more suitable fracture mechanics based ones.

REFERENCES

Bažant, Zdeněk, P. and Planas, Jaime (1997) *Fracture and size effect in concrete and other quasibrittle materials*, CRC Press.

Elices, M., Guinea, G.V. and Planas, J. (1992) 'Measurement of the fracture energy using three-point bend tests Part 3-Influence of cutting the P-u tail.pdf', *Materials and Structures*, 25, pp. 327–334.

Gaedicke, Cristian, Roesler, Jeffery and Shah, Surendra (2009) 'Fatigue Crack Growth Prediction in Concrete Slabs', *International Journal of Fatigue*.

Hillerborg, A., Modeer, M. and Petersson, P.E. (1976) 'Analysis of crack formation and crack growth in concrete by means of fracture mechanics and finite elements', *Cement and Concrete. Research*, 6.

Ioannides, Anastasios, M. and Peng, Jun (2006) 'International Journal of Pavement Engineering', *International Journal of Pavement Engineering*, 7(4).

Miner, A. (1945) 'Cumulative damage in fatigue', *Journal of Applied Mechanics*, 12, pp. A159–164.

NCHRP (2004) *Guide for Mechanistic-Empirical Design of New and Rehabilitated Pavement Structures*, http://www.trb.org/mepdg/guide.htm.

Olesen, J.F. and Stang, H. (2006) 'Nonlinear fracture mechanics and plasticity of the split cylinder test', *Materials and Structures*, 39(4), pp. 421–432.

OpenSees (2008) *Open System for Earthquake Engineering Simulation v1. 7.5*, Berkeley, Pacific Earthquake Engineering Research Center, University of California.

Reagel, F.V. and Willis, T.F. (1931) 'The effect of dimensions of test specimens on the flexural strength of concrete', *Public roads*, 12, pp. 37–46.

Rocco, C, Guinea, G.V., Planas, Jaime and Elices, M. (1999) 'Size effect and boundary conditions in the Brazilian test: Experimental verification', *Materials and Structures*, 32, pp. 210–217.

Sancho, J.M., Planas, J., Cendón, D.A., Reyes, E. and Gálvez, J.C. (2007) 'An embedded crack model for finite element analysis of concrete', *Engineering Fracture Mechanics*, 74, pp. 75–86.

Strauss, P.J., Slavik, M., Kannemeyer, L. and Perrie, B.D. (2007) 'Updating cncPave: inclusion of ultra thin continuously reinforced concrete pavement (UTCRCP) in the mechanistic, empirical and risk based concrete pavement design method', In *International Conference on Concrete Roads*, Midrand, South Africa.

Wu, R., Denneman, E. and Harvey, J.T. (2009) 'Evaluation of an embedded discontinuity method for finite element analysis of hot mix asphalt concrete cracking', In *Transportation Research Board 88th annual meeting 2009*, Washington.

Advances in Cement-Based Materials – van Zijl & Boshoff (eds)
© 2010 Taylor & Francis Group, London, ISBN 978-0-415-87637-7

Modified local bond stress—slip model for self-compacting concrete

P. Desnerck, G. De Schutter & L. Taerwe
Magnel Laboratory for Concrete Research, Department of Structural Engineering, Ghent University,
Ghent, Belgium

ABSTRACT: The force transfer in reinforced concrete is provided by the concrete-to-steel bond. This phenomenon has widely been studied for conventional vibrated concrete (CVC). For self-compacting concrete (SCC) however less test results are available.

To fill in this lack and to develop adapted standards for predicting the bond of reinforcement in SCC, an experimental program has been set up. The bond strength of reinforcement bars with different diameters has been tested by means of "beam-test" specimens. During testing the bond stress-slip response was recorded.

From the test results it can be seen that the maximum bond strength of SCC is slightly higher than for CVC when small bar diameters are studied. For larger bar diameters the difference becomes smaller. Comparison of the test data with bond models indicated that the bond clauses underestimated the bond strength for SCC as well as for CVC. Therefore an adjustment of the bond model has been made.

1 INTRODUCTION

Reinforced concrete is far from homogeneous. It is built up of steel and concrete and the concrete itself is not homogeneous neither. Reinforced concrete elements are basically designed so that the concrete can carry the compressive stresses and the steel can resist the tensile stresses. Therefore a good force transfer between the two materials is necessary which can only be achieved by an interaction between both materials, which is provided by bond between the reinforcement bars and the concrete.

The bond has an important influence on the behaviour of reinforced elements in the cracked stage. Crack widths and deflections are influenced by the distribution of bond stresses along the reinforcement bars and by the slip between the bar and the surrounding concrete.

Due to the importance of the interaction between steel and concrete a lot of research has been done in the past. In all these projects the main focus was on the reinforcement bar, its geometrical characteristics and how these characteristics influence the bond strength. With the appearance of new concrete types, such as steel-fibre reinforced concrete and high strength concrete, questions arose about the bond strength achieved with these concrete types, and the main focus of the research on bond shifted to the concrete and its composition (Martin 2002).

The same questions can be formulated for self-compacting concrete. A concrete type which, in fresh state, has the ability to flow under its own weight, fill the required space or formwork completely and produce a dense and adequately homogeneous material without a need for compaction (De Schutter et al. 2007). The advantages are clear: no need for vibration of the concrete, a higher quality of the finished element, reduced construction times,…. The self-compactability is achieved by adding a superplasticizer to the mixture and by reducing the amount of coarse aggregates. Although self-compacting concrete (SCC) is a relatively new material, already a lot of research has been done on the durability and the workability of the concrete type (De Schutter et al. 2008). Less studies have been focussing on the mechanical properties and more in particular the bond aspects.

Some programs have been carried out to determine the force transfer between concrete and reinforcement in self-compacting concrete. These studies show that the bond strength of steel in SCC is not lower than for conventional vibrated concrete (CVC), and may be even higher in some cases (Almeida et al. 2008, Chan et al. 2003, Zhu et al. 2004, Dehn et al. 2004). Nevertheless there is a great scatter in the results.

To get a better insight in the difference in bond strength between conventional vibrated concrete and self-compacting concrete and to develop modified models describing the bond stress-slip behaviour, this research program has been set up.

2 EXPERIMENTAL PROGRAM

The common way to test the bond strength of reinforcing bars in concrete is by means of pull-out tests (RILEM 1973). The behaviour of these types of specimens is quite different from that in reinforced elements subjected to bending: the reinforcing bar is under tension while the surrounding concrete is subjected to compression.

The beam test specimen suggested by RILEM recommendation RC6 part 1 (RILEM 1973) are more suitable to evaluate the bond strength of reinforced elements subjected to bending. The specimen, consisting of 2 half-beams connected on top by a hinge and at the bottom by a reinforcing bar, is loaded on top introducing bending moments in the beam. In this way a more realistic stress distribution inside and around the steel bar is created. Therefore in this study beam test specimens are used.

2.1 Materials

Three types of concrete have been used: 2 powder-type self-compacting concretes and one conventional vibrated concrete. Due to the fact that self-compacting concrete mixtures regularly have higher compressive strengths than conventional vibrated concrete for the same water to cement ratio (De Schutter 2007) and the fact that to achieve a low compressive strength (range 20–40 N/mm²) a higher W/C-ratio (often higher than allowed by the codes) is needed for SCC mixtures, the reference mixture CVC1 has been chosen to have a compressive strength of about 60 MPa. The first self-compacting concrete, SCC1, has the same water to cement ratio, resulting in a higher compressive strength. The second mixture, SCC2, has been designed to achieve a compressive strength comparable to the conventional vibrated concrete by reducing the amount of cement, resulting in a higher W/C-ratio.

All concretes were made with the same Portland cement type: CEM I 52,5 N. A natural sand 0/4 mm and 2 types of gravel (2/8 mm and 8/16 mm) have been chosen as aggregates. The amount of each aggregate type is the same for both self-compacting concretes, but the fraction of large aggregates was substantially lower than for the conventional vibrated concrete (only 56.9%). The amount of fine materials (cement and filler) is kept constant at 600 kg/m³.

As superplasticiser a polycarboxylic ether hyperplastiziser (Glenium 51) at a concentration of 35% has been used. The amount of superplasticizer is varied until the slump flow of the fresh concrete reached a value of about 750 mm.

The mix proportions are summarized in Table 1. More details about the mixing procedure and the used materials can be found in (Desnerck 2008).

The properties of the fresh concrete are determined immediately after mixing. For the self-compacting concrete the slump flow and V-funnel time are measured and for the conventional vibrated concrete only the slump is measured. No sign of segregation was detected during the slump flow tests for the self-compacting concrete. The averages of the measurements are summarized in Table 2.

The compressive and tensile strength of the concrete are determined at 28 days. The mean results of all tests are summarized in Table 2 as well.

The self-compacting concrete SCC2 has a comparable compressive and tensile strength as the conventional vibrated concrete CVC1, as was intended. The first self-compacting concrete SCC1, with the same W/C ratio, has a significantly higher strength.

Besides the concrete type, the steel bar diameter has been varied. In this research program, 5 different nominal diameters of the embedded reinforcement bars were chosen: 12, 20, 25, 32 and 40 mm. The nominal diameter ϕ, yield stress f_y and tensile strength f_u of the different high-strength hot-rolled reinforcing bars, are measured in the laboratory and are summarized in Table 3.

All reinforcing bars have the same rib pattern which consist of two longitudinal ribs at opposite

Table 1. Mix design for SCC and CVC mixes.

Materials (kg/m³)	CVC1	SCC1	SCC2
CEM I 52,5 N	360	360	300
Sand 0/4 mm	640	853	853
Gravel 2/8 mm	462	263	263
Gravel 8/16 mm	762	434	434
Limestone filler	–	240	300
Water	165	165	165
Superplasticizer	–	3.6	3.0

Table 2. Properties of fresh and hardened concrete.

	CVC1	SCC1	SCC2
Slump (mm)	36	–	–
Slump flow (mm)	–	745	745
V-funnel (s)	–	12.3	12.9
f_{ccub} (N/mm²)	58.4	71.7	62.1
f_c (N/mm²)	51.8	63.7	57.5
$f_{ct,fl}$ (N/mm²)	6.2	7.2	6.8
$f_{ct,sp}$ (N/mm²)	4.1	5.0	4.4

Table 3. Mechanical and geometric characteristics of the tested reinforcing bars.

ϕ [mm]	f_y [N/mm²]	f_u [N/mm²]	Max rib height [mm]	f_R [–]
12	629	740	0.99	0.0473
20	641	750	1.90	0.0717
25	578	674	1.59	0.0454
32	542	659	2.54	0.0602
40	570	681	2.70	0.0665

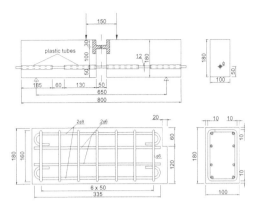

Figure 1. Size and reinforcement of beam test specimen type I (dimensions in mm).

sides of the bars and two series of transverse ribs. On one side of the bar, the transverse ribs are parallel to each other. On the other side 2 alternating series of parallel ribs can be seen. The rib pattern is measured in the laboratory using a laser measuring device. From the measured rib profiles, the maximum rib height and relative rib area f_R are derived.

2.2 Specimen details

To test the bond strength of reinforcing bars in the different concrete types, the standard "beam-test" geometry, as described in the RILEM recommendation, is used. The specimen dimensions depend on the bar size. For the smallest diameter (12 mm) a specimen type I (Figure 1) is used and for diameters 20 and 25 mm a specimen type II is used. For the largest diameters 32 and 40 mm an even larger specimen (type III) has been cast to avoid brittle failure by splitting of the concrete cover. More information on the dimensions and the reinforcement of the specimen can be found in (Desnerck 2008).

All beams are composed of 2 half-beams connected at the bottom by the tested reinforcing bar and at the top by a steel hinge. This hinge has been secured to the beam 14 days before testing by using a traditional mortar.

The prescribed bond length is 10 times the bar diameter ϕ. However this leads to yielding, and in some cases even rupture, of the reinforcement bar before reaching the ultimate bond strength. Therefore most of the specimens are cast with a bond length of 5 times the bar diameter. This has been achieved by putting plastic tubes over the remaining parts of the bar.

In all cases the actual bond length started at 230 mm from the centre of the beam (200 mm inside the concrete) except for the type I specimen where the bond length started at a distance of 160 mm from the centre of the specimen. To avoid an influence of the bar position, all bars are placed in the same way. the longitudinal ribs at mid-height.

2.3 Testing procedure

During the tests, the specimens were loaded at a constant rate corresponding to an increase in steel stress of 30 N/mm² per minute. For all specimen types, the actuator was positioned in the centre of the specimen and the total load P was transferred by means of a spreader steel profile to each half-beam (Figure 2). A load cell measured the load applied to the specimen during the test.

The slip of the bar, at its free end, was recorded using 3 linear variable differential transducers (LVDT) on both sides of the specimen.

Loading continued until the slip at one end of the specimen reached 3 mm. For the half-beam with 3 mm slip the bar was fixed in a clamping device so that the test could be continued without further slip at this side of the specimen. Loading continued until the slip at the second half of the specimen exceeded 3 mm as well.

2.4 Test results

From the obtained test results, values of the bond stress along the surface of the bonded reinforcing bar can be derived. The formulas to calculate the total force acting in the reinforcing bar depends on the geometry of the specimen used.

The mean bond stress can be calculated by assuming the force F_s in the reinforcing bar to be transferred to the concrete in the cylindrical zone of the embedment length l_d:

$$\tau_d = \frac{F_s}{l_d \cdot \pi \cdot \phi} = \frac{\sigma_s}{4 \cdot k} \tag{1}$$

by writing l_d as $k \cdot \phi$ and σ_s the tensile stress in the reinforcing bar. This stress is, as mentioned earlier,

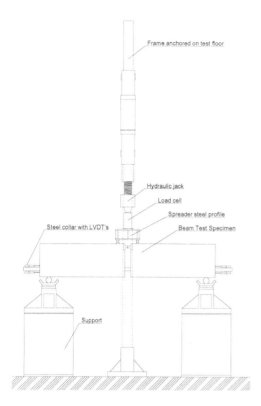

Figure 2. Test set-up for beam test specimen.

a function of the applied total load P and the geometry of the specimen.

$$\sigma_s = \beta \cdot \frac{P}{A_s} \qquad (2)$$

The factor β can be determined from the specimens dimensions and has a value of 1.25 for specimen type I (diameter 12 mm), 1.50 for specimen type II (diameters 20 and 25 mm) and 1.75 for specimen type III (diameters 32 and 40 mm).

Two values are of major interest: the ultimate bond strength τ_R and the so-called characteristic bond strength τ_M. The ultimate bond strength is defined as the bond stress corresponding to the ultimate load recorded during testing. The characteristic bond strength is calculated as the mean value of the bond stresses corresponding to a slip of 0.01 mm, 0.10 mm and 1.00 mm. Both values differ for the two halves of the specimen.

2.4.1 Influence of the concrete type

The main goal of the study is to compare bond strengths for self-compacting concrete with those for conventional vibrated concrete. In Figure 3 one

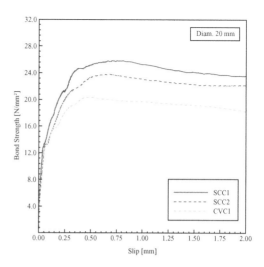

Figure 3. Bond stress—slip diagram for bar diameters 20 mm.

Table 4. Test results for beam tests with bond length of 5ϕ.

	τ_M [N/mm²]	DEV [N/mm²]	τ_R [N/mm²]	DEV [N/mm²]
SCC1–12	18.13	0.99	27.82	3.17
SCC1–20	14.94	0.77	24.07	1.84
SCC1–25	12.80	0.81	19.39	1.27
SCC1–32	11.24	0.59	20.49	1.07
SCC1–40	9.71	0.55	19.86	0.93
SCC2–12	15.77	1.47	25.70	2.93
SCC2–20	13.31	0.25	21.54	1.56
SCC2–25	12.10	0.29	18.60	2.03
SCC2–32	10.65	0.28	19.77	0.85
SCC2–40	8.81	0.29	17.48	0.44
CVC1–12	13.45	0.73	19.88	0.75
CVC1–20	12.96	0.53	19.46	0.82
CVC1–25	11.14	0.94	16.28	1.50
CVC1–32	9.67	0.28	18.10	1.00
CVC1–40	8.13	1.22	16.61	2.09

of the recorded bond stress—slip curves (mean of 4 measurements) is plotted for bar diameter 20 mm and different concrete compositions. In table 4 the values of the characteristic bond strength and the ultimate bond strength are given for all specimen types as well as the standard deviation (DEV).

Comparing the different types of concrete for the same bar diameter, CVC1 and SCC2 (which have almost the same compressive strength) have comparable values for the characteristic bond stress τ_M, except for bars diameter 12 mm for which a significant difference between the 2 concretes is noticed. The difference for the ultimate bond stress τ_R is somewhat larger. For all tests on SCC2, τ_R is above the ultimate bond stress of CVC1.

When the bond stress-slip relations of the different concrete types are plotted for tests on specimen with a reinforcing bar of the same diameter, it can be seen that the bond strength of SCC1 is larger than those of SCC2 and CVC1 (as was expected due to the higher compressive strength) at all stress levels, resulting in a steeper curve. For bar diameters of 40 mm the curves for SCC2 and CVC1 are almost identical for small slip values, while the bond stress level for SCC1 for the same slip is higher. For all other diameters, the bond stresses for the SCC2 specimens are higher for the same slip as recorded for the CVC1 specimens. In some cases the stresses even approach the values for SCC1.

2.4.2 *Influence of the bar diameter*

When the results of all test are compared, it can be seen that an increase in the bar diameter results in a decrease of τ_M and τ_R. As the concrete strength influences the bond properties of the concrete, the bond stress is normalised by the root of the compressive strength:

$$\tau_{R,n} = \frac{\tau_R}{\sqrt{f_c}} \qquad (3)$$

In Figure 4, $\tau_{R,n}$ is plotted for all concrete mixes and tested bar diameters.

The differences in the normalized ultimate bond strength for the conventional vibrated concrete and the self-compacting concrete is largest for bar diameters of 12 mm. The difference becomes smaller for higher bar diameters, but the results for self-compacting concrete are higher in all cases. There

are no significant differences between the normalized ultimate bond strength of SCC1 and SCC2, except in case of a 40 mm reinforcement bar.

By increasing the bar diameter, the slip at ultimate bond stress s_u is increasing in all cases. No significant difference can be noticed between the results for self-compacting concrete and the results for conventional vibrated concrete.

3 PREDICTION MODELS

In literature a lot of models to predict the ultimate bond strength, corresponding slip and equations to describe the bond stress-slip behaviour can be found. In these models several parameters such as bar diameter, concrete cover, concrete compressive strength, ... are incorporated.

All equations have been established by linear or non-linear regression on obtain test results with varying parameters, but most for conventional vibrated concretes with compressive strengths in the range between 20 and 50 MPa. Few tests have been done on high strength concretes with compressive strengths above 60 MPa. The compressive strengths of the concretes used in this research project are all around 60 MPa or higher for the SCC1 mixture.

3.1 *Bond stress—slip relation*

A model for the bond stress-slip behaviour is required to be able to make calculations of the crack pattern, crack widths,.... Therefore a relationship has been proposed in the Ceb-Fib Model Code 1990 (Figure 5). It consists of an increasing first branch up to the ultimate bond stress. This branch is followed by a plateau during which slip is increasing for constant bond stress, after which bond stress starts to decrease for increasing slip values. Finally a constant residual bond strength is reached which is due to pure friction between the reinforcing bar with the cracked concrete lugs and the surrounding concrete:

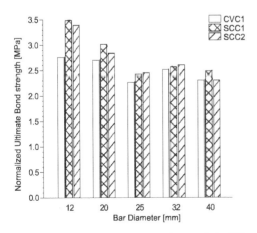

Figure 4. Normalized ultimate bond strength for different diameters and concrete compositions.

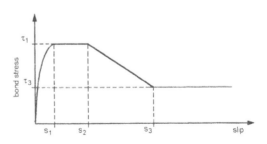

Figure 5. Prediction model for the bond stress—slip relationship according to MC90 and Huang et al.

$$\tau = \tau_1 \cdot \left(\frac{s}{s_1} \right)^{\alpha} \qquad \text{for} \quad 0 \le s \le s_1 \qquad (4)$$

$$\tau = \tau_1 \qquad\qquad\qquad s_1 < s \le s_2 \qquad (5)$$

$$\tau = \tau_1 - \left(\tau_1 - \tau_3 \right) \cdot \left(\frac{s - s_2}{s_3 - s_2} \right) \qquad s_2 < s \le s_3 \qquad (6)$$

$$\tau = \tau_3 \qquad\qquad\qquad s_3 < s \qquad (7)$$

The parameters in this model have been pre-scribed in the code for confined and unconfined normal strength concrete with good or other bond conditions. Huang et al (1993) proposed values for the parameters for normal and high strength con-crete under good bond conditions (Tables 5 and 6).

Both models (Ceb-Fip confined concrete and Huang high strength concrete) are compared with the measured bond stress-slip behaviour of the specimen (Figures 6 to 8). It can be seen that the predicted values calculated with the Huang-model for high strength concrete overestimate the bond strength in all cases. For the MC90-model, the values are underestimated for small diameters and in the range of the measured values for larger bar diameters.

Table 5. Values for the prediction equations according to Ceb-Fip MC90 (1999) for good bond conditions.

	Confined concrete	Unconfined concrete
s_1	1.0 mm	0.6 mm
s_2	3.0 mm	0.6 mm
s_3	distance betw. ribs	1.0 mm
α	0.4	0.4
τ_1	$2.5\sqrt{f_c}$	$2.0\sqrt{f_c}$
τ_3	$0.4\,\tau_1$	$0.15\,\tau_1$

Table 6. Values for the prediction equations according to Huang et al. (1996) for good bond conditions.

	High strength concrete	Normal strength concrete
s_1	0.5 mm	0.6 mm
s_2	1.5 mm	0.6 mm
s_3	distance betw. ribs	1.0 mm
α	0.3	0.4
τ_1	$0.4 \cdot f_{cm}$	$0.4 \cdot f_{cm}$
τ_3	$0.4\,\tau_1$	$0.4\,\tau_1$

The bar diameter has also an influence on the slip corresponding with the ultimate bond strength as can be seen on the graphs. Both models however have fixed values of this slip value regardless the bar diameter.

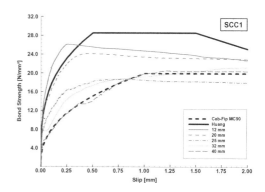

Figure 6. Comparison of measured bond stress—slip behaviour for SCC1 with prediction models.

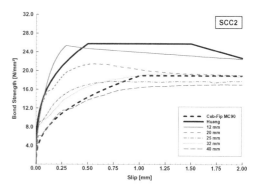

Figure 7. Comparison of measured bond stress—slip behaviour for SCC2 with prediction models.

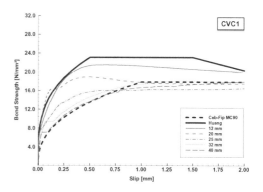

Figure 8. Comparison of measured bond stress—slip behaviour for CVC1 with prediction models.

Out of this it can be concluded that the models should be modified to take in to account the effect of the bar diameter.

3.2 *Maximum bond strength*

To get a better prediction of the ultimate bond strength more sophisticated models are necessary. Some of the models to predict the ultimate bond strength are show in Table 7 with the used unit system. Key parameters in these equations are the ratio of the concrete cover c to the bar diameter ϕ, the ratio of the bar diameter to the bond length l_d and the root of the compressive strength f_c.

In Figures 9 to 11 a comparison is made between the obtained values and the predicted ones for the different concrete mixes in a bond stress versus bar diameter diagram. It shows an underestimation of the ultimate bond strength in all cases, except for the Huang model (1996) which is developed for high strength concrete. The predicted value is

sometimes only 40% of the recorded one. Some models give bond strengths independently from the bar diameter, others seems to predict the trend quite good but give values that are too low, e.g. the models by Orangun (1977) and Harajli (1994). Therefore the coefficients out of the Orangun-model have been re-determined by linear regression based on the obtained results.

Due to the fact that the bond length has been kept constant in these test, the term with l_d becomes constant. The determined coefficients for both self-compacting concretes were almost identical, but the coefficients for the conventional vibrated concrete differed. The new equations are given in table 8. The good correlation between the

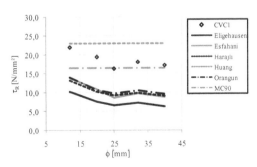

Figure 10. Bond strength versus bar diameter—comparison of models and test results for SCC2.

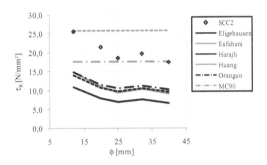

Figure 11. Bond strength versus bar diameter—comparison of models and test results for CVC1.

Table 7. Prediction models for bond strength.

Author	Equation	Units
Eligehausen	$\tau_R = 0.75 \cdot \sqrt{c/\phi} \cdot \sqrt{f_c}$	SI
Esfahani*	$\tau_R = 4.73 \cdot [(c/\phi) + 0.5]/[(c/\phi) + 5.5] \cdot \sqrt{f_c}$	SI
Harajli	$\tau_R = [1.2 + 3 \cdot (c/\phi) + 50 \cdot (\phi/l_d)] \cdot \sqrt{f_c}$	Psi
Huang**	$\tau_R = 0.45 \cdot f_{cm}$	SI
Orangun	$\tau_R = [1.22 + 3.23 \cdot (c/\phi) + 53 \cdot (\phi/l_d)] \cdot \sqrt{f_c}$	Psi
MC 90	$\tau_R = 2.5 \cdot \sqrt{f_{ck}}$	SI

* Equation is valid for $f_c > 50$ MPa.
** Equation is valid for f_c between 60 and 120 MPa.

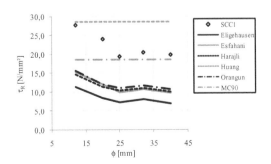

Figure 9. Bond strength versus bar diameter comparison of models and test results for SCC1.

Table 8. Modified prediction models for bond strength.

Concrete type	Equation	Units
SCC	$\tau_R = [1.77 + 0.49 \cdot (c/\phi)] \cdot \sqrt{f_{cm}}$	SI
CVC	$\tau_R = [1.87 + 0.35 \cdot (c/\phi)] \cdot \sqrt{f_{cm}}$	SI

predicted and the measured bond strengths can be noticed in figure 8 (for SCC mixtures).

4 CONCLUSIONS

Based on the obtained results, the following conclusions can be made:

a. The bond strength of self-compacting concrete is as high as the bond strength for conventional vibrated concrete when large bar diameters are studied. For smaller bar diameters, the bond strength of SCC is slightly higher, with the largest difference occuring for the smallest bar diameters.
b. For equal water to cement ratio the compressive strength of the powder-type self-compacting concrete is higher (due to the limestone filler content), and so are the maximum and characteristic bond strengths.
c. The slip corresponding to the maximum bond strength is increasing for increasing bar diameters.

Considering the bond models, the following can be concluded:

a. The bond stress-slip model out of MC90 does not implement the influence of the bar diameter, resulting in a model that underestimates the ultimate bond strength for small bar diameters and approaches the values for τ_R for large bar diameters.
b. The bond stress—slip model does not take into account the influence of the bar diameter on the slip corresponding with the ultimate bond strength.
c. Almost all existing models for predicting the ultimate bond strength underestimate the actual value for SCC as well as for CVC
d. A modification can be made to the models to get a better prediction of the ultimate bond strength, as discussed.

REFERENCES

CEB-FIP. 1990. *Model Code 1990*. Comité Euro-International du Béton, Lausane, France
Chan, Y., Chen, Y. & Liu, Y. 2003. Development of bond strength of reinforcement steel in self-consolidating concrete, *ACI Structural Journal*, 100 (4).

De Schutter, G. & Boel, V. (eds) 2007. *Proceedings of the 5th International RILEM Symposium on Self-Compacting Concrete*, RILEM Proceedings PRO54, RILEM Publications.
De Schutter, G., Bartos, P., Donome, P. & Gibbs, J. 2008. *Self-Compacting Concrete*, Caithness, Whittles Publishing.
Dehn, F., Holschemacher, K. & Weiβe, D. 2000. Self-Compacting Concrete (SCC) Time Development of the Material Properties and the Bond Behaviour, *LACER*, 5.
Desnerck, P., De Schutter, G. & Taerwe, L. 2008. Bond Strength of Reinforcing Bars in Self-Compacting Concrete: Experimental Determination, *3rd North American Conference on the Design and Use of Self-Consolidating Concrete (SCC2008)*, Chicago, USA.
Harajli, M.H. 1994. Development/splice strength of reinforcing bars embedded in plain and fiber reinforced concrete, *ACI Structural Journal*, 91 (5), 511–520.
Harajli, M.H. 2007. Numerical bond analysis using experimentally derived local bond laws: A powerful method for evaluating the bond strength of steel bars. Journal of Structural Engineering. 133 (5), 695–705.
Huang, Z., Engström, B. & Magnusson, J. 1996. *Experimental investigation of the bond and anchorage behaviour of deformed bars in high strength concrete*. Report 94:4, Chalmers University of Technology, Chalmers.
Martin, H. 2002. Bond Performance of Ribbed Bars (Pull-Out-Tests)—Influence of Concrete Composition and Consistency, *Proceedings of the 3rd International Symposium: Bond in Concrete—from Research to Standards,* 289–295, Budapest.
Menezes de Almeida Filho, F., et al. 2008. Bond-slip Behavior of Self-Compacting Concrete and Vibrated Concrete using Pull-out and Beam Tests, *Materials and Structures*, 41, 1073–1089.
Orangun, C.O., Jirsa, J.O. & Breen, J.E. 1977. A reevaluation of test data on development length and splices. *ACI Journal*, 74(3), 114–122.
Rilem. 1973. Technical recommendations for the Testing and Use of Construction Materials: RC6, Bond Test for Reinforcing Steel: 2. Pull-out Test, *Materials and Structures*.
Rilem. 1973. Technical recommendations for the Testing and Use of Construction Materials: RC6, Bond Test for Reinforcing Steel: 1. Beam Test, *Materials and Structures*, 96–105.
Zhu, W., Sonebi, M. & Bartos, P.J.M. 2004. Bond and interfacial properties of reinforcement in self-compacting concrete, *Materials and structures*, 37, 442–448.

Advances in Cement-Based Materials – van Zijl & Boshoff (eds)
© *2010 Taylor & Francis Group, London, ISBN 978-0-415-87637-7*

Pullout of microfibers from hardened cement paste

Carsten Rieger & Jan G.M. van Mier
ETH Zurich, Institute for Building Materials, Zurich, Switzerland

ABSTRACT: In order to enhance properties of hybrid fiber reinforced cementitious composites, microfibers can be added to bridge microcracks. The pullout behaviour of such fibers, and with that the bridging capacity, depends on the geometry and the properties of the fiber and matrix material. In the present investigation straight fibers with a smooth surface were cut from a finewire and embedded in a pure cement paste. Steel wires with a diameter of 50 µm were embedded 2, 3 and 4 mm in a cement paste and were pulled out. Pullout tests were performed to understand the interaction between microfibers and cement matrix. The pullout-strength, the pullout force-displacement diagrams and the debonding energy were determined. Two different types of wire were used. One wire type was annealed, the other one was unannealed. The annealed wire shows a lower strength but a much higher strain capacity than the unannealed wire. The wire was pulled out of cement pastes with different water/cement ratios. The pullout tests were performed using a tensile loading stage that was developed especially for these experiments. With this loading stage it is possible to perform the pullout tests in an Environmental Scanning Electron Microscope (ESEM), which allows for optical measurements. The results of the pullout tests, the different behaviour of unannealed and annealed wire in different cement matrices and ESEM pictures of the pullout behaviour are presented in this paper. With the help of energy considerations the pullout-mechanisms are further clarified.

1 INTRODUCTION

In recent years hybrid fibre concretes have been developed with flexural strengths exceeding 45–50 MPa (Rossi and Renwez 1996), (Markovic, Walraven, and van Mier 2003), (Staehli and van Mier 2004) and others. It is well known that the strength of fibre concrete depends on the fibre distance (Romualdi and Batson 1963). High fibre contents reduce the fibre distance, and with that the size of the cracks.

Long fibres tend to ball at higher fibre volumes, and the production of self-levelling/self-compacting materials becomes difficult. Short fibres are in that respect much better and can be added at larger volumes. This results in the development of hybrid fibre concrete with a full hierarchy of fibre sizes, from the micro to the meso-scale. Steel fibres or high-modulus fibres seem still preferential as they can much better take over the carrying capacity of the brittle cement matrix after it fractures.

Investigations on steel micro-fibre reinforced cement composites show that the strength of the material can be improved significantly (Banthia and Sheng 1991). The constraint of micro-cracks due to the addition of micro-fibres was observed by (Lawler, Wilhelm, Zampini, and Shah 2003). The pre-preak performance and strength was improved by preventing the formation of macro-cracks. The durability of steel micro-fiber reinforced mortars was also assessed (Pigeon, Pleau, Azzabi, and Banthia 1996).

This study deals with micro-fibres that were cut from a steel wire (see figure 1(b)). This ensures a straight geometry and a smooth surface of the fibre. Micro-steel fibres that are available on the market are usually cut from a steel block and show an irregular geometry with a rough surface (see figure 1(a)).

Due to the smooth surface the steel fibres can be pulled out of the matrix whereas the fibres with the rough surface cannot be pulled out and break. The pulled out micro-fibres in a crack are shown in figure 2. Three-point bending tests show the different behaviour of cement paste containing

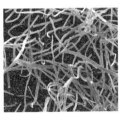
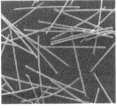

(a) Industrially produced steel micro-fibres

(b) Cut fibres from a steel finewire

Figure 1. Different types of steel micro-fibres.

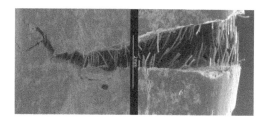

Figure 2. Crack opening and pulled out micro-fibers.

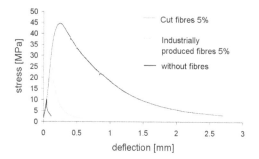

Figure 3. Three-point-bending tests on cement specimens with different steel-fibre types.

industrially produced or cut steel fibres (see figure 3).

This study focuses on single fibre pullout tests with two different steel fibres embedded in cement matrices with different water-cement ratios. There are many studies about pullout tests with different fibre types in cement matrices using different techniques for producing and testing (for example (Pinchin and Tabor 1978), (Stang and Shah 1986), (Singh, Shukla, and Brown 2004), (Kim, El-Tawil, and Naaman 2008) and others). In this study the pullout tests were performed with a special tensile testing stage in an environmental scanning electron microscope. With this the fibre pullout could be observed at high magnification with a high accuracy.

2 MATERIALS

The pullout tests were performed on specimens made of cement paste with an embedded steel fine-wire with a diameter of 50 µm. The specimens were cylindrical-shaped with a diameter of 10 mm. The thickness is varied with varying embedded length. The cement used is an Ordinary Portland Cement (CEM I 52,5 R). The chemical composition is listed in table 1.

According to the small diameter of the fine-wire with 50 µm the grains of the used cement had to be small too. For this reason a cement with a Blaine value of 5060 cm^2/g was used. The grain size distribution is listed in table 2. Almost all grains are smaller

Table 1. Chemical composition of the cement used for the specimens.

Clinker phase	Fraction [%]
CaO	61.5
SiO_2	19.8
Al_2O_3	4.9
Fe_2O_3	3.3
MgO	2.1
K_2O	0.83
Na_2O	0.31

Table 2. Grain size distribution of the cement used for the specimens.

Grain size [µm]	Fraction [%]
16	70.8
32	94.8
45	99.4
63	100
90	100
200	100
d(10%)	1.6 µm
d(50%)	9.5 µm
d(90%)	26.6 µm

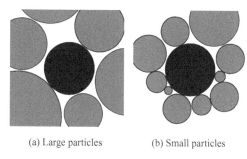

(a) Large particles   (b) Small particles

Figure 4. Steel fibre (black) between different sized cement particles (grey) before hydration.

than 45 µm. After the grains hydrate the CSH structure has a much smaller size. The initial packing before hydration is very important; too much hollow space should be prevented (see figure 4), since this will after hydration lead to large porosity, which may impair the quality of the bond.

The fine-wire is made of stainless steel (AISI 304 L). Two different types of wire were used for the pullout tests. One type was annealed (temperature treated) and the other one was unannealed. The annealed wire (Type A) shows a much higher strain capacity but a lower tensile strength than the unannealed wire (Type B) (see table 4). The chemical composition of the two wire types is listed in table 3.

Table 3. Chemical composition of the two different wire types.

| Elements | Fraction [%] | |
	Annealed	Un-annealed
C	0.023	0.02
Si	0.39	0.38
Mn	1.38	1.38
P	0.03	0.03
S	0.0005	0.0006
N	0.024	0.023
Cr	18.39	18.18
Cu	0.46	0.41
Mo	0.43	0.35
Ni	9.75	9.71

Table 4. Mechanical properties of the two different wire types (Type A = annealed, Type B = un-annealed).

Type	Ult. stress R_m [MPa]	$R_p 0.2$ [MPa]	Elongation A100 [%]	D [mm]
A	919	449	26.9	0.0498
B	2361	2231	0.55	0.0503

3 SAMPLE PREPARATION

3.1 Casting

Accuracy of sample preparation is very important to get reliable results. Because of the small dimensions of wire and specimen small deviations can have large influences on the results. To remove possible residues of fibre production or pollution on the surface, the wire was cleaned with acetone before using it for specimen preparation. Furthermore it had to be ensured that the wire is aligned in the center-line of the cement cylinder. For this purpose the wire was clamped in the mould and aligned. After that the mould was closed and the cement was filled in. A similar technique for producing pullout specimens with micro-fibres can be found in (Katz and Li 1996).

3.2 Curing

The wet cement paste was stored for 1 day in a climate chamber with 95% R.H. and 20°C. After 1 day the specimens were demoulded and stored in tap-water for 27 days.

3.3 Preparing for testing

After 28 days the samples were grinded on the topside to get a cylindrical specimen with two flat sides (one moulded side and the grinded side). The

Figure 5. Schematic of a pullout specimen with support.

(lower) grinded side was glued to a steel support which was used to mount the specimen in the testing device, see figure 5.

4 TESTING DEVICE

The testing device was a special development for the single fibre pullout test. The testing device had to fit in the available environmental scanning electron microscope (ESEM). With these boundary conditions the micro tensile testing stage was developed and produced at the Institute for Building Materials at ETH Zurich. The device consists of a stepper motor with a gear that can move a cross-beam over threaded rods. The force is measured with a 20 N load cell and the displacement of the cross-beam can be measured with a displacement transducer (see figure 6). For exact measurements of the fibre pullout the experiments can be performed in the ESEM and the relative movement between fibre and matrix can be observed under high magnification. With the loading stage pullout tests can be performed in displacement control, i.e. the rotation of the stepper-motor is controlled. Different testing velocities can be programmed, for example, a testing rate and a much higher rate for moving the cross-beam in the start position. All pullout tests were performed with a rate of 1 μm/sec.

5 TESTING SERIES

The two types of fine-wire were pulled out of cement pastes that were produced with different water/cement-ratios. These tests were performed with different embedded length (l_e) of the fibres. In table 5 an overview of all parameter variations is shown.

For each series 9 specimens were produced. 5 were tested outside the ESEM and one was tested in the ESEM. In the remainder of the paper

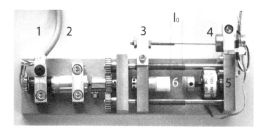

Figure 6. Tensile loading stage with 1: stepper motor, 2: gear (1:900), 3: cross-beam, 4:2 mm displacement transducer, 5:20 N load-stage, 6: specimen.

Table 5. Testing series.

Wire-type	w/c-ratio	l_e[mm]
Un-annealed (UA)	0.2	2/3/4
or	0.3	2/3/4
Annealed (AN)	0.4	2/3/4

UA2, UA3, UA4 refer to un-annealed wire tests, UN2, UN3, UN4 to annealed wire tests, where the number refers to the embedded length.

6 RESULTS

From the pullout tests the maximum pullout force (F_{max}), the force at the end of the linear-elastic behaviour (F_{el}) and the area under the pullout curve from the maximum pullout force (G_{po}) were determined, see figure 7. G_{po} was considered because from the peakload the pullout process starts. The elastic part of the curve can be considered as the wire behaviour. The wire behaviour depends on the free wire length in the testing device (l_0). Dividing the displacement by l_0, the strain of the wire can be calculated. The most accurate method to correct the curves from the free wire behaviour is to perform the test in the ESEM, because the pullout length of the wire can be measured with more accuracy than with just the testing device.

The largest pullout force was reached with the un-annealed wire and with the lowest water/cement-ratio, which was expected. The influence of w/c-ratio on the maximum pullout force is much larger with 2 mm embedded length than at 3 mm and 4 mm.

It was expected that with increasing water/cement-ratio the maximum pullout strength decreases. This effect was only observed with the un-annealed wire. The tests with the annealled wire show a nearly constant maximum pullout force with increasing w/c-ratio. The following figures show the mean-values and the standard deviation.

The specimens with $l_e = 4$ mm and $l_e = 3$ mm show nearly the same pullout force.

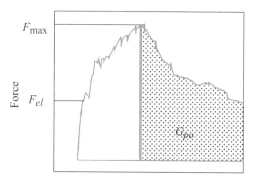

Displacement

Figure 7. Pullout force-displacement diagram with description of various parameters.

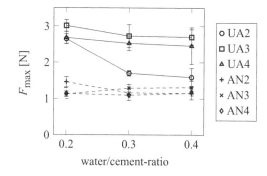

Figure 8. Maximum pullout force with different w/c-ratios, solid lines = un-annealed wire, dashed lines = annealed wire.

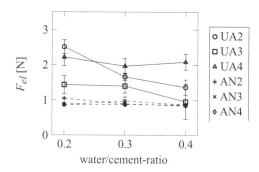

Figure 9. Force at end of linear-elastic regime with different w/c-ratios, solid lines = un-annealed wire, dashed lines = annealed wire.

The force at the end of the linear elastic regime (F_{el}) shows only differences for the un-annealed wire. The influence of the w/c-ratio is again with $l_e = 2$ mm the largest. With $l_e = 3$ mm a decrease of F_{el} with w/c = 0.4 was observed, whereas with $l_e = 4$ mm F_{el} stays nearly constant with different w/c-ratios on a high level.

The constant F_{el} with the annealed wire can be explained from the wire behaviour. The annealed wire shows at a force of about 1 N a clear bend-over in the force-displacement diagram. The pullout specimen behaves the same because the debonding and pullout-process starts later, as indicated in figure 10.

Next the pullout energy G_{po} was considered. The pullout energy was calculated as the area under the pullout curve from F_{max} to the end (F = 0). In order to compare the values for different embedded lengths G_{po} was divided through the w_{max} (original embedded length).

Figure 11 shows G_{po} for specimens with un-annealed wire. A large influence of the embedded lengths can be observed, although the values are normalised. With 2 mm embedded length the energy is lower than with 3 and 4 mm embedded length. With increasing w/c-ratio there is a tendency for the energy to increase, too.

A possible explanation for the increasing pullout energy with larger l_e can be found in longitudinal cross-sections of the specimens, see figure 12. The figure shows a polished cross-section through the fibre and the interface. One can see that the interface shows a higher porosity at the right side near the surface where the fibre is pulled out. Surface effects like this can be responsible for a reduced bond between fibre and matrix. This effect is relatively larger with smaller embedded lengths.

The specimens with annealed wire show a nearly constant pullout energy for all embedded lengths, but on a much lower level than specimens with un-annealed wire, see figure 13.

The pullout energy tends to increase with increasing w/c-ratio (for specimens with un-annealed wire) whereas the maximum pullout force decreases. This shows that with low w/c-ratios a higher strength of the material can be achieved, but the strain capacity is, as expected, lower in comparison to larger w/c-ratios.

This phenomenon can also be seen by looking at the pullout curves for different embedded lengths and is described for example in (Kim, Baillie, and Mai 1991). The maximum pullout force changes not that much, but the pullout energy changes a lot with increasing embedded length, see figure 14.

The lower the w/c-ratio the less water is available for hydration. This results in a dense matrix containing large un-hydrated cement grains. With increasing w/c-ratio the degree of hydration increases, but also the porosity increases because of the larger amount of redundant water (see figures 15(a) to 15(c)).

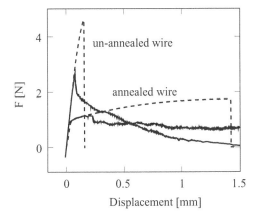

Figure 10. Force-Displacement curves of un-annealed and annealed wire (dashed) with typical pullout results (full lines).

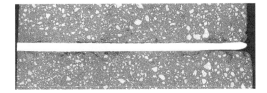

Figure 12. Longitudinal polished cross-section through fibre (white in the middle $d = 50$ µm) and surrounding matrix.

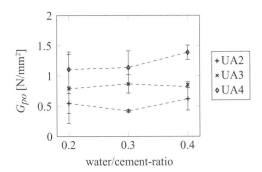

Figure 11. Pullout energy per mm embedded length for un-annealed wire.

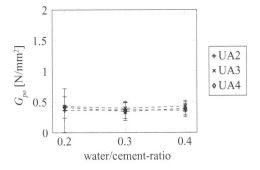

Figure 13. Pullout energy per mm embedded length for annealed wire.

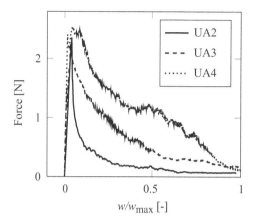

Figure 14. Pullout curves for different embedded lengths UA w/c = 0.2.

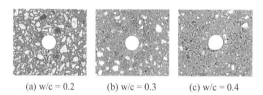

(a) w/c = 0.2   (b) w/c = 0.3   (c) w/c = 0.4

Figure 15. ESEM pictures of polished cross-sections (black is porosity, light gray hydrated cement and irregularly shaped white grains are un-hydrated cement particles).

The higher porosity in the interface between fibre and matrix leads to a lower adhesion or static friction. This explains the decrease of F_{max} and F_{el} for specimens with un-annealed wire. Figures 15(a) to 15(c) show ESEM pictures of polished cross-sections with different w/c-ratios. The wire diameter $d = 50 \mu m$ is equal in all specimens.

With pullout tests in the ESEM the movement of the wire was observed. This also showed a difference between w/c-ratios. With w/c = 0.2 only the blank wire was pulled out, where as with w/c-ratios of 0.3 and 0.4 cement particles bonded to the wire were also pulled out. That would indicate that with higher w/c-ratios there is not only an adhesive crack along the fibre-matrix interface, but also a crack in the matrix interfacial zone. This can be an explanation for the larger pullout energy at higher w/c-ratios. With increasing porosity the interlock between interface and particles bonded to the fibre gets important for the pullout behaviour.

7 PULLOUT-TESTS IN THE ESEM

Of each series one pullout-test was performed in the ESEM. In the ESEM the pullout tests can

be observed at high magnification. A further advantage is that during the pullout test pictures can be taken. With these pictures the relative displacement of the wire close to the matrix can be measured (see figure 16).

The particles on the free wire are observed and with that the fibre-pullout can be measured close to the matrix.

Using this method the influence of the free wire deformation, that is measured with the displacement transducer mounted on the testing device (i.e. #4 in figure 6) disappears. Comparing the values of the displacement transducer and the ESEM measurements leads to figure 17.

The Δw values show a linear increase until they stay on a constant level. This function can be used to correct the pullout curves for the free wire behaviour. Figure 18 shows a part of a corrected and an initial pullout curve.

Figure 18 shows that the pullout curve has only to be corrected up till the peak-load. There is no need to correct the pullout behaviour after the first peak.

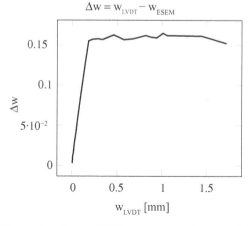

Figure 16. Fibre pullout viewed in ESEM; the position of cement particles at the fibre surface is a measure for the fibre pull-out.

Figure 17. From ESEM measurements corrected displacement.

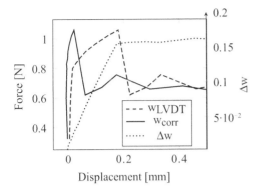

Figure 18. Corrected pullout curve.

8 CONCLUSIONS

This study shows the in fluence of wire type, w/c-ratio and embedded length on the pullout behaviour of micro-fibres with a diameter d = 50 μm. The specimens with annealed wire show a much lower pullout resistance and a lower pullout energy. The influence of w/c-ratio and embedded length is negligible.

The pullout tests with un-annealed wire show differences in embedded lengths and w/c-ratio. With 2 mm embedded length the influence of w/c-ratio is much higher than with larger embedded length. The lower the w/c-ratio, the higher the pullout resistance of the wire. The reason for this behaviour is the denser matrix with lower w/c-ratios. By contrast the energy absorbed during pullout tends to be larger with higher w/c-ratios. This can be explained with the higher porosity in the fibre-matrix interface. Due to this, there is no clear adhesive failure between fibre and matrix. The wire is pulled out and also some cement particles bonded to the fibre are pulled out. This leads to particle interlock during pullout.

The specimens with 3 mm and 4 mm embedded lengths showed the highest pullout resistance. The maximum pullout force of $l_e = 3$ mm is close to those with $l_e = 4$ mm. The pullout energy is the largest with 4 mm embedded length, caused by the larger regime of frictional pullout. The pullout energy per mm embedded length increases with increasing l_e, but only for specimens with un-annealed wire. This can possibly be explained from surface effects as depicted in figure 12. These effects do not occur in specimens where the fibres are completely embedded (for example dog bones for tensile tests). For this reason the results of G_{po} for $l_e = 4$ mm are considered more representative. Furthermore a method to correct the pullout behaviour from the free wire behaviour is introduced. When performing pullout tests in the ESEM, the fibre-pullout can be observed with high magnification. The fibre-pullout close to

the matrix can be observed and the pullout length can be measured. Also the pulled out wire can be observed and analysed due to particles bonded on the pulled out fibre. With this method the different pullout behaviour with varying w/c-ratios can be explained.

ACKNOWLEDGMENTS

The support with specimen preparation, ESEM measurements and ESEM pictures by Gabriele Peschke and Marcel Oberer is kindly acknowledged.

REFERENCES

Banthia, N. and J.K. Sheng (1991). Micro-reinforced cementitious materials. *Fiber-Reinforced Cementitious Materials 211*, 25–32.

Katz, A. and V.C. Li (1996). A special technique for determining the bond strength of micro-fibres in cement matrix by pullout test. *Journal of Materials Science Letters 15*, 1821–1823.

Kim, D.J., S. El-Tawil, and A.E. Naaman (2008). Loading rate effect on pullout behaviour of deformed steel fibers. *ACI Materials Journal* (105-M65), 576–584.

Kim, J.-K., C. Baillie, and Y.-W. Mai (1991). Interfacial debonding and fibre pull-out stresses. *Journal of Materials Science 27*, 3143–3154.

Lawler, J.S., T. Wilhelm, D. Zampini, and S.P. Shah (2003). Fracture processes of hybrid fiber-reinforced mortar. *Materials and Structures 36(257)*, 197–208.

Markovic, I., J.C. Walraven, and J.G.M. van Mier (2003). Development of high performance hybrid fibre concrete. *A.E. Naaman and H.W. Reinhardt, editors, High Performance Fiber Reinforced Cement Composites (HPFRCC4), Proceedings of the Fourth International RILEM Workshop, Ann Arbor, USA*, 277–300.

Pigeon, M., R. Pleau, M. Azzabi, and N. Banthia (1996). Durability of microfiber-reinforced mortars. *Cement and Concrete Research 26(4)*, 601–609.

Pinchin, D.J. and D. Tabor (1978). Interfacial phenomena in steel fibre reinforced cement ii: Pull-out behaviour of steel wires. *Cement and Concrete Research 8(1)*, 15–24.

Romualdi, J. and G. Batson (1963). Mechanics of crack arrest in concrete (2nd edn.). *Proc. Am. Soc. Civil Engrs. Vol 89(No. EM3)*

Rossi, P. and S. Renwez (1996). High performance multi-modal fibre reinforced cement composites (hpmfrcc). *Proceedings of the 4th International Symposium on High Strength/High Performance Concrete, Paris*, 525–535.

Singh, S., A. Shukla, and R. Brown (2004). Pull-out behavior of polypropylene fibers from cementitious matrix. *Cement and Concrete Research 34(10)*, 1919–1925.

Staehli, P. and J.G.M. van Mier (2004). Three-fibre-type hybrid fibre concrete. *V.C. Li, C.K.Y. Leung, K.J. Willian and S.L. Bllington, editors, Proceedings 5th International Conference on Fracture of Concrete and Concrete Structures (FraMCoS-5), Vail, Colorado, USA*, 1105–1112.

Stang, H. and S.P. Shah (1986). Failure of fibre-reinforced composites by pull-out fracture. *Journal of Materials Science 21*, 953–957.

Advances in Cement-Based Materials – van Zijl & Boshoff (eds)
© 2010 Taylor & Francis Group, London, ISBN 978-0-415-87637-7

Shear behavior of reinforced Engineered Cementitious Composites (ECC) beams

I. Paegle
Department of Structural Engineering, Riga Technical University, Latvia

G. Fischer
Department of Civil Engineering, Technical University of Denmark, Denmark

ABSTRACT: This paper describes an experimental investigation of the shear behavior of beams consisting of steel reinforced Engineered Cementitious Composites (ECC). Based on the strain hardening and multiple cracking behavior of ECC, this study investigates the extent to which ECC can improve the shear capacity of beams loaded primarily in shear and if ECC can partially or fully replace the conventional transverse steel reinforcement in beams. However, there is a lack of understanding of how the fibers affect the shear carrying capacity and deformation behavior of structural members if used either in combination with conventional transverse reinforcement or exclusively to provide shear resistance. The experimental investigation focuses on the influence of fibers on the shear caring capacity and the crack development in ECC beams subjected to shear. The experimental program consists of ECC with short randomly distributed PVA (polyvinyl alcohol) fiber beams with different stirrup spacing and reinforced concrete (RC) beams for comparison. Displacement and strain measurements taken using the ARAMIS photogrammetric data acquisition system by means of processing at high frame rate captured images of applied a high contrast speckle pattern to the beams surface. The multiple micro cracking resulting from the strain-hardening response of ECC in tension develop in a diagonal between the load and support point. The formation of multiple micro cracks is highly dependent on the tensile stress-strain behavior of the ECC. The shear crack formation mechanism of ECC is investigated and found to be characterized by an opening of the cracks prior to sliding. Several analytical models on shear design of ECC and concrete beams are evaluated and compared to the experimentally obtained results. The provisions of the Eurocode and ACI Code are found to be over-conservative but can be modified by utilizing the tensile strength of ECC. An expression for the load carrying capacity is proposed by expressing the ECC shear strength in terms of the crack angle.

1 INTRODUCTION

Engineered Cementitious Composites (ECC) belong to a class of materials known as High Performance Fiber-Reinforced Cement composites (HPFRCC), which are materials that exhibit pseudo strain-hardening behavior under uniaxial tension (Naaman & Reinhardt 2003). Cementitious materials with these characteristics exhibit a tensile post-cracking strength grater than the first cracking strength. To achieve this goal, the brittle cementitious matrix is reinforced by randomly distributed fibers such that the resulting composite undergoes damage in the form of multiple cracking and the resulting deformations are uniform in the material and therefore considered as strain.

ECC is characterized by a high tensile strength and ductility caused by a distinct cracking pattern consisting of multiple micro cracks. This behavior is achieved with a moderate amount of short randomly distributed synthetic fibers on the order of 2% by volume. ECC typically has an ultimate tensile strength between 4 and 6 MPa, a first crack strength 3–5 MPa, a tensile strain capacity 2–5%, a compression strength of 20–90 MPa and Young's modulus of 18–34 MPa. The tensile stress-strain behavior of ECC is illustrated in Figure 1 and compared to concrete and fiber reinforced concrete (FRC). ECC's tensile stress-strain behavior is characterized by three stages. The elastic stage, where matrix and fibers deform elastically as the load is shared by the matrix and fibers in combination relative to their stiffness and volume. The second stage is called pseudo strain hardening—characterized by an increase in load capacity after first crack accompanied by formation of multiple cracks, the strain hardening behavior continues until the bridging strength of fibers is reached.

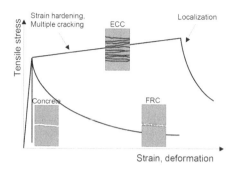

Figure 1. Comparison of stress-strain behavior of ECC, FRC and concrete (Li & Fischer 2002).

At this point a localized crack opening occurs. The third stage is the strain softening stage, which is similar to that of FRC.

The micro cracking behavior of ECC is a result of high tensile strength and a particular pullout behavior of the fibers, which form a bridge across the crack and prevent large cracks to develop and propagate. The limited crack widths of ECC are desirable due to reduced risk of infiltration of water and other materials which can be harmful and cause corrosion of the reinforcement.

2 SHEAR BEHAVIOR OF FRC

In many civil engineering and building structures, particular structural members have to resist shear loading, such as coupling beams, columns, beam ends, short cantilevers, etc. Traditionally in reinforced concrete structures, these members are reinforced with steel shear reinforcement in form of discrete stirrups. The shear failure mode of these members is typically brittle and should be avoided to prevent rapid collapse and consequently a higher safety factor is assumed for this failure mode. The shear strengths predicted by various currently used design codes for beam sections can vary by factors of more than 2. In contrast, the flexural strength predicted by the same codes is unlikely to vary by more than 10% (Bentz et al. 2006). This disparity is not only due to the lack of a rational theory for calculating the shear strength for all variations of RC beams, but also because of different safety factors used for materials and structures in shear to prevent brittle collapse.

Another critical aspect of shear resistance is related to reinforcement congestion and lack of concrete quality and compressive strength. Shear reinforcement, such as stirrups which are placed too closely in an element, may result in poor concrete compaction, causing a reduction in compressive strength and shear resistance of the member. Therefore, if the shear strength and ductility of

concrete as a material could be improved, shear failure of RC beams could be avoided and the mode of failure could be changed from brittle to ductile.

For the past decades the use of fiber reinforced concrete (FRC) or fiber reinforced cement composites (FRCC) as a means to enhance the shear resistance in RC flexural members subjected to monotonic loads has been extensively investigated. In many cases, these investigations come to the conclusion that the addition of fibers to the concrete can enhance the shear capacity and ductility of RC members.

In general, the experimental findings reported in the reviewed literature include multiple crack formation of ECC beams lead to equal or higher loading capacity than the comparable concrete beams in shear. The formation of multiple micro cracks is highly dependent on the tensile stress-strain behavior of the ECC. Other experimental results indicated that the addition of a relatively low volume friction of fibers to the concrete may allow a reduction of transverse reinforcement requirements in flexural members (Parra-Montesinos & Chompreda 2007).

Several approaches have been proposed to determine shear capacity of reinforced FRC (R/FRC) beams. The analytical model for predicting the shear capacity of R/HPFRCC beams is based on a truss and arch model (Shimuzu et al. 2004). The model relies on the theory of plasticity and is computed in order to describe the load transfer and to predict the shear capacity of reinforced concrete beams. The shear design provisions of the ACI 318 code provisions and the Eurocode EN 1992 (EC2) are also based on the truss model. The analytical model for R/HPFRCC was developed using expressions from the Architectural Institute of Japan (AIJ) for design of RC structures; the existing formulas were modified by including a term which represents the tensile stress carried by a HPFRCC material in a similar effect as the effect of transverse reinforcement (Kabele 2006). The following expression for the shear capacity of fiber reinforced concrete beams was developed based on this model:

$$V_u = bjd(p_w f_{yw} + \sigma_t)\cos\phi + \tan\theta(1-\beta)bhvf_c/2 \tag{1}$$

where b is the width of the beam, p_w is the transverse reinforcement ratio, jd is the distance between upper and lower reinforcement bars, f_{yw} is the yield strength of stirrups and σ_t is the tensile stress carried by ECC. The amount of the tensile capacity of the ECC matrix, which can be utilized in resisting the applied shear force was defined in four expressions proposed by several researchers. Those four expressions are dependent on either the ratio of the direct tensile strength or first cracking strength in direct tension. The results obtained from this model are highly depended on the amount of utilized tensile strength of the ECC, which varies by a factor of more than two depending on the assumptions made in these calculations.

In some experiments described in the literature it can be found that the assumptions and calculation details are designed to determine a maximum shear capacity of HPRCC beams in a relatively precise manner for a small number of experiments. The application of these calculations to other examples requires additional experimental work and further investigation of the various test configurations. This study is intended to contribute to this effort and adds to the currently available information.

3 EXPERIMENTAL PROGRAM

For this study of the shear behavior of reinforced ECC beams, experimental tests were performed to determine the influence of fibers on the shear carrying capacity and crack development in ECC beams subjected to shear.

The experimental program consists of ECC with two kinds of short randomly distributed PVA (polyvinyl alcohol) fiber beams with different stirrup spacing and reinforced concrete (RC) beams for comparison. The properties of two types of PVA fiber are listed in Table 1.

To determine the tensile stress-strain behavior and compressive strength of ECC, direct tensile and compression tests were carried out. For all ECC beams, the same mortar composition was used, consisting of fly ash, cement, water, sand, quartz powder and 2% by volume of fibers. The concrete mix used for the reference beams was made with standard components using cement, water and several sizes of aggregates. The concrete mix was intended to have the same compressive strength as the ECC used in the companion beam specimens.

3.1 Material properties

3.1.1 Compression tests
The compression tests on both the ECCs and concrete were conducted on standard cylinders with a diameter 100 mm and a height of 200 mm. The test machine with a capacity of 3000 kN was used for the compression tests. The cylinders were loaded to failure with a loading rate of 6.28 kN/s.

3.1.2 Tensile tests
The tensile stress-strain response of ECC was determined with specimens known as "dogbones". These specimens had a cross section of 25 mm × 50 mm in the representative section. The tests were conducted in a deformation controlled test machine with a capacity of 250 kN. A photogrammetric data acquisition system was used to take displacement measurements by means of processing images captured at high frame rate of the specimen surface with previously applied speckle pattern. The specimens were loaded with a loading rate of 0.5 mm/min in a configuration shown in Figure 3.

The results from the compression and tension tests for ECC and RC are shown in Table 2, where f_c is ultimate compression strength, f_t is the ultimate tension strength, f_{cr} is the first crack strength in direct tension and ε_u is the ultimate strain capacity. All results are shown as mean values. The tensile stress–strain relationship of ECCs is shown in Figure 2.

3.2 Shear tests

The shear beam tests were conducted to examine the shear behavior ECC beams. Both ECC and concrete beams with different amount of transverse reinforcement were tested in order to evaluate and compare the results.

Table 2. Test results from compression and tensile tests.

Material	f_c MPa	f_t MPa	f_{cr} MPa	ε_u %
ECC0	53.6	4.5	4.0	2.5
ECC1	38.7	4.5	4.0	0.9
Concrete (RC)	51.0			

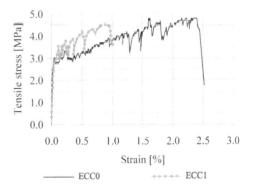

Table 1. Properties of two types of PVA fibers.

Fibre	ECC-matrix	Type	Ø µm	L mm	f_t MPa	E GPa	Strain capacity %
REC 15	ECC0	PVA	40	8	1560	40	6.5
RFS 602	ECC1	PVA	26	6	1183	28.4	9.2

Figure 2. Tensile stress—strain relationship of ECC.

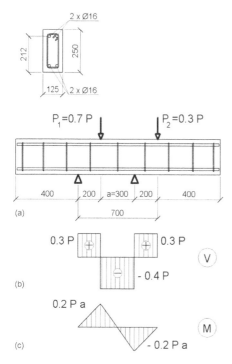

Figure 4. Test set-up and area of interest.

Figure 3. Shear beam test configuration (a), shear force distribution (b), moment distribution (c).

Table 3. Dimensions and materials of test beams.

| Beam | Material | Stirrup | |
		∅ mm	Spacing mm
ECC1	ECC1	–	–
ECC0	ECC0	–	–
ECC 1 d	ECC0	6	200
ECC ½ d	ECC0	6	100
ECC ¼ d	ECC0	6	50
RC 1 d	RC	6	200
RC ½ d	RC	6	100
RC ¼ d	RC	6	50

Table 4. Test results from shear beam tests.

Beam	First shear crack stress, τ_{cr} MPa	Mean ultimate shear stress, τ_u MPa	Mean ultimate shear strain, γ_u rad × 10^3	Main shear crack angle ϕ deg
ECC1	1.4	3.3	6.7	–
ECC0	2.1	4.6	6.9	–
RECC 1 d	1.2	5.8	7.7	44.2
RECC ½ d	1.4	6.7	9.1	43.0
RECC ¼ d	1.2	7.5	12.5	43.5
RC 1 d	1.3	4.3	7.8	42.5
RC ½ d	1.4	5.3	9.0	43.3
RC ¼ d	1.3	6.2	12.4	42.5

3.2.1 Test set up

The test set-up for the shear beam tests is shown in Figure 3. The load configuration was designed to comply with the Ohno shear beam test used in previous investigations (Li 1994, Kanda et al. 1998).

The actual test set-up is shown in Figure 4. The load was applied through a secondary load beam and rollers with a diameter 60 mm. The beams were loaded in a displacement controlled procedure with a loading rate of 1.2 mm/min.

All beams were reinforced with longitudinal reinforcement bars ∅16 mm, two of them were provided in the top and two in the bottom of beam. Part of the ECC beams and all reinforced concrete (RC) beams were reinforced with transverse reinforcement ∅6 mm. Yield strength f_y of both longitudinal and transverse reinforcement was 550 MPa. A summary of the amount of transverse reinforcement in the beams and the material is shown in Table 3.

The photogrammetric data acquisition system was used to take displacement measurements of the front surface of the specimens in the area of interest (shown in Figure 4). The stochastic speckle pattern was applied to the beam surface prior to the test.

3.2.2 Test results

The test results in terms of ultimate strength and strain are shown in Table 4 with respect to first crack strength τ_{cr}, ultimate shear stress τ_u, ultimate shear strain γ_u, and shear crack angle ϕ.

4 DISCUSSION

4.1 The ultimate shear strength

Based on the comparison of specimens RC and RECC, the ultimate shear strength of ECC beams is higher than that of the concrete beams with the same stirrup spacing and similar compression strength. The shear stress-strain relationship of the RECC beams with different amount of transverse reinforcement is shown in Figure 5. The shear capacity of ECC beams without shear reinforcement is comparable to the concrete beams with stirrup spacing d. Comparing the ultimate shear stress in the ECC beams and with that in the concrete beams with the same amount of shear reinforcement, it can be seen that the proportionally highest contribution of ECC is found for lower amount of transverse reinforcement. At a stirrup distance d, the ECC beams can take 34% more of shear stress, at distance d/2—27%, but at distance of d/4, ECC beam can take just 10% higher shear stresses than the conventional concrete beam.

In the provisions of EC2 the shear resistance of the beams with vertical shear reinforcement is based on the governing failure being in either yielding of the stirrups or crushing of concrete. Therefore the design shear resistance of the beams is taken as the smaller value of the shear resistance limited by yielding of the shear reinforcement or the crushing of the concrete compression struts. As a result, the design shear resistance of concrete beam increases by a factor of two if the distance between stirrups is decreased by a factor of two (assuming that the capacity of the concrete strut is not exceeded). From the experimental data in Figure 5 it can be seen that increasing the amount of transverse reinforcement by a factor of two increases the ultimate shear capacity only by about 20%.

In Figure 6, various differences become apparent between experimentally obtained data and EC2 calculated results. At the lower amount of transverse reinforcement (stirrup distance d), the experimentally received ultimate shear load exceeds the design value by more than two times. At a stirrup spacing of d/2, the experimentally

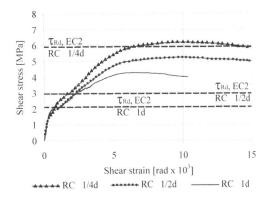

Figure 6. Average shear stress strain relationship and design ultimate shear stress according to Eurocode EN 1992 for RC beams.

obtained value exceeds calculations by approximately 70–80%, but at stirrup spacing d/4 only by approximately 5%. Furthermore, the EC2 calculations are more precise with increasing amounts of transverse reinforcement.

4.2 Crack formation

The images obtained from the image acquisition system shown in Figure 7 (a)–(f) illustrate the development of cracks in ECC beam specimen with stirrup distance d

The images (a)–(f) represent the characteristics of the crack pattern in the ECC beams in general. Image (a) illustrates the major strains in the beam at 20% of the ultimate shear strength. The first flexural cracks appear in the regions where the flexural moment in the shear span is the largest. The first diagonal shear cracks appear in image (b) which corresponds to 25% of ultimate shear strength. The flexural cracks develop further into the beam and the cracks in the lower right corner start to develop due to shear with an inclination toward the right support point. Shear cracks continue to develop further into the beam in image (c) which corresponds to 35% of ultimate shear strength. The cracks propagate at an angle of approximately 44° between the load points. A second shear crack develops parallel to the first crack. The two cracks seen in the images are in reality several micro cracks, which cannot be differentiated due to limited resolutions of the cameras used to capture the images at the given distance to the specimen. Increased micro cracking occurs parallel to the first developed shear crack band in image (d) and (e). The crack widths increase and cracks start to orientate towards the right support point as they reach the top longitudinal reinforcement. More flexural cracks develop further into the beam and the widths of the cracks increase. Multi-

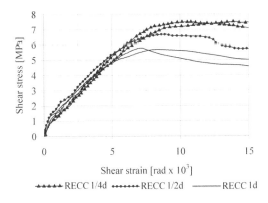

Figure 5. Average shear stress strain relationship for ECC beams.

ple shear cracks develop (Fig. 8) directly from the load point to the support point (f). The subsequent ultimate failure occurs in a localized fracture plane parallel to the first developed shear crack as the two beam parts are separated perpendicular to the fracture plane.

4.3 Crack development

Previous investigations on the shear behavior of reinforced ECC or HPFRCC beams have not reported details of the crack formation process and opening vs. sliding mechanism. To investigate the crack development of the different beam configurations in terms of crack opening and sliding, a non-contact image acquisition system is used. The analysis is performed by fixing two virtual points on the acquired image of the beam surface on both sides of a single crack and subsequently obtaining

the relative displacement between them to determine the crack width and sliding distance. The line between these two points is set to be perpendicular to the crack as illustrated in Figure 9.

In Figure 10 (a) and (b), a crack opening/sliding vs. shear stress relationship is shown for an ECC beam without shear reinforcement (ECC0) and a concrete beam (RC d) with stirrup spacing of d.

The two vertical lines shown in the plots represent the starting of crack opening and crack sliding, respectively. From these plots it can be seen that in all cases crack opening precedes crack sliding. The particle with the greatest grain size in the ECC is the quartz sand with a maximum diameter of 0.18 mm. Beams with ECC start the crack sliding process at the point when the opening of the crack is roughly 0.1 mm, which is equivalent to approximately ½ of the maximum grain size contained in the ECC material. The cracks in the ECC beams start to open at an increased rate when they exceed a width of 0.2 mm and sliding exceeds 0.08 mm. In contrast to that, the primary crack opening in the concrete beams evolves a lot faster. It should be noted that despite a largest particle size of 16 mm in the concrete specimens, the crack opening rate increased rapidly at a crack opening of approximately 0.25 mm.

Both, fibers and transverse reinforcement crossing the shear cracks, affect crack sliding and opening. Therefore, at the same shear stress in the RECC and RC beams, the crack opening in the RECC beam is approximately half of that observed in the RC beams. The maximum crack opening and sliding values at the stage of maximum applied load are shown in Table 5. From the comparison of these values it can be concluded that the crack widths measured in the RECC beams are smaller than those in the RC beams at similar or larger applied shear stresses and deformations.

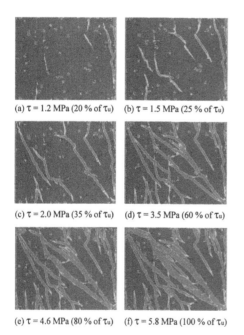

(a) τ = 1.2 MPa (20 % of τu)   (b) τ = 1.5 MPa (25 % of τu)

(c) τ = 2.0 MPa (35 % of τu)   (d) τ = 3.5 MPa (60 % of τu)

(e) τ = 4.6 MPa (80 % of τu)   (f) τ = 5.8 MPa (100 % of τu)

Figure 7. Crack formation in ECC beam specimen at different load stages.

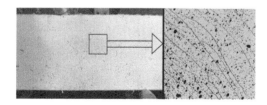

Figure 8. Multiple shear cracks in ECC beam.

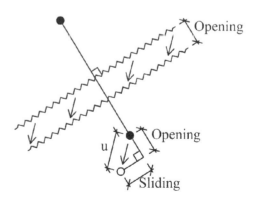

Figure 9. Illustration of analysis method for crack opening/sliding.

4.4 Derivation of expression for effective shear strength

The failure modes observed in the tested specimens are found to have similar characteristics as observed and documented in concrete compression specimens, such as cubes or cylinders, where typically a crack forms due to tension perpendicular to the major principal compressive stress. In the compression test, the crack is suppressed until sliding occurs due to compression, whereas the shear cracks in the tested beam specimens are opening before sliding. In the cylinder the crack line develops into a diagonal sliding crack that fails as the frictional strength of the crack surface is exceeded. In the shear beams the crack opening is restrained by the fibers until they allow a sufficiently large opening for the interlock between the aggregates to be overcome, which

is consequently followed by sliding. Furthermore, it can be seen that the ultimate strength of a member is a function of the angle at which the load is applied relative to the crack direction, where the contributing factors to the shear capacity of the section is the matrix cohesion, the aggregate interlock, which itself depends on the crack opening, and the tensile strength of the fibers. This shear strength f_{sh} is translated to the force which is demanded to deform the crack surface and allows sliding without any tension or compression of the crack. As the crack opens, f_{sh} is reduced by the cohesive strength of the matrix f_{co}, which is equal to the first crack strength in pure tension. The angle between load direction and crack orientation θ_{load}, see Figure 11 (Johansen & Svare 2009), is set to 90° for pure tension, −90° for pure compression, and 0° for pure shear.

An expression for the ultimate strength that can be applied to the ends of the modeled specimen, σ_u, can then be described by the bridging strength of the crack. See equations 2 and 3.

$$\sigma_{u1} = -\frac{f_{sh}}{f_1(\theta_{load})}, \quad \theta_{load} \in [-90^\circ, 0] \tag{2}$$

$$\sigma_{u2} = \frac{f_{sh} - f_{c0}}{f_2(\theta_{load})}, \quad \theta_{load} \in (0, 90^\circ] \tag{3a}$$

$$\sigma_{u3} = \frac{f_t}{f_3(\theta_{load})}, \quad \theta_{load} \in (0, 90^\circ] \tag{3b}$$

where

$$f_1(\theta_{load}) \quad \text{gives} \quad \begin{cases} \sigma_u \to \infty \ \text{for}\ \theta_{load} \to -90^\circ \\ \sigma_u = f_{sh} \ \text{for}\ \theta_{load} \to 0^\circ \end{cases} \tag{4}$$

$$f_2(\theta_{load}) \quad \text{gives} \quad \begin{cases} \sigma_u = f_{sh} - f_{co} \ \text{for}\ \theta_{load} \to 0^\circ \\ \sigma_u = f_t \ \text{for}\ \theta_{load} = 90^\circ \end{cases} \tag{5}$$

$$f_3(\theta_{load}) \quad \text{gives} \quad \begin{cases} \sigma_u = 0 \ \text{for}\ \theta_{load} = 0^\circ \\ \sigma_u = f_t \ \text{for}\ \theta_{load} = 90^\circ \end{cases} \tag{6}$$

From equations (4), (5) and (6) functions in equation (2) and (3) follow:

$$f_1(\theta_{load}) = \cos(\theta_{load}) \tag{7}$$

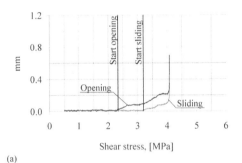

(a)

(b)

Figure 10. Crack opening and sliding to shear stress relationship for (a) ECC0 and (b) RC d beams.

Table 5. The maximum values of crack opening and sliding.

Beam	Max opening mm	Max sliding mm
ECC1	0.63	0.21
ECC0	0.70	0.32
RC 1 d	1.0	0.37
RC ½ d	1.4	0.51

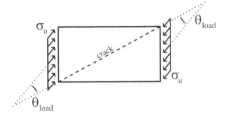

Figure 11. Geometric relationship between applied load and direction of failing surface.

$$f_2(\theta_{\text{load}}) = (\cos(\theta_{\text{load}}))^{-\sqrt{\theta_{\text{load}}}} \qquad (8)$$

$$f_3(\theta_{\text{load}}) = \frac{90}{\theta_{\text{load}}} \qquad (9)$$

The load carrying capacity in tension σ_{ut} is seen to be a function of both aggregate interlock σ_{u2} and the strength of the fibers contribution σ_{u3}. In compression, the fibers are assumed not to contribute when the sliding is initiated without opening of the crack, hence the load carrying capacity in compression $\sigma_{uc} = \sigma_{u1}$. The equations for the load carrying capacity then yield:

$$\sigma_{uc} = -\frac{f_{sh}}{\cos(\theta_{\text{load}})}, \qquad \theta_{\text{load}} \in [-90^\circ, 0] \qquad (10)$$

$$\sigma_{ut} = \frac{f_{sh} - f_{c0}}{(\cos(\theta_{\text{load}}))^{-\sqrt{\theta_{\text{load}}}}} + f_t \frac{\theta_{\text{load}}}{90}, \qquad \theta_{\text{load}} \in [(0, \ 90^\circ]$$

$$\qquad (11)$$

Typically, compression cylinders fail in diagonal sliding at a loading direction vs. crack angle $\theta_{\text{load}} = 26.6^\circ$, which results in $f_{sh} = f_c \cdot \cos(26.6^\circ)$. The dependence derived from the equations derived above is plotted in Figure 12. The input values used in this model for different materials are stated in Table 2.

5 CONCLUSIONS AND REMARKS

Comparing the ultimate shear strength of reinforced ECC and concrete beams with the same amount of shear reinforcement, the highest contribution of ECC is found for lower amounts of transverse reinforcement. At stirrup distance d, the ECC beams can take 34% more shear load, whereas at stirrup spacing d/2 and d/4, the increase in shear capacity compared to the companion RC beams is 27 and 10%, respectively.

Comparing the experimental results with design values from EC2, it can be seen that for the lower amount of transverse reinforcement, the experimentally obtained ultimate shear capacity exceeds the design values by a factor of two or more. The EC2 calculations are more precise with increasing amounts of transverse reinforcement.

Comparing the ultimate shear strain for concrete and ECC beams with the same amount of shear reinforcement, the shear strain is similar for both materials, while the corresponding shear capacity for ECC beams is higher. From the comparison of crack development in the beams, it can be concluded that the crack widths measured in the ECC beams are smaller than those in the RC beams at similar of larger applied shear stresses and deformations. The smaller crack opening is a result of the multiple cracking behavior of ECC. The shear crack development can be characterized by crack opening

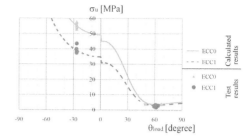

Figure 12. Calculated load carrying capacity for ECC0 and ECC1 beams with results from compression tests and shear beam tests.

and crack sliding and it was observed in all tests that crack opening occurs before crack sliding. The maximum crack opening and crack sliding of the failure crack at the maximum load stage differs more than two times, for both ECC and concrete beams.

REFERENCES

Bentz, E.C., Vecchio, V.J. & Collins, M.P., 2006. Simplified Modified Compression Field Theory for Calculating Shear Strength of Reinforced Concrete Elements, Structural Journal, V103: 614–624.

Eurocode 2: Design of concrete structures—Part 1-1: General rules and rules for buildings (EN 1992-1 - 1:2004), 2004, CEN.

Johansen, S. & Svare, P.-A. 2009. Shear behaviour of PVA-ECC and reinforced concrete beams, DTU report.

Kabele, P., 2006. Fracture Behavior of Shear-Critical Reinforced HFRCC Members, Proceedings on the 49th International RILEM Workshop on High Performance Fiber Reinforced Cementitious Composites in Structural Applications, Honolulu, USA: 383–392.

Kanda, T., Watanabe, S. & Li, V.C., 1998. Application of Pseudo Strain Hardening Cementitious Composites to Shear Resistant Structural Elements, Fracture Mechanics of Concrete Structures, Proceedings on FRAMCOS-3, AEDIFICATIO Publishers, Freiburg, Germany: 1477–1490.

Li, V.C., Mishra, D.K., Naaman, A.E., Wight, J.K., LaFave, J.M., Wu, H.C. & Inada, Y., 1994. On the Shear Behavior of Engineered Cementitious Composites, Advanced Cement Based Materials vol. 1 is. 3: 142–149.

Li, V.C. & Fischer, G., 2002. Reinforced ECC—An Evolution from Materials to Structures, Proceedings of the First FIB Congress, Osaka, Japan: 105–122.

Naaman, A.E. & Reinhardt, H.W., 2003. High Performance Fiber Reinforced Cement Composites, RILEM Publications S.A.R.L., Proceedings PRO 6, France.

Parra-Montesinos, G.J. & Chompreda, P., 2007. Deformation Capacity and Shear Strength of Fiber Reinforced Cement Composite Flexural Members Subjected to Displacement Reversals, Journal of Structural Engineering, ASCE, Vol. 133, No. 3: 421–431.

Shimuzu, K., Kanakubo, T., Kanda, T. & Nagai, S., 2004. Shear Behavior of Steel Reinforced PVA-ECC Beams, 13th World Conference on Earthquake Engineering, Vancouver, Canada.

Advances in Cement-Based Materials – van Zijl & Boshoff (eds)
© 2010 Taylor & Francis Group, London, ISBN 978-0-415-87637-7

Mechanical interaction of Engineered Cementitious Composite (ECC) reinforced with Fiber Reinforced Polymer (FRP) rebar in tensile loading

L.H. Lárusson, G. Fischer & J. Jönsson
Department of Civil Engineering, Technical University of Denmark, Denmark

ABSTRACT: This paper introduces a preliminary study of the composite interaction of Engineered Cementitious Composite (ECC), reinforced with Glass Fiber Reinforced Polymer (GFRP) rebar. The main topic of this paper will focus on the interaction of the two materials (ECC and GFRP) during axial loading, particularly in post cracking phase of the concrete matrix. The experimental program carried out in this study examined composite behavior under monotonic and cyclic loading of the specimens in the elastic and inelastic deformation phases. The stiffness development of the composite during loading was evaluated as well as crack widths and crack distributions in the ECC. Results indicate that the interaction of the ductile ECC together with the elastic brittle behavior of the GFRP make a highly compatible ductile composite. The combination of multiple cracking and limited crack width of ECC insures good stain distribution which in terms results in less mechanical deterioration during loading.

1 INTRODUCTION

1.1 Background

The experimental program presented in this paper is part of a study that focuses on research and development of a prefabricated continuous expansion joint using composite materials such as Engineered Cementitious Composites (ECC) and Glass Fiber Reinforced Polymer (GFRP) reinforcing bars to improve both the performance as well as fabrication and implementation methods of current expansion joints.

The mechanical interaction of conventional concrete matrices and steel reinforcement in tensile loading has been a subject of previous investigations (Bischoff & Paixao 2004, Fischer & Li 2002), which indicate that the three main contributing mechanisms in the composite behavior are the tensile loads carried over transverse cracks in the concrete, the shear forces between the cementitious matrix and the reinforcement, and the forces carried by the steel reinforcement at the crack locations. The brittle nature of conventional concrete however prevents it from carrying any significant tensile forces across the crack faces. The use of a fiber reinforced cement composite to substitute the concrete matrix was intended to maintain a force transfer over the cracks and furthermore utilize the ductile nature of ECC in particular to alter the relative deformation behavior between cementitious matrix and reinforcement towards compatible deformations between both materials and thus reducing or eliminating interfacial bond stresses. Furthermore, the use of FRP reinforcement in place of conventional steel was intended to provide the resulting reinforced ECC composite member with a relatively low stiffness and elastic deformation behavior to minimize residual deformations.

The multiple cracking characteristics of ECC results in reduced crack width and increased deformation capacity of the reinforced ECC member, which is the objective of the targeted application, while the corrosion resistance of the FRP reinforcement results in increased durability as well as elastic deformation behavior.

Continuous expansion joints in road and bridge structures, so-called link slabs (Caner & Zia 1998), are typically constructed of regular steel reinforced concrete (R/C), which in many cases is cast in place. In these steel reinforced concrete link slabs, the concrete forms transverse cracking as deformations from the adjacent bridge spans are imposed. Upon cyclic reversal of these imposed deformations, the steel-concrete interface degrades and the steel reinforcement is exposed to surface water and possibly deicing salt, which can lead to corrosion and loss of composite integrity.

1.2 Materials

Engineered cementitious composite (ECC) is a subclass of high performance fiber reinforced cementitious composites (HPFRCC). The ability of ECC to increase its tensile strength after first crack formation is realized by a micromechanically designed interaction between fibers, cementitious matrix and the interfacial bond between fibers and cementitious matrix. As a result ECC can exhibit the strain capacity of 3–5% by developing multiple cracking during tensile loading process, over 300 times that of normal concrete (Li & Lepech 2006). The strain hardening attribute of ECC is realized by an intricate relation between a limited crack width and the resulting multiple cracking. This ability distinguishes ECC from regular brittle concrete and conventional tension softening fiber reinforced concrete (FRC) as is illustrated in figure 1.

ECC used in this study is composed of a fine grained cementitious mortar consisting of cement, fly-ash, quarts powder, fine sand, water, admixtures and fiber reinforcement. The lack of aggregates or coarser sand in the cementitious matrix is due to their adverse effect on the fiber distribution and a required limitation on the fracture energy of the mortar to obtain the strain hardening and multiple cracking properties.

The ability of ECC to exhibit high strain capacity while keeping its structural integrity is an important benefit of ECC utilized in this experimental program with the objective to obtain a structural member with large reversible deformation capacity.

A study carried out on the interaction of ECC and steel reinforcement (R/ECC) in comparison to regular concrete and steel reinforcement (R/C) revealed a significant difference in the stress distribution of cementitious matrix and steel reinforcement after transverse cracking formed in the composite (Fischer & Li 2002). In the R/ECC specimen, the presence of multiple cracks in ECC along with the fiber bridging between the transverse cracks allows for a uniform stress distribution in the cracked and un-cracked parts of the ECC, as opposed to a localized crack formation in

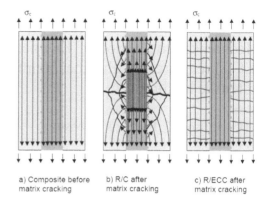

Figure 2. Crack formation and stress distribution in R/C and R/ECC (Fischer & Li 2002).

conventional concrete and consequently a stress concentration in the steel reinforcement of the R/C. This is illustrated in figure 2.

As an alternative to steel reinforcement, the use fiber reinforced polymer (FRP) rebars introduces two major benefits: 1) enhanced corrosion resistance or even elimination of damage by reinforcement corrosion and 2) a low effective stiffness of the member with elastic deformation behavior and reduced residual deformations. Although the FRP reinforcement possesses high tensile strength and a good fatigue and damping response, the relatively low elastic modulus, ranging from 40 to 150 GPa, results in relatively large deformations (Bischoff & Paixao 2004), which is undesirable in typical applications as it can result in large deflections. In the particular application investigated in this study, however, the low stiffness and large deformability are design goals and therefore very attractive material characteristics.

In this study, glass fiber reinforced polymer (GFRP), a subclass of FRP, was used as a substitute for steel reinforcement. The rebars are made from glass fibers impregnated with epoxy resin and coated with coarse sand to improve adhesion.

GFRP is a linear elastic, brittle material, i.e. it exhibits a linear elastic behavior in tensile loading up failure while steel reinforcement has a ductile yielding and strain hardening phase before failing.

The GFRP rebars used in this study are a commercially available product, with tensile strength of 655 MPa and a relatively low elastic modulus of 40.8 GPa.

2 CONCEPT AND DESIGN

Continuous expansion joints in bridge structures and elevated roads can be composed of steel reinforced concrete in so-called link slabs, which are

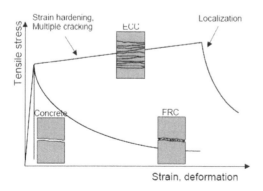

Figure 1. Schematic tensile stress-strain behavior of cementitious materials.

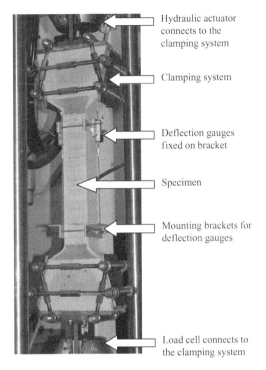

— Hydraulic actuator connects to the clamping system

— Clamping system

— Deflection gauges fixed on bracket

— Specimen

— Mounting brackets for deflection gauges

— Load cell connects to the clamping system

Figure 3. Test configuration for the large dog bone specimens with clamping system in place.

designed to accommodate relative displacements of the adjacent spans. By using ductile cementitious material like ECC instead of brittle concrete and elastic, non corrosive GFRP rebar instead of conventional steel reinforcement in theses joints it is expected that the structural performance and durability of these joint can be improved. The aim of the work presented in this paper is to evaluate the mechanical interaction of a ECC and GFRP composite bar element in monotonic and cyclic loading. Furthermore the effective stiffness along with the transverse cracking behavior of the GFRP reinforced ECC elements are of particular interest.

The test setup shown in figure 3 was designed to evaluate the composite elements under reversed cyclic axial loading. The purpose of the dog bone shaped specimen geometry and their fixation in the loading machine is to confine and hold the specimen at both ends in a prestressed state to secure the load transfer zone and to prevent slipping of the specimen at the supports during cyclic loading.

3 EXPERIMENTAL PROGRAM

3.1 Test specimens

In this experimental program, three large dog bone shaped specimens were cast and tested (LDB-1,

LDB-2 and LDB-3 please renumber the specimens in the following). The GFRP reinforced ECC elements were each 1000 mm long with a narrower 500 mm long middle stretch and a constant cross section where strain measurements were taken. Over the whole length, a single GFRP rebar was positioned in the center extending from one end to the other in the longitudinal direction.

The first dog bone specimen (LDB-1) had a cross section of 150×150 mm^2 and Ø13 mm GFRP rebar, resulting in a reinforcement ratio of $\rho = 0.6\%$. The second and the third element (LDB-2 and LDB-3) had the same cross section, 100×110 mm^2 and a single Ø13 mm GFRP rebar, resulting in reinforcement ratio $\rho = 1.2\%$.

After the elements were taken out of their casting moulds, they were air-cured at a constant temperature of 20°C and relative humidity between 50–60%. Some drying shrinkage induced cracking was observed in element LDB-1, both in the longitudinal and transverse direction, whilst no such cracks were observed for elements LDB-2 or LDB-3.

To measure the tensile strength of ECC in this study a number of tension test (small dog bones) were prepared. These test specimens were subsequently cured in the same conditions as the large dog bone elements to ensure a similar curing environment.

3.2 Test setup and procedure

On each end of the large dog bone specimens a clamping system was used to provide a sturdy connection between the specimen and the applied load. Furthermore the clamping system enhances and secures the load transfer zone between the ECC and the GFRP. By confining the ECC surrounding the GFRP rebar, the load transfer zone becomes more effective and failure by pullout is less likely. Finally, steel cylinders filled with epoxy glue were used to secure the protruding ends of the GFRP to further prevent failure by rebar pullout.

A schematic of the test setup is shown in figure 4. The elements were subjected to axial tensile and compressive loading by applying the load

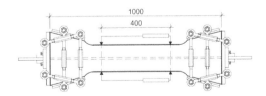

Figure 4. Schematic test configuration for the large dog bones elements with clamping system in place. Loading was applied to the clamping system at the ends.

Figure 5. Small dog bone specimen used to derive tensile strength and behavior of ECC.

directly at the clamping mechanism. Two linear displacement gauges were positioned on opposite sides of the representative cross-section of the element to measure longitudinal displacements over a 400 mm gauge length.

Additionally to these conventional measurement devices element LDB-3 and a number of small ECC dog bone specimens were equipped with a non-contact image analysis system to measure global and local deformations to subsequently deduce strain measurements. This system was used to evaluate and determine the strain, crack with, crack spacing and overall crack development during testing. For elements LDB-1, LDB-2 and LDB-3 as well as the small dog bone tests, physical observations and remarks were also made on the crack distribution and spacing.

For element LDB-1, five different loading sequences were applied by deforming the specimen up to a certain strain and then cycling it back to zero deformation before the setup was loaded up to failure.

For elements LDB-2 and LDB-3, loading was applied up to approximately 60% of the assumed ultimate tensile loading capacity (f_u) of the cross section. Subsequently from that position a displacement controlled cyclic load was applied as sinus type cycles at a frequency of 0.04 Hz.

Accompanying these large dog bone tests on the GFRP reinforced ECC, the tensile stress-strain response of the ECC matrix was evaluated by testing small dog bone (SDB) specimens in tensile loading. The SDB specimens were 500 mm long with a 210 mm long narrower representative cross section of 50 × 25 mm, see figure 5. The specimens were firmly clamped at both ends and a tensile load was applied with a load rate of 0.5 mm/min.

4 RESULTS

The first element (LDB-1) was a pilot test where the results were used to refine both the test setup and test procedures. Elements LDB-1, LDB-2 and LDB-3 were all tested in monotonic tensile loading, followed by a cyclic loading sequences. The results focus on the composite behavior and crack formation of each element while comparing it to the measured material response of ECC and GFRP.

4.1 Stiffness behavior

The response of element LDB-1 during monotonic loading is shown in figure 6, for comparison the GFRP response is plotted.

The first load sequence shows the behavior of the un-cracked element upon initial loading, reaching about 75 kN before the ECC matrix starts to yield, equivalent to a nominal composite tensile stress of 3.3 MPa. From the material testing of LDB-1, where small ECC dog bone specimens were tested in tensile loading, the first crack formation occurred on average at a stress value of 4.8 MPa, but measurements ranged from 3.1 to 6.2 MPa.

The initial tension-stiffening effect of the un-cracked ECC matrix surrounding the GFRP rebar decreases at each subsequent loading sequence, while ultimately converging at an elevated level parallel to the elastic response of the bare GFRP rebar.

The response of elements LDB-2 and LDB-3 during the initial monotonic loading up to approximately 60% of the assumed ultimate limit capacity of the cross section (f_u) plus the first 10 cycles are shown in figure 7. For comparison the bare GFRP bar response is also plotted.

Elements LDB-2 and LDB-3 both reached about 46 kN tensile load before first crack formation was observed, equivalent to a nominal composite tensile stress of 4.2 MPa. In comparison, the average first crack tensile strength of ECC obtained in the corresponding small dog bone tests was 3.5 MPa. Furthermore both elements follow the same response path and reached approximately the same 0.8% strain at the target load of 76 kN.

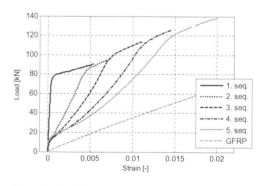

Figure 6. Load-strain responses of element LDB-1 during five different loading sequences. The response of the GFRP is shown in reference to the composite response.

The stiffness response of the composite elements LDB-2 and LDB-3 are equivalent to the stiffness of the bare GFRP rebar while the difference in elevation corresponds to the tensile strength of ECC in addition to GFRP rebar. By

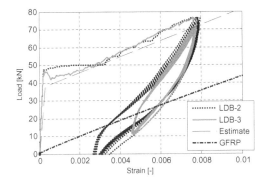

Figure 7. Load-strain response of elements LDB-2 and LDB-3 during loading up to approximately 60% of ULS followed by the first ten cycles of the cyclic loading sequence. The GFRP response and the superimposed composite response are shown in reference to the actual data.

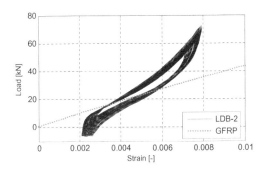

Figure 8. Cyclic load-strain response of element LDB-2 during 1000 cycles.

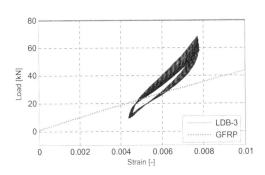

Figure 9. Cyclic load-strain response of element LDB-3 during 1000 cycles.

superimposing the contribution of the GFRP rebar and the contribution of the ECC matrix to estimate the composite response of the GFRP reinforced ECC element, a bilinear response was obtained based on the tensile behavior of ECC and the tensile response of the bare GFRP bar. The resulting estimate is seen in comparison with LDB-2 and LDB-3 in figure 7.

The cyclic response of LDB-2 and LDB-3 are shown in figures 8 and 9 respectively. The cyclic loading sequence for LDB-2 was initiated at 0.78% strain (76 kN) and unloaded to 0.29% strain (−6 kN compression), corresponding to total deformation of 2.1 mm of the representative section. During the loading sequence, the hysteresis loops sagged slightly at its upper crest while the strain shifted slightly to the left, both at the upper and lower crest.

For LDB-3 the cyclic loading sequence was initiated at 0.77% strain (76 kN) and unloaded to 0.46% strain (11 kN), corresponding to a total deformation of 1.3 mm of the representative section.

As with LDB-2, the hysteresis loops of LDB-3 sagged at its upper crest but held its position relatively steadily at the lower end.

The strain hardening behaviors of three ECC SDB specimens during tensile loading are seen in figure 10. All the specimens have a first cracking strength of approximately 3.4 MPa, ultimately reaching a tensile strength of 3.9 MPa (SDB-1) and 5.0 MPa (SDB-2 and 3) at strain levels of 4.5%, 2.6% and 3.3% respectively.

4.2 Cracking behavior

From physical observations, the average crack spacing and distribution for all LDB elements were evaluated, while for element LDB-3 and selected SDB specimens an optical image acquisition system was used.

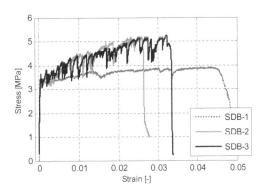

Figure 10. Three examples of the stress-strain response of small ECC dog bone specimens in tensile loading.

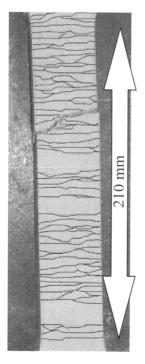

Figure 11. Example of crack distribution in small ECC dog bone specimen after tensile loading. The specimen corresponds to the material used in specimen LDB-3. Cracks are highlighted for visualization.

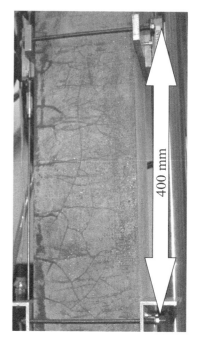

Figure 12. Crack formation in element LDB-1, flexural cracks on left edge along with transverse and longitudinal shrinkage induced cracks at the representative section.

For LDB-1 the average crack spacing was observed to be 17 mm with a fairly even distribution over the measured middle section. Prior to testing, shrinkage induced cracking was observed over the whole specimen, both in longitudinal and transverse direction. Furthermore it was noted that some flexural cracks had formed on the element due to a slight misalignment of the specimen in the testing setup, see figure 12. This was also clear from the offset in the measurements of the two strain gauges located on each side of the representative cross section.

The average crack spacing for element LDB-2 was found to be 14 mm with a fairly even distribution over the representative middle section as can be seen in figure 13.

From the physical observations of element LDB-3 an average crack spacing of 18 mm was measured, beside a minor localization the distribution was found to be even over the element, similar to LDB-2.

Results of the strain and crack width analysis of element LDB-3 with the optical image system are shown in figures 14 and 15.

The crack widths were found to be in the range from 90–250 µm, forming at a strain of 0.01–0.45%.

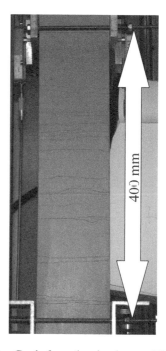

Figure 13. Crack formation in element LDB-2 over the representative section. Cracks are highlighted for visualization.

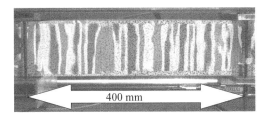

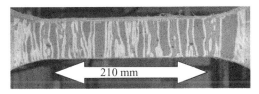

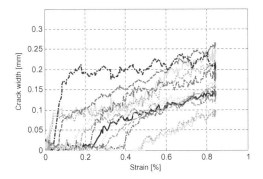

Figure 14. Strain distribution in element LDB-4 during tensile loading, visualized with the image analysis system. Example shows an overlay of the strain contours on the image of the tested specimen.

Figure 15. Development of crack opening of selected cracks on element LDB-3 during loading up to 76 kN. Crack opening plotted against the tensile strain imposed on the specimen.

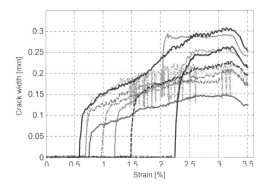

Figure 16. Example of crack opening in 9 randomly selected cracks on a small dog bone specimen as a function of tensile strain.

At the approximate 0.8% strain, none of the cracks shown in figure 15 seem to have fully leveled out, i.e. reached a stable crack width.

The same crack width detection method was used for the small dog bone specimens of the ECC material and crack widths ranging from 60–300 μm were observed. Crack width development in a

Figure 17. Strain distribution shown for SDB specimen during tensile loading, visualized with the image analysis system.

SDB specimen as a function of strain is shown in figure 16.

The strain distribution in a SDB specimen obtained from image analysis is presented in figure 17.

5 DISCUSSION

5.1 Composite load-deformation behavior

For element LDB-1, the decrease in tension stiffening effect of the composite (as seen in figure 6) at repeated load applications is most likely due to interfacial bond degradation between the GFRP rebar and the ECC matrix. In addition, an increase in the de-bonded length of the fibers bridging the transverse cracks also leads to a loss of fiber bridging stiffness and consequently in a reduction of the tension stiffening effect of the cracked ECC matrix.

However, for all sequences, the initial stiffness is the same (up to about 10 kN load), subsequently a converging trend is observed towards to the bare GFRP rebar stiffness at a slightly elevated level due to the stiffening effect of the un-cracked ECC segments between the cracks.

When crack saturation along the length of the representative section is achieved, the response of the element levels out and converges in a line almost parallel to that of the GFRP. This indicates that the composite behavior of the element follows the superimposed tension behavior of the ECC and the GFRP rebar quite well.

Unlike the other specimens, LDB-2 was loaded up to approximately 60% of f_u (76 kN) mark in a load controlled loading as opposed to displacement controlled loading. As a result, the load—strain response for LDB-2 was governed by sudden strain gains rather than sudden drops in load due to forming of cracks.

The overall tension stiffening behavior of elements LDB-2 and LDB-3 are almost identical, converging in a line slightly stiffer than the response of the bare GFRP rebar.

When comparing the response of elements LDB-2 and LDB-3 to the estimated superimposed response shown in figure 7, it should be noted that the measured elastic response of the specimen is stiffer than the estimated value. Furthermore the actual post-crack response is also observed to be stiffer than the estimated response, which indicates that the composite stiffness is slightly greater than the superposition of ECC matrix and GFRP rebar.

From the cyclic response of LDB-2 shown in figure 8, it can be observed that the load response drops by about 20% as cyclic loading is applied, equivalent to a softening of the composite. Furthermore, the shape of the hysteresis loops becomes increasingly narrow, i.e. the distance between the loading and unloading trajectory is reduced due to crack saturation. In the current study, the focus is on accommodating deformations rather than resisting load and therefore, the loss in load response and the reduction of energy dissipated in the deformation process are not considered as a loss in performance.

Element LDB-3 loses about 15% of its loading response during cyclic load application, see figure 9. Similar to LDB-2, LDB-3 also exhibits decreasing energy dissipation however, to a smaller extent.

5.2 Crack behavior

From physical inspection of the specimens after testing, the crack spacing of elements LDB-1, LDB-2 and LDB-3 are fairly consistent with spacing ranging from 14 mm to 18 mm. Furthermore it was observed that the number of cracks formed after the initial monotonic loading did not increase during the subsequent cyclic loading procedure.

Using the image analysis system, it was observed in specimen LDB-3 that at about 0.8% strain, none of the cracks have reached the maximum crack widths observed in the ECC tensile tests, see figure 15. This is consistent with the crack formation in ECC as found in the SDB tests, where the crack widths started to form a plateau after approximately 2.5% strain.

Most of the cracks seem to form with approximately the same approximate stiffness, although the initial instant opening in the SDB specimen is occasionally found to be larger than that observed in element LDB-4, see figure 16. The crack widths for LDB-3 and the SDB specimens were in the same range, although forming at a lower strain in LDB-4.

The average value of the crack width is in good comparison with previous tension tests carried out on steel reinforced ECC where crack widths stabilized at approximately 200 μm (Fischer, 2002). However the crack spacing in the steel reinforced ECC elements was 10 mm, lower than that found in this study.

6 CONCLUSIONS

The combination of the multiple cracking and strain hardening characteristics of Engineered Cementitious Composites (ECC) with the relatively low stiffness of GFRP resulted in a relatively large elastic composite deformation capacity and formation of cracks with a maximum crack width of 0.25 mm. Furthermore, compatible deformations between ECC and GFRP in tension enabled repeatable deformation cycles in the elements while maintaining composite integrity due to reduced interfacial bond stresses between ECC matrix and GFRP rebare.

For all composite elements the tension stiffening behavior was shown to be directly related to the tension response of the bare GFRP rebar after cracking in the ECC had initiated. Furthermore the composite response of the each of the elements can be reasonably estimated by superimposing the strain hardening response of ECC and elastic response of the bare GFRP rebar.

REFERENCES

Bischoff, P.H. & Paixao, R. 2004. Tension stiffening and cracking of concrete reinforced with glass fiber reinforced polymer (GFRP) bars. *Canadian Journal of Civil Engineering*, No. 31. pp. 579–588.

Caner, A. & Zia, P. 1998. Behavior and Design of Link Slab for Jointless Bridge Decks. *Precast Concrete Institute Journal*. May–June. pp. 68–80.

Fischer, G. & Li, V. C. 2002. Influence of Matrix Ductility of Tension-Stiffening Behavior of Steel Reinforced ECC. ACI Structural Journal. V. 99. No. 1 January–February pp. 104–111.

Li, V.C. & Lepech, M. 2006. General Design Assumptions for ECC. Proceedings of the International RILEM Workshop on HPFRCC in Structural Applications. Published by RILEM SARL. pp. 269–277.

Naaman, A.E. & Reinhardt, H.W. 1995. Characterization of HPFRCC. In Proceedings of HPFRCC 2. pp. 1–23.

Advances in Cement-Based Materials – van Zijl & Boshoff (eds)
© *2010 Taylor & Francis Group, London, ISBN 978-0-415-87637-7*

Comparing the behaviour of high strength fibre reinforced and conventional reinforced concrete beams

E.P. Kearsley & H.F. Mostert
Department of Civil Engineering, University of Pretoria, Pretoria, South Africa

ABSTRACT: Modern admixtures make it possible to manufacture high performance concrete with compressive strengths in the region of 100 MPa. It should thus theoretically be possible to reduce the required size of elements in a concrete structure. Researchers have also proven that steel fibres can be used to increase the tensile and shear strength of concrete. For example it is possible to reduce the stirrups required as shear reinforcing in structures when using fibre reinforced concrete. Design engineers however, are reluctant to use fibre reinforcing to replace shear stirrups as the structural design codes used in most countries do not make provision for the use of fibres to increase the shear resistance of concrete elements. In this paper, experimental results will be given to show that significant material saving is possible, through the use of high strength fibre reinforced concrete.

1 INTRODUCTION

Internationally there has been a trend towards using high strength concrete in the pre-cast and pre-stressed concrete industry to increase product output and reduce product loss during handling. The use of higher strength can result in thinner sections and longer beam spans and thus a reduction in dead load. Higher strength concrete does however fail in a more brittle manner but steel fibres can be added to the concrete to ensure non-brittle failure.

The minimum dimensions of stirrups required as shear reinforcing in concrete beams often limits the minimum dimensions of reinforced concrete beams. The use of fibre reinforcing in stead of shear stirrups makes it possible to significantly reduce the size of pre-cast concrete beams.

In South Africa the relatively low cost of aggregate and labour has historically limited the financial viability of using high strength pre-cast pre-stressed elements in reinforced concrete structures. In the recent past the cost of building materials and labour has increased significantly and the aim of this project was to compare the structural behaviour and cost of 30 MPa reinforced concrete to that of 90 MPa fibre reinforced pre-stressed concrete.

2 MIX COMPOSITION AND MATERIAL COST

A 30 MPa and a 90 MPa mix were designed and the mix composition is indicated in Table 1. A pure Portland cement was used and extended with fly

Table 1. Mix composition and material cost.

Material	30 MPa mix kg/m³	90 MPa mix kg/m³	Cost $/t
Cem I 42.5R	207.6	419.2	$191.25
PFA	35.3	69.9	$18.75
CSF	–	34.9	$187.50
Water	173.0	153.7	$0.63
Sand	988.5	978.2	$13.45
Stone	1136.8	866.4	$14.38
Superplasticizer	1.8	9.2	$1875.00
Steel fibres	–	90.8	$2000.00
Density	2543.1	2622.4	

ash (PFA) and Condensed Silica fume (CSF). The aggregate used was 9.5 mm dolomite stone and crushed dolomite sand. Premia P100, as supplied by Chryso was used as superplasticizer and the steel fibre was 30 mm long, hooked-ended hard drawn wire fibres with a diameter of 0.5 mm.

The June 2009 cost of the different materials used are indicated in Table 1 (for an exchange rate of R8 for $1) and it is clear that the materials used to increase the strength of the concrete cost significantly more than the materials used in the conventional 30 MPa concrete. The concrete material cost per cubic meter was calculated as $73.52 for the 30 MPa concrete and $312.71 for the 90 MPa concrete. The 325% increase in material cost has historically discouraged engineers and contractors from using high strength fibre reinforced concrete in structures. High strength fibre reinforced concrete can only be economically viable if the element dimensions and reinforcing can be reduced to the extent where the total structural cost is no longer negatively affected by the increased material cost per cubic meter.

3 EXPERIMENTAL SETUP

To establish whether it could be financially viable in South Africa to use high strength fibre reinforced concrete in structures in stead of normal 30 MPa concrete a comparative test was set up, where two different beams were designed to fail at the same load.

A conventional reinforced concrete beam was designed and the material properties of the high strength fibre reinforced concrete were used to reduce the beam size and the volume of steel in the high strength beam.

3.1 Contribution of steel fibres

The use of steel fibres can enhance the flexural and the shear strength of reinforced concrete beams. Vanderwalle et al (2000) indicated that the contribution of the steel fibres towards the bending strength of the beam can be calculated as:

$$F_{sf} = 0.67 \, V_f \, f_f \, b \, (h{-}x) \tag{1}$$

where:

F_{sf} = maximum force in steel fibres
V_f = volume fraction of fibres
f_f = design yield strength of steel fibres
$h{-}x$ = height of cross sectional tension zone.

Vanderwalle et al (2000) indicated that the shear resistance of beams can be calculated as follows:

$$V_{Rd3} = V_{cd} + V_{fd} + V_{wd} \tag{2}$$

with:

V_{Rd3} = design shear resistance of beam
V_{cd} = the shear resistance of the beam without shear reinforcement
V_{fd} = contribution of the steel fibre shear reinforcement
V_{wd} = contribution of shear stirrups.

$$V_{cd} = [0.12 \, k \, (100 \, \rho_l \, f_{ck})^{1/3}] \, b \, d \tag{3}$$

$$k = 1 + \sqrt{200/d} \tag{4}$$

$$\rho_l = A_s/(b \, d) \tag{5}$$

$$V_{fd} = k_l \, \tau_{fd} \, b \, d \tag{6}$$

$$k_l = (1600 - d)/1000 \tag{7}$$

$$\tau_{fd} = 0.12 \, f_{eqk,3} \tag{8}$$

$$V_{wd} = (A_{sw}/s) \, 0.9 \, d \, f_{ywd} \tag{9}$$

where:

f_{ck} = compressive cylinder strength (assumed to be 85% of cube strength)
τ_{fd} = design value of the increase in shear strength due to steel fibres
s = spacing between shear reinforcement
A_{sw} = area of yield reinforcement
f_{ywd} = design yield strength of shear reinforcement.
$f_{eqk,3}$ = equivalent flexural strength

The equivalent flexural strength can be determined by testing Modulus of Rupture (MoR) beams in deflection control. Beams with a cross section of 150 mm × 150 mm were loaded at third points while spanning over a distance of 450 mm. The bending stress plotted as a function of the midspan deflection for beams from both the 30 MPa and the 90 MPa mixture are indicated in Figure 1. The 30 MPa had a flexural strength of 5.88 MPa. The effect of the fibre reinforcing can clearly be seen, with the high strength concrete cracking at a stress of 15.8 MPa but the post-cracked stress reaching a value as high as 18.6 MPa.

The equivalent flexural strength was calculated as the average flexural stress for a deflection of up to 3 mm. The equivalent flexural strength for the 90 MPa concrete was calculated as 13.1 MPa. The shear resistance as calculated using Equation (2) was calculated as 118.0 kN for the 30 MPa beam and 116.8 kN for the 90 MPa beam.

3.2 Design of 30 MPa beam

A 30 MPa beam with dimensions as indicated in Table 2 was reinforced with high yield steel bars with theoretical yield strength of 450 MPa. Two Y16 bars were used as main reinforcing and for shear reinforcing Y10 stirrups at 150 mm spacing

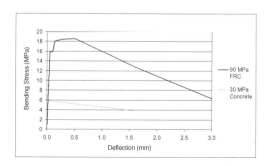

Figure 1.   Flexural strength of beams.

Table 2. Beam designs.

	30 MPa beam	90 MPa beam	Cost $/m
Beam size			
Length	4.2 m	4.2 m	
Depth	300 mm	300 mm	
Width	150 mm	100 mm	
Beam volume	189 l	126 l	
Reinforcing			
Y16	2 × 4.7 m		$1.91
Y10	28 × 0.9 m		$0.76
Y6	2 × 4.1 m		$0.45
7 mm wire		3 × 4.8 m	$0.51

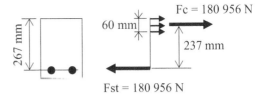

Figure 2. Design principle for 30 MPa beam.

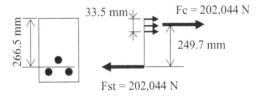

Figure 3. Design principle for 90 MPa beam without pre-stress or fibre contribution.

was used. The stirrups had a 15 mm concrete cover and two Y6 bars were used in the top of the beam to keep the stirrups in position. The total length and unit cost for each diameter of the bar used in the 30 MPa beam can be seen in Table 2.

Theoretically the two 16 mm bars will yield at a tensile force of 180.96 kN. In the limit state, just before the beam fails in flexure, the maximum stress in the concrete compression block can be assumed to be 67% of the concrete strength. When the steel in the 30 MPa concrete beam yields, the concrete stress in the compression zone would be 20.1 MPa over an area 150 mm wide and 60 mm deep (see Figure 2). The total bending moment that can be resisted by the 30 MPa beam would be 42.88 kNm. This moment can be applied to the beam in a four point bending test where the beam is supported on rollers to span 3.9 m and the load is applied at third points. The 30 MPa beam should resist a load of 65.98 kN before the main reinforcing bars starts yielding. The maximum shear force in the beam at this load would be 32.99 kN, which is significantly less than the shear resistance of the beam.

3.3 Design of 90 MPa beam

High strength fibre reinforced concrete can best be used in a factory environment where pre-cast members are manufactured and cured in a controlled environment. The concrete volume of the 90 MPa beam was reduced by a third by reducing the width of the beam to 100 mm as indicated in Table 2. To reduce the volume of steel used in the high strength concrete is was decided to use pre-stressing steel wires as reinforcing. The wires have a theoretical strength of 1750 MPa and the cost of the wires can be seen in Table 2. If three 7 mm wires are used in a 90 MPa beam the force required to yield the wires would be 202 kN.

If the wires are not pre-stressed we can use the principles as illustrated in Figure 3. The wires will

yield when a concrete stress of 60.3 MPa is applied over an area 100 mm wide and 33.5 mm deep. Two of the wires were placed 25 mm from the bottom of the beam but for practical reasons the third wire was placed 40 mm from the bottom of the beam as can be seen in Figure 3. The maximum bending moment that the 90 MPa beam should be able to resist when the steel starts yielding is 50.46 kNm. In this calculation the contribution of the steel fibres towards the flexural strength has not yet been taken into account.

In a four point bending test this moment will be the result of an applied load of 77.63 kN. The beam will have to resist a shear force of at least 38.82 kN without the use of shear reinforcing.

The wires in the 90 MPa been were however pre-stressed to 50% of the theoretical capacity (875 MPa) and therefore the stress in the beam can be calculated as follows:

$$f_{top} = \frac{P}{A} - \frac{Pe}{z} + \frac{M}{z} \qquad (10)$$

$$f_{bot} = \frac{P}{A} + \frac{Pe}{z} - \frac{M}{z} \qquad (11)$$

where:

f_{top} = stress in the top of the section
f_{bot} = stress in the bottom of the section
P = pre-stressing force
A = cross sectional area
e = eccentricity of the stressed wires
$z = \dfrac{bh^2}{6}$
M = the externally applied moment

The pre-stressing of the three wires should cause a compressive stress of 3.37 MPa across the whole concrete cross section. The eccentricity of the wires results in an additional compressive stress of 7.85 MPa in the bottom of the beam cross section but an equivalent tensile stress in the top of the beam cross section. Before any external loads are applied to the beam the resultant stresses in the extreme top and bottom fibres of the beam would be 4.48 MPa (tension) and 11.21 MPa (compression) respectively.

When an external load is applied to the beam the bottom fibres of the beam would eventually fail when the flexural capacity of the beam, as determined in Figure 1 is exceeded. The flexural stiffness of the pre-stressed concrete beam will decrease as soon as the concrete cracks. A cracking stress of 15.8 MPa will be reached when the bending moment is 40.52 kNm. This moment will result from an applied load of 62.3 kN.

The high fibre content does however result in a flexural capacity higher than the cracking strength of the concrete, and as can be seen from Figure 1, the fibres limit the reduction of stiffness after cracking. The flexural stress capacity of the fibre reinforcing is 18.6 MPa and this stress will be reached when the moment caused by external loading is 44.72 kNm. A load of 68.8 kN will be required to exceed the flexural capacity of the fibre reinforced concrete.

4 EXPERIMENTAL RESULTS

4.1 Material properties

The compressive strength of each concrete mix was determined from three 100 mm cubes that were cured in water at 25°C and tested 28 days after casting. The average 28-day compressive strengths for the two mixes were 31.5 MPa and 114.9 MPa.

The measured strength of the reinforcing used in this experiment can be seen in Figure 4. The 16 mm high yield steel bars had a yield strength of 537 MPa while a 0.1% proof stress of 1636 MPa was recorded for the 7 mm wire. The tensile strength of the wire was 1828 MPa. The modulus of elasticity (slope of the stress-strain diagram) was recorded as 189 GPa for the high yield steel and 202 GPa for the wire.

The actual strength of the reinforcing and the concrete can be used to try and predict more accurately what the failure load of the beams would be. The force that will cause the two 16 mm bars to actually yield should be 216.1 kN and for a concrete compressive strength of 31.5 MPa it can be assumed that the stress in the compression block is limited to 21.1 MPa. The beam should fail in

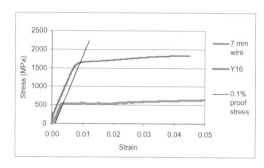

Figure 4. Strength of steel reinforcing.

Figure 5. Test setup for testing beams.

bending at a moment of 50.3 kNm resulting from a load of 77.4 kN.

4.2 Reinforced concrete beams

The experimental setup used to test the 4.2 m long beams can be seen in Figure 5. The beams were supported by rollers to span 3.9 m and the load was applied with a closed loop Materials Testing System (MTS). The tests were conducted in deflection control and the applied load as well as the midspan deflection was recorded.

The load deflection behaviour of the 30 MPa and the 90 MPa reinforced concrete beams can be seen in Figure 6. The 30 MPa concrete beam behaves in a linear elastic manner up to a load of 68.3 kN. Thereafter the main reinforcing bars seem to yield and the beam fails in tension. This value compares well with the theoretically predicted value of 65.98 kN.

The reduced cross-sectional area of the 90 MPa fibre reinforced concrete beam should result in a lower stiffness and thus higher deflections. The beam was pre-stressed to reduce the deflection and from Figure 6 it can clearly be seen that for equal loads the high strength beam deflects less than the 30 MPa beam for deflections up to 10 mm. The effect of the pre-stressing is eliminated when

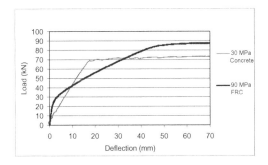

Figure 6.   Load deflection behaviour of beams.

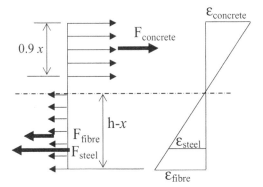

Figure 7.   Fibre reinforced concrete stesses.

Figure 8.   Strain measurement.

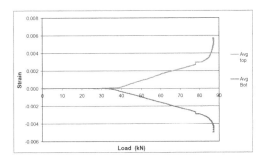

Figure 9.   Strain in high strength beam.

the stress in the bottom of the beam changes from compression to tension. At a load of 25.9 kN the bottom of the pre-stressed beam starts experiencing tensile stresses and from Figure 6 it can be seen this is the load where the stiffness of the 90 MPa beam changes.

When the load increases above 25.9 kN, the beam clearly has a lower stiffness than the 30 MPa beam. The 90 MPa beam acted in a linear elastic manner up to a load in the region of 85 kN. This load is significantly higher than the load predicted in Figure 3 by not taking the fibres into account. The contribution of the fibre reinforcing towards the flexural strength can be calculated using the actual measured values for the concrete and steel strengths.

The 3 wires will yield at a load of 188.9 kN and the maximum compressive stress in the concrete is assumed to be 77 MPa (67% of the measured cube strength). To be conservative it can be assumed that the maximum tensile stress in the fibre reinforced concrete, as indicated in Figure 7, is 13.1 MPa.

The position of the neutral axis can be determined by measuring the strain in midspan. The setup used to measure strain can be seen in Figure 5. More detail of the frame is visible in Figure 8. A measuring frame was attached with pins to the

centre of the beam (35 mm from the top of the beam and 75 mm from the bottom of the beam). Four LVDT's were set up to measure the change in length over a 300 mm distance. Two LVDT's (one on each side of the beam) were attached 10 mm from the top of the beam and two 50 mm from the bottom of the beam.

The strain calculated from the LVDT readings is plotted as a function of applied load in Figure 9. Up to a load of about 35 kN very little strain was recorded but thereafter the movement measured by both the top and the bottom LVDT's increased linearly with load up to a load of approximately 85 kN. Thereafter the strain increased without any increase in load.

By assuming that planes remain plane during bending, the recorded readings can be used to calculate not only the position of the neutral axis but also the strain at the extreme fibres in the top and the bottom of the beam. When the load reaches 85 kN the strain was calculated as 0.0038 in the extreme top fibre and 0.0049 in the extreme bottom fibre. The strain in the wires was calculated to be 0.0039,

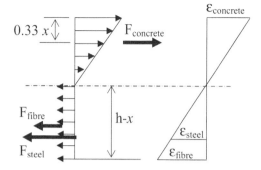

$0.33\,x$

$F_{concrete}$

$\varepsilon_{concrete}$

F_{fibre}

$h\text{-}x$

ε_{steel}

F_{steel}

ε_{fibre}

Figure 10. Stress distribution at elastic limit.

Table 3. Beam costs.

	30 MPa beam	90 MPa beam
Beam volume	189 l	126 l
Concrete cost	$13.90	$39.40
Y16	$17.94	
Y10	$19.16	
Y6	$3.65	
7 mm wire		$7.29
Total cost	$54.65	$46.69
Total weight	512 kg	335 kg

which relates to a stress of 795 MPa, which is significantly lower than the yield stress of the steel. If the pre-stressing stress of 875 MPa is added to this stress, it is clear that the steel does yield. The neutral axis was calculated to be 131.26 mm from the top of the beam. If we assume a stress block for the steel fibres as indicated in Figure 7, we can use equation 1 to calculate the contribution of the steel fibres towards the bending strength of the beam.

The fibre volume is 1.15% and the yields strength of the fibres can be assumed to be 2000 MPa. The maximum tensile stress in the fibre reinforced concrete was calculated as 15.36 MPa and this stress can be multiplied by the area of the beam below the neutral axis to determine the tensile force in the fibre reinforced concrete. The contribution of the fibres towards the tensile force is 259.2 kN.

For equilibrium to exist in the cross section the tensile forces should be equal to the compressive force. The total tensile force was calculated as 448 kN, but the measured position of the neutral axis is such that the assumption of a limit stress state in the concrete can not be correct. The stress distribution in the compression zone is more likely to be linear elastic, as indicated in Figure 10. A triangular stress distribution as indicated in Figure 10 would result in a compressive force of 454.7 kN in the concrete. This is about the same than the tensile force. The bending resistance of the section can be calculated by using the tensile forces and the total bending resistance can be calculated as 86.6 kNm with the steel fibres contributing 44.6 kNm of the total moment resistance.

5 COST OF BEAMS

The total material cost of the two beams was compared using the information as indicated in Table 3. These results clearly indicate that the use of fibre reinforced high strength concrete can not only result in a 35% reduction in own weight but also in a 15% reduction in cost.

6 CONCLUSION

The materials used in high strength fibre reinforced concrete are significantly more expensive than the materials used in normal 30 MPa concrete. The results of this investigation however proves that it is possible to reduce both the own weight and the cost of structural elements by using high strength fibre reinforced concrete efficiently. High strength concrete can only be financially viable if the volume and type of materials used is optimized.

The cross sectional area of concrete beams can be significantly reduced if fibres are used as shear reinforcing in stead of stirrups. Design codes of practice should include design guidelines for the use of fibre reinforced concrete. The design guidelines as published by Vandewalle et al in 2000 can be used to predict the strength of high strength fibre reinforced concrete beams.

ACKNOWLEDGEMENTS

The authors would like to express appreciation for the research grants made by the Duraset business unit of Aveng and the THRIP-program of the South African National Research Foundation that made this research possible.

REFERENCES

Kearsley, E.P. & Mostert H.F. 2004. The effect of fibres on the shear strength of reinforced concrete beams. *6th RILEM Symposium on Fibre Reinforced Concretes (FRC)—BEFIB 2004*, Varenna, Italy, 955–964.

Mindess, S. & Young, J.F. 1981. *Concrete*, USA: Prentice-Hall.

Vanderwalle, L et al. 2000. 'Recommendations of RILEM TC162-TDF: Test and Design Methods for Steel Fibre Reinforced Concrete: σ-ε design method (recommendations)', *Materials and Structures*. 33: 75–81.

Vanderwalle, L et al. 2003. RILEM TC 162-TDF: 'Test and design methods for steel fibre reinforced concrete' σ–ε design method. Final Recommendation. *Materials and Structures*. 36: 560–567.

Structural mechanics and applications

Advances in Cement-Based Materials – van Zijl & Boshoff (eds)
© *2010 Taylor & Francis Group, London, ISBN 978-0-415-87637-7*

Self-compacting lightweight aggregate concrete for composite slabs

A.-E. Bruedern & V. Mechtcherine
Institute of Construction Materials, TU Dresden, Germany

W. Kurz & F. Jurisch
Department of Steel and Composite Steel Construction, TU Kaiserslautern, Germany

ABSTRACT: In the construction of high-rise buildings, very often lightweight structures are used in order to minimize costs. Because of the placement of trapezoidal steel sheets independently of the crane and because of their simultaneous function as formwork, working platform, reinforcement and horizontal stiffeners, the application of composite slabs can be an economically advantageous construction method. In the construction of new buildings, and especially in the renewal of existing buildings, lightweight structures are required. Composite slabs of lightweight concrete combine the advantages of lightweight, rapid construction and extensive industrial prefabrication. This paper describes the development of a pumpable, self-compacting, lightweight aggregate concrete (SCLC) which has optimized properties for use in composite slabs. In the first step, a few variations of SCLC with a bulk density class D1.6 were developed. These materials varied in their mechanical properties, e.g., compressive strength, tensile strength, and Young's modulus. Furthermore, different shrinkage reducing admixtures (SRA) were added to the mix of one chosen SCLC, which accordingly showed favourable mechanical properties. With the addition of SRA, shrinkage deformations in the concrete could be reduced considerably. In the next step SCLC compositions for the bulk density classes D1.4 and D1.8, respectively, were developed. Finally, the mechanical performance of composite slabs made with the developed SCLC compositions was studied by means of four-point bend tests. Particular attention was directed at interactions between the concrete and sheet steel as well as at the failure mechanisms inherent in the slab system. The test results clearly demonstrated the relevance of concrete properties to the improvement of the performance of the composite slabs.

1 INTRODUCTION

Composite structures represent one of the most economic solutions for flooring in structural engineering. The advantages of crane-independent, rapid installation of steel decking sheets combined simultaneously with their formability, their making a ready working platform available, and their reinforcement and stiffening elements for horizontal loads are responsible for the success of this technology. To achieve minimal dead weight in such structures, the use of lightweight concrete is necessary. In particular, self-compacting, pumpable lightweight aggregate concrete can simplify the construction process and optimize quality as well as minimize noise generation and erection time.

The aim of the ongoing project is to develop self-compacting lightweight concrete well suited to use in composite floors and to investigate the load-bearing behaviour of the composite slabs.

Since the complex interaction of the composite members (SCLC and steel) in their adhesion, friction, sheet deformation, and clamping effect is influenced by profile geometry and the particular properties of the SCLC, several parameter combinations have been investigated experimentally and by means of finite element analysis.

2 DEVELOPMENT OF SELF-COMPACTING LIGHTWEIGHT CONCRETE

The first SCLC compositions with commercially produced lightweight aggregates (LWA) were developed approximately 10 years ago (Mueller and Mechtcherine 2000). In the following years intensive research was undertaken in order to improve the rheological behaviour of SCLC and make it pumpable (Mechtcherine et al. 2003; Haist et al. 2003).

Although some research was also directed toward improving the ductility of SCLC by the addition of short fibres (Mechtcherine et al. 2003), generally speaking the optimisation of the properties of this new material in its hardened state with regard to particular applications has so far never been a topic of investigation. In the framework of this project, a purposeful further development of SCLC technology has begun to take into account the special demands of construction using light composite slabs. Apart from excellent flowability and robust pumpability, such application demands high tensile strength, high stiffness, low brittleness of the concrete, and low shrinkage and creep deformations as well. Four of the SCLC compositions developed, belonging to the bulk densities classes from D1.4 to D1.8 are presented in this paper (see Table 1).

For all compositions expanded clay was used as LWA. The coarse LWA had a grain size of 2 mm to 10 mm, the grain size of lightweight sand ranged from 0 mm to 4 mm. Since the porous expanded clay can absorb high amounts of mixing water, which may considerably affect the consistency of fresh concrete, the expanded clay aggregates were pre-wetted with water (18.5% by mass of coarse LWA and 21% by mass of lightweight sand). This additional water did not influence the strength of SCLC negatively. However, it may have had an effect on its shrinkage behaviour.

An ELBA laboratory mixer with a capacity of 60 litres was used at the TU Dresden for

the concrete development. The mixture SCLC-1 (cf. Table 1) was developed in an earlier project (Mechtcherine et al. 2003) and used in this study as reference concrete and the basis for a further optimisation.

The total mixing time was 3 minutes in the preliminary tests. The reference concrete SCLC-1 showed a slump flow of 680 mm, a flow time t_{500} of 3.8 s and the V-funnel flow time of 6.3 s. During first mixings of SCLC-1 it was observed that a part of the coarse lightweight aggregates was ground or ruptured due to the intensive mixing process. To minimize this, the mixing intensity was reduced, but the total mixing time was still limited to 3 minutes. Since the super plasticizer on the polycarboxylate ether basis could not develop its full effect during this short time period at the applied low mixing intensity, the consistency of the fresh concrete mixture became stiffer when tested directly after mixing: the slump flow decreased to 610 mm. However, in the process of the subsequent "post-mixing" with very low intensity (which simulated the post-mixing in a ready-mix concrete lorry) the SCLC became more workable. Figure 1 shows the change of the concrete consistency over time, which is appropriate to the application of this material as ready-mix concrete in the production of composite slabs at the construction site.

SCLC-1 was used for the production of a first series of slabs at the TU Kaiserslautern using a concrete plant mixer with a volume of 1 m³. Testing of characteristic concrete properties in the fresh and hardened state provided nearly the same values of the characteristic properties as those obtained at the TU Dresden.

Primary based on the results of the first four-point bend tests (cf. Chapter 4), a few material parameters of SHCC were selected to be improved in order to enhance the mechanical performance

Table 1. SCLC compositions developed.

Concrete Density class	SCLC-1 D1.6	SCLC-4 D1.4	SCLC-6 D1.8	SCLC-7 D1.6
Cement*	10.3	10.9	10.3	10.7
Fly-Ash	9.5	9.9	5.0	9.9
Silica Fume (solid)	–	–	0.6	0.7
Water**	15.9	16.5	17.7	15.4
SP***	0.95	0.64	0.8	1.0
VA****	0.15	0.30	–	–
LSP*****	–	–	5.1	–
Sand 0/2	23.7	–	–	21.5
Split 5/8	–	–	22.3	–
LWA 2/10	37.4	33.0	–	39.2
LWA 0/4	–	26.2	36.3	–

All components are given in % by volume, except SP and VA.
* CEM II/A-LL 32.5 R
** Without pre-wetting Water
*** Super plasticizer (SP) in % by mass of cement
**** Viscosity agent (VA) in % by mass of cement
***** Limestone powder (LSP)

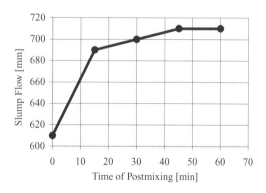

Figure 1. Change of the slump flow during the low-intensity post-mixing.

of the composite slabs subsequently. The following changes were the objective:

– extension of the range of bulk densities classes to D1.4 and D1.8,
– increase in the tensile strength,
– increase in the Young's modulus,
– reduction of creep and shrinkage deformations

To reduce the bulk density of SCLC to the density class D1.4 (mixture SCLC-4), the normal-weight quartz sand utilised in the reference mixture was substituted by lightweight sand (cf. Table 1). The mixture SCLC-6 was designed to meet the bulk density class D1.8 and simultaneously to increase the Young's modulus and the tensile strength in comparison to the reference composition. It was attained by replacing the coarse LWA by basalt split and by adding silica fume, while the quartz sand was replaced by lightweight sand (cf. mixture SCLC-6 in Table 1). The silica fume was used in form of a colloid suspension with 50% by mass of solid material. To improve the grain size distribution, which was disturbed by the introduction of the split aggregates, some limestone powder was used additionally in the mixture SCLC-6. Finally, mixture SCLC-7 was developed with the goal of reducing the bulk density by a small amount (in order to be more secure in reaching the density class D1.6) and to increase the tensile strength. The first goal was achieved by increasing the content of LWA slightly and simultaneously by lowering the portion of quartz sand slightly. An increase in the tensile strength should be attained, similar to the SCLC-6 composition, by adding silica fume suspension (cf. Table 1).

The properties of the hardened SCLC were tested at a concrete age of 28 days according to the German standard DIN 1048. Table 2 gives the results of the mechanical tests performed. As expected, SCLC-4 (bulk density class D1.4) showed lower strength and stiffness in comparison to the reference SCLC-1 (bulk density class D1.6). The reduction of the compressive strength was moderate; however, as a result of using LWA for all aggregate grain fractions, there was a pronounced decrease in the values of the most relevant parameters, Young's modulus and the splitting tensile strength. In contrast the testing of the mixture SCLC-6 (bulk density class D1.8) revealed a considerable increase in tensile strength, leaving the perceptual increase in the compressive strength and Young's modulus well behind. Finally, SCLC-7 showed, again in comparison to the reference SCLC-1, a slightly lower dry density, while comparing better to the bulk density class D1.6. The compressive strength and Young's modulus of this mixture were somewhat below the reference values, but a significant

Table 2. Mechanical properties for different SCLC mixtures.

Concrete bulk Density class	SCLC-1 D1.6	SCLC-4 D1.4	SCLC-6 D1.8	SCLC-7 D1.6
Dry density*	1630	1390	1765	1543
Compression strength**	40/43	34/36	46/48	39/41
Young's modulus	20000	14000	21000	18700
Splitting tensile strength	2.4	1.4	3.4	2.8

* The dry density is given in kg/m³, all mechanical properties are given in MPa.
** Compressive strength measured on cylinder/cube.

Figure 2. Equipment used for shrinkage measurements.

perceptual increase in the splitting tensile strength could be attained.

Since the SLCL compositions developed have a high proportion of fines and increased total water content due to the pre-wetting of LWA, high shrinkage deformation can be expected. In the composite slab, concrete shrinkage deformations are constrained by the profiled steel sheet, which leads to tensile stresses in concrete. This effect is particularly pronounced at the upper chord of the profiled sheet, a region, where stress concentrations from mechanical loading occur as well. To decrease the shrinkage deformations and therefore to counteract early cracking, two different shrinkage reducing admixtures (SRA) and one concrete sealant agent (SA) were used as concrete additives. Their effect on shrinkage behaviour was tested on prisms made of the reference concrete SCLC-1 using the Graf-Kaufmann method (cf. Figure 2).

Figure 3 shows the test results for total shrinkage, including the autogenous and drying components. The addition of SRA resulted in a

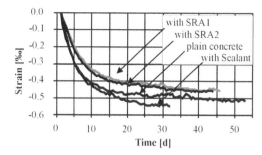

Figure 3. Development of total shrinkage strain over time for SCLC without and with shrinkage reducing additives (SRA) and sealant agent (SA), respectively.

Figure 4. Instrumented ring test for estimating concrete cracking due to restraint of shrinkage deformations.

clear decrease in shrinkage deformations in comparison to the reference mixture without such additive. In contrast, the use of concrete sealant agent not only was not beneficial, but even caused some increase in shrinkage.

To estimate the effect of the additives on cracking behaviour of SCLC instrumented ring test were performed. The setup of this test is shown in Figure 4. It consists of two steel rings of different diameter. Fresh SLCL was placed between the rings. At a concrete age of one day, the outer steel ring was removed and the circumference of the concrete annulus was exposed to desiccation at the standard laboratory climate (20°C, 65% RH). The inner steel ring was equipped with strain gauges for measuring deformations of this ring induced by the shrinkage of the concrete annulus. From the deformation values measured the stress in the SCLC annulus could be estimated using formula by Hossain & Weiss (2004).

Figure 5 shows typical cracks as observed in the ring test right after the crack formation in SCLC. In their tendency, the test results agree well with the

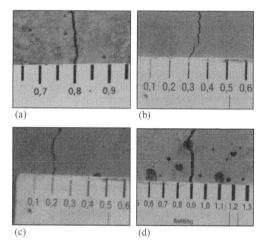

Figure 5. Crack widths observed in ring tests right after crack formation:

a) Reference composition SCLC-1, cracking age 15 days
b) SCLC-1 with SRA1, cracking age 27 days
c) SCLC-1 with SRA2, cracking age 39 days
d) SCLC-1 with sealant, cracking age 19 days.

findings of the shrinkage measurements. However, in the ring tests the differences were much more pronounced: while the reference concrete cracked at the age of 15 days and showed a crack width of 0.8 mm, the mixes with the additives SRA1 and SRA2 displayed considerably smaller crack openings of 0.2 mm and 0.3 mm, respectively. The cracking occurred also considerably later in SCLC-1 with SRA, at a concrete age of 27 days (SRA1) and 39 days (SRA2), respectively. The largest crack width was registered for concrete with the sealant: it was 0.9 mm right after crack formation at a concrete age of 19 days.

3 TESTING THE PUMPABILITY OF SCLC

One of the key properties required for an efficient application of SLCL for the production of the composite floors is robust pumpability. This property was tested by producing 1 m³ of fresh SCLC in a regular concrete plant and pumping it in a circle for a while and subsequently casting of composite slabs with the pumped concrete. Here one experiment, in which SCLC-7 with addition of SRA1 was pumped will be described.

The lightweight aggregates were pre-wetted in the 1 m³-mixer. Subsequently, the concrete was mixed with the same mixer. The slump flow of fresh SLCL right after mixing was 650 mm.

A truck-mounted concrete pump M28-4 from Putzmeister was used in the test. The walls of the

Figure 6. Casting a composite slab with pumped SCLC.

Table 3. Test results for composite slab specimens with SCLC.

Specimen		A	B
f_{yp}	[MPa]	337.8	337.8
$0.9\,f_{cm,cube}$	[MPa]	38.43	38.43
Length	[m]	3.4	2.2
Width	[m]	0.70	0.695
M_u	[kNm]	37.2	29.1
η	[%]	79	54
τ	[kN/m²]	512.3	525.2

pipe system were first lubricated by pumping an amount of a premix made of cement, sand and water. Subsequently, the pump started to deliver SCLC. After the premix and the very first portion of SCLC were pumped into the waste, the pumping process was changed to the cycle regime.

The velocity of pumping was set to 8–10 m³/h. After 10 min of pumping it was interrupted for 10 min to investigate the condition of the mix, then the pumping process continued. The velocity was increased to 70 m³/h. The testing of the mix during the interruption showed that no segregation occurred. A significant decrease in slump flow was observed to 490 mm indicating some loss in the workability of the material. This could likely be traced back to the surplus of water absorption by LWA due to the high pumping pressure. With an addition of super plasticizer (0.2 % by mass of cement) and some water (approximately 3 litres) the desired consistence was reached again. The pumping test was continued for another 5 minutes and then finished by casting composite slabs with the pumped SCLC (cf. Figure 6).

4 TESTS WITH COMPOSITE SLABS

4.1 Test setup and main results

Preliminary tests with two composite slab specimens (denoted here as specimen A and specimen B) were performed in order to obtain initial references regarding their applicability to and behaviour of self-compacting lightweight in composite floors. The experiment setup was designed similar to previous tests on composite slabs with lightweight concrete by Faust at Leipzig University (Faust 2002) and Kessler at the TU Kaiserslautern (Kurz & Kessler, unpubl.). According to Koenig & Faust (1997) and Kurz & Kessler (2007), the composite slabs showed ductile bearing behaviour with longitudinal shear failure in 4-point-bend tests. The specimens had approximately

same size; the tests were carried out according to Eurocode 4.

The specimens were made using a profiled sheet steel type Super-Holorib SHR51 with a nominal thickness of 1.00 mm. They had a width of 70 cm and a height of 14 cm. The length of the one specimen type was 3.40 m ("long" shear length of 90 cm) and that of the other specimen type was 2.20 m ("short" shear length of 60 cm). At the points of load application, crack-initiating sheets were installed. The same dimensions were chosen for the new specimens A and B, to be tested in this project with SCLC. The concrete composition SCLC-1 was used (cf. Chapter 2). During the testing of specimen A (with a "long" shear length of 90 cm), the first slip appeared at the load of 58.8 kN (25.8 kNm). The maximum load was 87.2 kN (37.2 kNm), which was 48% above the load of the first slip. Thus the load bearing behaviour can be classified as "ductile". Specimen B (with a "short" shear length of 60 cm) reached a peak load of 112.3 kN (29.1 kNm), while the first slip appeared at 71.4 kN (18.9 kNm). With a load increase of 57% after the first slip the load bearing behaviour can be classified as well as "ductile". The abort criterion of mid-span deflection of L/50 was insignificant for both tests.

From these maximum loads shear strengths of 525.2 kN/m² and 512.3 kN/m², respectively, could be obtained for A and B. The calculation was done according to the regulations of Eurocode 4, without taking into account the support pressure, and under assumption of a triangular compression stress distribution (Faust 2002). The primary test results are given in Table 3.

4.2 Deformation and failure behaviour of the composite slabs

While the maximum loads achieved were within the expected range, the failure mode and crack pattern in part showed unusual features, which were significantly different to the well known behaviour of normal weight concrete specimen. First cracks

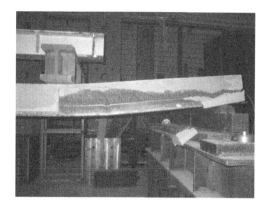

Figure 7. Specimen A at the ultimate limit state.

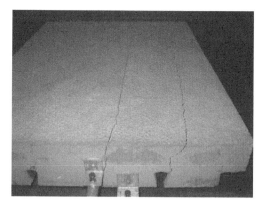

Figure 9. Longitudinal splitting cracks in specimen A.

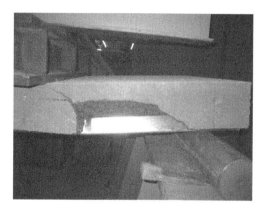

Figure 8. Specimen B at the ultimate limit state.

appeared, as expected, at the crack-initiating sheets and on the longitudinal edges of the specimen (cf. Figure 7). These horizontal, longitudinal cracks split the concrete cover at the height of the composite sheet's upper chord. In contrast to composite slabs made of ordinary concrete, an additional crack appeared with further load increases and opened within a distance of approximately 20 cm from the crack-initiating sheet, while the initial crack at the crack-initiating sheet was closed. For specimen A this crack was vertical in the beginning and split later. One branch headed towards the load application area and the other grew between the load application and the support. With further increases in load another crack formed in the same way at a distance from the first equal approximately the height of the slab. This crack ran along the compression trajectories toward load application as well. Specimen B showed only one considerably curved crack heading toward load application (see Figure 8).

The more the cracks opened, the more their edges shifted against each other. For Specimen B a vertical displacement of the crack edges could be recognised and local buckling of the composite sheet was observed. After cutting specimen B it could be observed that the curved crack ran through the entire width. The concrete at the lower crack edge was clearly damaged. Burls of the composite sheet left abrasion marks on the concrete, and the angle of the concrete rib was sheared off. These marks have gone undetected in tests of ordinary concretes of normal weight.

Specimen A with its "long" shear length and specimen B with its "short" shear length were both split longitudinally after reaching maximum load. Each crack started at an upper chord angle of the Super-Holorib-sheet and ran inclined through the entire specimen thickness. Specimen A was split at each rib of this undercut profile (see Figure 9). The test results observed by Koenig & Faust (1997) and Kurz & Kessler (2007) with conventional lightweight concrete showed similar behaviour.

5 SUMMARY

Several SCLC compositions suitable for the production of composite slabs were developed in this investigation. The bulk density classes ranged from D1.4 to D1.8. To secure a target workability of SCLC over the delivery and casting time LWA were pre-wetted by a defined amount of water. Furthermore, it was shown that the mixing intensity and duration play a significant role with regard to concrete workability. Shrinkage measurements showed a clear, positive effect of the addition of shrinkage reducing additives, while the addition of a sealing agent led to no improvement. These results were confirmed in the instrumented ring tests: the cracking of the specimens with SRA occurred at a higher concrete age, and crack widths were considerably lower in comparison to the specimens made without the addition of SRA. Furthermore, the pumpability of SCLC was validated by pumping tests.

Composite slabs made of SCLC showed a ductile failure in the bend tests. The crack pattern was very different to that known in the corresponding tests on composites slabs made with ordinary concrete. The observed failure mode was not only the well known and assumed longitudinal shear failure, but also a combined failure from transverse force and longitudinal shear. The investigations are in progress.

ACKNOWLEDGEMENTS

This research has been sponsored by the *Bundesamt für Bauwesen und Raumordnung* with funding from the initiative "*Zukunft Bau*" (reference number Z 6-10.08.18.7-07.9/II 2-F20-07-19). The authors are responsible for the content of this report.

REFERENCES

Faust, T. (ed.) 2002. *Leichtbeton im konstruktiven Ingenieurbau.* Berlin: Ernst & Sohn.

Haist, M., Mechtcherine, V., Beitzel, H. & Mueller, H.S. 2003. Retrofitting of building structures using pumpable self-compacting lightweight concrete. In Wallevik, O. & Nielsson, I. (eds.) *Self-Compacting Concrete; Proc of the 3rd Intern. Symp. Reykjavik, 17–20 August 2003*: 776–785. RILEM Publications S.A.R.L.

Hossain, A.B. & Weiss, J. 2004. Assesing residual stress development and stress relaxation in restrained concrete ring specimens, Cement & Concrete Composites, 26, pp. 531–540.

Koenig G. & Faust T. 1997. *Abschlussbericht zu Verbunddecken aus Leichtbeton.* Leipzig:Universität Leipzig, unpublished.

Kurz, W. & Kessler, C. 2007. *Bericht zu Tastversuchen an Holorib Verbunddecken mit Leichtbeton, Nr. 104/07.* Kaiserslautern: Technische Universität Kaiserslautern, unpublished.

Mechtcherine, V., Haist, M., Staerk, L. & Mueller, H.S. 2003. Optimisation of the rheological and fracture mechanical properties of lightweight aggregate concrete. In Brandt, A.M., Li, V.C. & Marshall, I.H. (eds.), *Brittle Matrix Composites; Proc. of the Seventh Intern. Symp., Warsaw, 13–15 October 2003*: 301–310. Cambridge: Woodhead Publishing Ltd.—Warsaw: ZTUREK.

Mueller, H.S. & Mechtcherine, V. 2000. Selbstverdichtender Leichtbeton. Sachstandbericht Selbstverdichtender Beton (SVB) *Deutscher Ausschuss für Stahlbeton (DAfStb) Heft 516*: 74–84.

Advances in Cement-Based Materials – van Zijl & Boshoff (eds)
© *2010 Taylor & Francis Group, London, ISBN 978-0-415-87637-7*

Influence of fibre type and content on properties of steel bar reinforced high-strength steel fibre reinforced concrete

K. Holschemacher & T. Mueller
Leipzig University of Applied Sciences (HTWK Leipzig), Leipzig, Germany

ABSTRACT: The effect of fibres in combination with steel bar reinforcement is still widely unexplored. For this reason a research program was started, focused on the influence of steel fibre types and amounts as well as different bar reinforcements on flexural tensile strength, fracture behaviour and workability of steel bar reinforced high-strength steel fibre reinforced concrete (HSSFRC). The parameters which were investigated are in detail the influence of fibre geometry (straight with end hooks, corrugated), tensile strength of fibres (1100 N/mm^2, 1900 N/mm^2) and different steel bar reinforcements in combination with diverse fibre contents (20, 30, 40 and 60 kg/m^3). The bases for evaluating the efficiency of the different steel fibres together with the reinforcing steel are force-displacement curves of steel fibre reinforced concrete (SFRC) beams, gained in four point bending tests. According to the regulations, published by the German Concrete Association (DBV), beams of 150 × 150 mm cross section with a length of 700 mm were used as specimens. By summarising the test results the paper gives a support of the choice of the suitable fibre types and dosages that interacted most efficient in the combination with steel bar reinforcement and shows the influence of the reinforcing steel on the distribution of fibres in the cross section.

1 INTRODUCTION

High-strength concrete (HSC) or also called high performance concrete (HPC) is a construction material of which the cube compressive strength according to the German regulations ranges between 67 N/mm^2 and 115 N/mm^2. HSC is predominantly utilized in structural elements loaded by high compressive forces such as columns in building or bridge constructions. In comparison to normal performance concrete (NPC) HPC has a lot of advantages. Among other things this concrete type is characterised by a high density and strength as well as a good workability. The existing disadvantages like the shrinkage of concrete or the increasing brittleness can be improved by the addition of steel fibres. Compared to conventional HSC the most important benefits of HSSFRC are the hindrance of the development of macrocracks, the delay of the propagation of microcracks to macroscopic cracks and the better ductility after microcracks. HSSFRC is also tougher and demonstrates higher residual strengths.

2 TEST PROGRAM

The test program was divided into separate parts in which each one of them was studied in detail: influences of geometry and tensile strength of fibres as well as the effects of different bar reinforcements. The applied fibre contents were 20, 30, 40 and 60 kg/m^3. The experimental investigations were carried out according to the German regulations [2] for steel fibre reinforced concrete. Based on the large number of test results only three of four fibre types with the fibre contents of 20, 40 and 60 kg/m^3 are presented in this paper.

2.1 *Materials and mixture proportions*

The properties of three selected steel fibre types used for the HSSFRC mixtures are shown in Table 1. In total 4 steel fibres were tested. The illustrated fibre types have a length (l_f) of 50 mm, a diameter (d_f) of 1.0 mm and an aspect-ratio ($\lambda = l_f/d_f$) of 50. These fibre types mainly differ in their shape (straight and corrugated) as well as in their tensile strength (f_t). Fibre types F1 and F3 are normal-strength fibres with a tensile strength of circa 1100 N/mm^2. F2 is a high-strength fibre with a tensile strength of approximately 1900 N/mm^2. The value n_f represents the number of fibres per kg. In all cases the modulus of elasticity averages 200000 N/mm^2.

The following materials were used:

– CEM II/A-M(S-LL) 42,5R with conformity with EN 197-1 from CEMEX OstZement GmbH
– Fly ash EFA-Füller from BauMineral GmbH

Table 1. Overview of investigated steel fibres.

Geometry	Parameter		
Fibre type F1	$l_f =$	50	mm
	$d_f =$	1	mm
	$\lambda =$	50	-
	$f_t =$	1100	N/mm^2
	$f_t^* =$	1222	N/mm^2
	$n_f =$	3150	kg^{-1}
Fibre type F2	$l_f =$	50	mm
	$d_f =$	1	mm
	$\lambda =$	50	-
	$f_t =$	1900	N/mm^2
	$f_t^* =$	1762	N/mm^2
	$n_f =$	3100	kg^{-1}
Fibre type F3	$l_f =$	50	mm
	$d_f =$	1	mm
	$\lambda =$	50	-
	$f_t =$	1100	N/mm^2
	$f_t^* =$	925	N/mm^2
	$n_f =$	2850	kg^{-1}

* determined fibre tensile strength (average of 10 fibres).

Table 2. Overview of investigated concrete compositions.

Components		Concrete Compositions			
		R^0	F1$^{20/40/60}$	F2$^{20/40/60}$	F3$^{20/40/60}$
		Content in kg/m^3			
Cement	c	400.0			
Total Water	w	132.0			
w/c ratio		0.330			
Additives	Steel Fibres	0	20 / 40 / 60	20 / 40 / 60	20 / 40 / 60
	Fly Ash	100.0			
w/(c+FA) ratio		0.264			
Admixtures	PCE	10	10.4/11.6/12.4	10.4/11.6/11.6	10.4/11.6/11.6
	Retarder	0.8	0.8 / 2.0 / 2.4	2.0 / 2.0 / 2.4	2.0 / 2.0 / 2.4
Aggregates	0/2 Sand	696.9			
	2/8 Gravel	443.4			
	8/16 Gravel	638.4			

- Superplasticizer ISOLA BV/FM11 based on poly carboxylic ether (PCE) produced by CEMEX Admixtures GmbH
- Retarder ISOLA VZ 1 based on phosphate from CEMEX Admixtures GmbH
- Aggregates: 0/2 mm sand as well as 2/8 mm and 8/16 mm gravel

Table 2 shows the different concrete compositions of various HSC and HSSFRC mixtures.

The mixture design R^0 (reference concrete) represents a fibreless concrete composition. Based on this mixture for different fibre amounts (20, 40, 60 kg/m^3) HSSFRC compositions were produced with the fibres types F1, F2 and F3. The HSSFRC mixtures were essentially the same apart from the usage of different amounts of steel fibres in comparison with the HSC composition R^0. A slightly higher proportion of PCE was used as well as the addition of a retarder was required to maintain the workability with increasing fibre dosage over the working period. All concrete mixtures were mixed in a 0.5 m^3 single-shaft compulsory mixer. The yield for each mixture was 0.380 m^3. In order to reduce the influence of the moist aggregates for all concrete compositions dried aggregates were used.

2.2 Fresh properties tests

One workability test, as mentioned, was carried out for HSC and HSSFRC mixtures in accordance with the German standard DIN EN 12350-5. The test was carried out in the lapse of 5–10 minutes after emptying the concrete mixer. The *flow table test* consists of the determination of the mean diameter in the horizontal concrete spread on a table after lifting the cone-shaped mould with compaction. Hence the table top is raised until it meets the stop and then dropped freely 15 times. In this test the workability of the produced mixtures was investigated. The table flow test also gives an indication of the cohesion of the fresh concrete. That results in a possible determination of segregation or bleeding. The target slump for the investigations should have a range between 56 and 62 cm.

2.3 Hardened properties tests

The concrete compressive and splitting tensile strength was measured on three cubes with an edge length of 150 mm in each instance. The modulus of elasticity was investigated on three cylinders with a height of 30 cm and a diameter of 15 cm. Furthermore four-point bending tests were carried out for determining the post cracking behaviour. For this purpose HSC and HSSFRC beams with a cross section of 150 × 150 mm and a length of 700 mm were used as specimens (Figure 1).

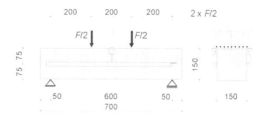

Figure 1. Experimental Set-up of the four point bending test.

The casting of the specimens, the curing procedure and the test set-up were chosen according to the German regulations [2] which are similar to those of RILEM. The beams were loaded orthogonal to the casting direction. The load was applied using displacement controlled method with a rate of 0.2 mm/min. The deflection was recorded by using one LVDT on each side of the beam.

Altogether 18 beams per test series were produced. In each case 6 beams with two different grades of steel bar reinforcements and 6 beams without reinforcing steel were investigated, as shown in Figure 2.

The distance between the bottom edge of the reinforcing steel and the concrete surface was approximately 2 cm. Longitudinal tensile reinforcement had hooks at the beam ends to ensure adequate anchorage. Additionally stirrups (diameter of 6 mm) were attached at the supports to hold the bar reinforcement in position. Deformed steel bars with a diameter of 6 and 12 mm and specified yield strength of 500 N/mm² (BSt 500) were used as bottom longitudinal reinforcement (Figure 3). The hardened properties of the concretes were tested 28 days after casting.

3 EXPERIMENTAL RESULTS

3.1 Properties of fresh concrete

Due to the low water/binder-ratio the concrete compositions were characterised by a high viscosity that results in an aggravated workability. However during and after the tests a uniform distribution and a random orientation of fibres without any signs of balling or clustering were observed. Table 3 presents the workability test results and

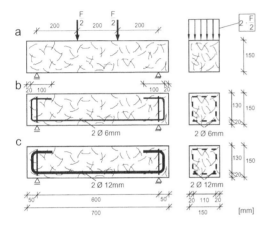

Figure 2. Test specimens with different bar reinforcements: a) no bar reinforcement; b) 2 Ø 6 mm; c) 2 Ø 12 mm.

Table 3. Properties of fresh concrete.

Concrete composition	Properties of fresh concrete		
	Flow table test	Concrete temperature	Ambient temperature
	[cm]	[°C]	[°C]
R ^{0}	62.5	22.1	7.2
F1 ^{20}	63.0	20.1	3.5
F1 ^{40}	55.5	21.9	6.5
F1 ^{60}	57.3	19.8	2.5
F2 ^{20}	57.0	21.3	6.3
F2 ^{40}	56.0	20.5	4.8
F2 ^{60}	62.8	22.3	11.0
F3 ^{20}	62.0	25.2	18.5
F3 ^{20*}	58.0	26.6	18.8
F3 ^{40}	64.5	33.6	26.8
F3 ^{60}	61.5	29.0	23.0

F3^{20*} repetition of mixture with bar reinforcement of 2 Ø 12.

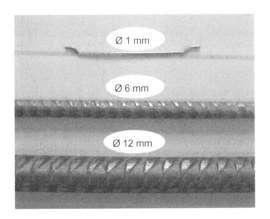

Figure 3. Comparison between sizes of used bar reinforcements and steel fibre.

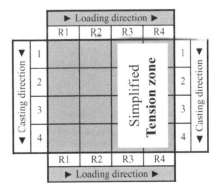

Figure 4. Discretization in cells of specimen's fracture surface.

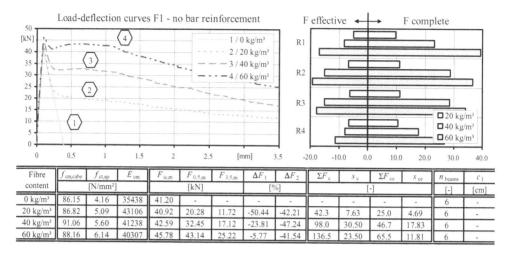

Fibre content	$f_{cm,cube}$	$f_{ct,sp}$	E_{cm}	$F_{ic,m}$	$F_{0.5,m}$	$F_{3.5,m}$	ΔF_1	ΔF_2	ΣF_c	s_c	ΣF_{ce}	s_{ce}	n_{beams}	c_1
	[N/mm²]			[kN]			[%]		[-]				[-]	[cm]
0 kg/m³	86.15	4.16	35438	41.20	-	-	-	-	-	-	-	-	6	-
20 kg/m³	86.82	5.09	43106	40.92	20.28	11.72	-50.44	-42.21	42.3	7.63	25.0	4.69	6	-
40 kg/m³	91.06	5.60	41238	42.59	32.45	17.12	-23.81	-47.24	98.0	30.50	46.7	17.83	6	-
60 kg/m³	88.16	6.14	40307	45.78	43.14	25.22	-5.77	-41.54	136.5	23.50	65.5	11.81	6	-

Figure 5. Average load-deflection curves for different fibre contents and distribution of effective and complete fibres in the loading direction without bar reinforcement (F1).

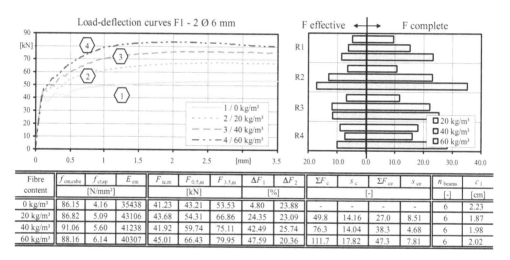

Fibre content	$f_{cm,cube}$	$f_{ct,sp}$	E_{cm}	$F_{ic,m}$	$F_{0.5,m}$	$F_{3.5,m}$	ΔF_1	ΔF_2	ΣF_c	s_c	ΣF_{ce}	s_{ce}	n_{beams}	c_1
	[N/mm²]			[kN]			[%]		[-]				[-]	[cm]
0 kg/m³	86.15	4.16	35438	41.23	43.21	53.53	4.80	23.88	-	-	-	-	6	2.23
20 kg/m³	86.82	5.09	43106	43.68	54.31	66.86	24.35	23.09	49.8	14.16	27.0	8.51	6	1.87
40 kg/m³	91.06	5.60	41238	41.92	59.74	75.11	42.49	25.74	76.3	14.04	38.3	4.68	6	1.98
60 kg/m³	88.16	6.14	40307	45.01	66.43	79.95	47.59	20.36	111.7	17.82	47.3	7.81	6	2.02

Figure 6. Average load-deflection curves for different fibre contents and distribution of effective and complete fibres in the loading direction with 2 Ø 6 mm (F1).

concrete temperatures of different HSC and HSS-FRC mixtures.

3.2 *Properties of hardened concrete*

In all cases the failed surfaces of HSC and HSS-FRC mixtures revealed uniform distribution of aggregates, confirming segregation resistance and stability of the mixtures produced. To evaluate the degree of the fibre segregation, the specimen's fracture surface was discretized in four rows and four columns of cells, presented in Figure 4. Generally there was a steady increase of the percentage of fibres in the casting direction related to the complete fibre distribution that was observed. In this connection similar fibre distributions under varying longitudinal reinforcement were obtained.

In Figures 5 to 7 it is exemplary shown the influence of fibre dosage (20, 40 and 60 kg/m³) for diverse reinforcement ratios (0.25% with 2 Ø 6 mm and 1.0% with 2 Ø 12 mm) on the pre- and post cracking behaviour of HSC and HSSFRC of fibre type F1. The plotted load-deflection curves were determined in four-point bending tests. Each

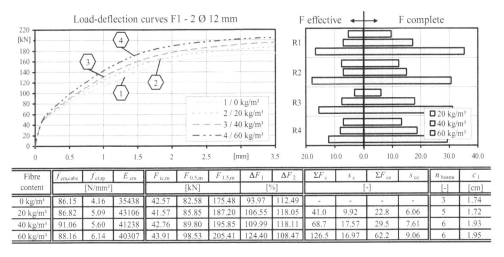

Figure 7. Average load-deflection curves for different fibre contents and distribution of effective and complete fibres in the loading direction with 2 Ø 12 mm (F1).

Fibre content	$f_{cm,cube}$	$f_{ct,sp}$	E_{cm}	$F_{ic,m}$	$F_{0.5,m}$	$F_{3.5,m}$	ΔF_1	ΔF_2	ΣF_c	s_c	ΣF_{ce}	s_{ce}	n_{beams}	c_1
	[N/mm²]			[kN]			[%]		[-]				[-]	[cm]
0 kg/m³	86.15	4.16	35438	42.57	82.58	175.48	93.97	112.49	-	-	-	-	3	1.74
20 kg/m³	86.82	5.09	43106	41.57	85.85	187.20	106.55	118.05	41.0	9.92	22.8	6.06	5	1.72
40 kg/m³	91.06	5.60	41238	42.76	89.80	195.85	109.99	118.11	68.7	17.57	29.5	7.61	6	1.93
60 kg/m³	88.16	6.14	40307	43.91	98.53	205.41	124.40	108.47	126.5	16.97	62.2	9.06	6	1.95

curve is the average relationship of three to six test beams (n_{beams}) as it is shown in the table under the graphs. In addition to the presented load-deflection curves the average numbers of complete and effective fibres (F complete, F effective) of the failed cross section are given in the opposite diagram as well as in the following table. The abbreviation ΣF_c refers to the number of fibres in the complete cross section of the fractured surface. The number of effective fibres is characterized with an additional index e (ΣF_{ce}). Effective fibres are those that are not located directly at an edge, are not orientated almost parallel to the failed cross-section and stick out more than 5 mm. Additionally the standard deviation (s) of the number of fibres is shown. An overview of selected material properties and average number of fibres in the failed cross-section is given in the table under the load-deflection curves. The strength values (cube compressive strength $f_{cm,cube}$, splitting tensile strength $f_{ct,sp}$ and the modulus of elasticity E_{cm}) are presented in the first three columns. In the middle part the load values (initial crack load $F_{ic,m}$, load corresponding to 0.5 mm ($F_{0.5,m}$) and 3.5 mm deflection ($F_{3.5,m}$)) based on test series as well as and the load differences between $F_{0.5,m}$ and $F_{ic,m}$ (ΔF_1) as well as between $F_{3.5,m}$ and $F_{0.5,m}$ (ΔF_2) as percentage were stated. The average concrete cover (c_l) of the longitudinal reinforcement is listed in the last column.

An overview of selected material properties and average number of fibres in the failed cross-section for different fibre types (F1, F2 and F3), bar reinforcements (2 Ø 6 mm, 2 Ø 12 mm) as well as for varying fibre contents (20, 40, 60 kg/m³) is given in

Table 4. For all test beams of one test series the fibre number and distribution were determined. The tension zone was defined as half of the cross section. Additionally, the strength values based on test series ($f_{ck,cube}$; $F_{max,m}$; $F_{0.5,m}$; $F_{3.5,m}$) and the load difference between $F_{0.5,m}$ and $F_{max,m}$ as well as between $F_{3.5,m}$ and $F_{0.5,m}$ were stated as load and percentage were stated.

Regarding the results of fibre type F1 without bar reinforcement, there was a steady increase of the load level for increased fibre content in the post cracking range, as depicted in Figure 5. By doubling or tripling the fibre content from 20 to 40 kg/m³ or 20 to 60 kg/m³ the load bearing capacity could not be doubled or tripled after the peak load. This behaviour can be justified by lower percentage of effective fibres at increased fibre amount. As it follows from Figure 5, for higher fiber contents the reduction in postcracking load is lower. At 40 kg/m³ fibre content or above the average load-deflection curves demonstrated a rise of the load after the matrix failure until a deflection of approximately 0.7 mm.

Figure 6 presents the average load-deflection curves of fibre type F1 in combination with reinforcing steel (2 Ø 6 mm). As it is shown with increasing fibre content distinct load improvements compared to curve R⁰ without steel fibres were observed. These averaged 25% for a fibre dosage of 20 kg/m³ and up to 49% for a fibre amount of 60 kg/m³ at a deflection of 3.5 mm. It was also noticed that for the fibre content of 60 kg/m³ from a deflection of approximately 2 mm the load reduced under increasing deflection. This can be explained by the fibre pull out, resulting in lower bond strengths. Flexural tensile failure was observed for all test beams independent of the fibre content.

Table 4. Material properties and average number of fibres for different HSC/HSSFRC mixtures.

Bar reinforcement	Fibre type	Fibre content	Compressive strength	Initial crack load	Load corresponding to 0.5 mm deflection	Load corresponding to 3.5 mm deflection	Load difference between $F_{0.5,m}-F_{max,m}$		Load difference between $F_{3.5,m}-F_{0.5,m}$		Average number of fibres in the cross section of the tested beams				Number of beams
											Complete		Tension zone		
		V_f	$f_{ck,cube}$	$F_{max,m}$	$F_{0.5,m}$	$F_{3.5,m}$	ΔF_1		ΔF_2		ΣF_c	ΣF_{ce}	ΣF_t	ΣF_{te}	
		(kg/m³)	(N/mm²)	(kN)	(kN)	(kN)	(kN)	(%)	(kN)	(%)	-	-	-	-	-
no Ø	R	0	81.15	42.27	-	-	-	-	-	-	-	-	-	-	6
	F1	20	81.82	42.53	20.30	11.73	-22.23	-52.3	-8.6	-42.2	42.3	25.0	21.7	13.0	6
		40	86.06	44.25	32.43	17.10	-11.82	-26.7	-15.3	-47.3	98.0	46.7	46.2	23.2	6
		60	83.16	47.27	43.10	25.20	-4.17	-8.8	-17.9	-41.5	136.5	65.5	60.7	29.2	6
	F2	20	86.36	43.53	21.07	18.40	-22.47	-51.6	-2.7	-12.7	43.8	23.8	21.8	12.0	6
		40	86.66	43.80	35.30	22.23	-8.50	-19.4	-13.1	-37.0	83.3	34.3	42.7	17.3	6
		60	87.76	44.30	40.03	29.57	-4.27	-9.6	-10.5	-26.1	116.8	49.5	57.3	26.2	6
	F3	20	81.16	44.53	16.27	1.87	-28.27	-63.5	-14.4	-88.5	32.0	18.0	18.5	10.3	6
		40	90.43	46.18	34.70	6.63	-11.48	-24.9	-28.1	-80.9	85.3	43.2	43.3	22.2	6
		60	88.33	46.20	41.17	9.50	-5.03	-10.9	-31.7	-76.9	122.8	61.8	62.5	33.3	6
2 Ø 6 mm	R	0	81.15	42.83	43.21	53.53	0.38	0.9	10.3	23.9	-	-	-	-	6
	F1	20	81.82	44.60	54.31	66.83	9.71	21.8	12.5	23.1	49.8	27.0	29.7	15.8	6
		40	86.06	44.23	59.74	75.13	15.50	35.0	15.4	25.8	76.3	38.3	38.2	19.2	6
		60	83.16	45.90	66.43	79.87	20.53	44.7	13.4	20.2	111.7	47.3	53.3	21.7	6
	F2	20	86.36	45.93	52.34	66.30	6.41	14.0	14.0	26.7	31.8	17.0	16.3	8.5	6
		40	86.66	45.47	57.59	74.93	12.13	26.7	17.3	30.1	76.5	32.2	34.5	15.5	6
		60	87.76	45.10	64.58	82.83	19.48	43.2	18.3	28.3	123.5	46.5	57.3	20.3	6
	F3	20	81.16	46.50	54.49	59.27	7.99	17.2	4.8	8.8	31.8	19.2	14.7	10.2	6
		40	90.43	45.33	57.60	66.90	12.27	27.1	9.3	16.1	71.0	39.5	33.0	18.0	6
		60	88.33	48.60	66.42	73.77	17.82	36.7	7.3	11.1	112.3	59.0	49.0	25.0	6
2 Ø 12 mm	R	0	81.15	44.87	82.58	175.48	37.72	84.1	92.9	112.5	-	-	-	-	3
	F1	20	81.82	42.36	85.85	186.96	43.49	102.7	101.1	117.8	41.0	22.8	19.2	10.0	5
		40	86.06	43.53	89.80	195.43	46.26	106.3	105.6	117.6	68.7	29.5	36.5	15.5	6
		60	83.16	45.57	98.53	204.83	52.96	116.2	106.3	107.9	126.5	62.2	60.5	27.7	6
	F2	20	86.36	43.90	88.35	186.37	44.45	101.3	98.0	110.9	41.5	19.5	19.7	10.0	6
		40	86.66	43.90	90.00	191.60	46.10	105.0	101.6	112.9	81.2	34.8	38.2	16.8	6
		60	87.76	42.83	93.84	201.00	51.01	119.1	107.2	114.2	119.8	42.8	58.3	21.5	6
	F3	20*	89.54	44.63	84.45	188.34	39.82	89.2	103.9	123.0	42.7	22.3	24.7	13.6	7
		40	90.43	46.27	90.76	196.87	44.50	96.2	106.1	116.9	82.3	45.0	42.0	25.3	3
		60	88.33	44.67	93.24	200.33	48.57	108.7	107.1	114.9	111.0	45.2	54.8	22.2	6

Under increasing fibre volume in conjunction with a bar reinforcement of 2 Ø 12 mm a continuous rise of the load levels in the post cracking range was obtained for fibre type F1 (Figure 7). On average the load increase ranged from 7% (20 kg/m³ fibre amount) to 17% (60 kg/m³ fibre dosage) at a deflection of 3.5 mm. For beams with a reinforcement ratio of 1% (2 Ø 12 mm) two different kinds of failure modes were noticed. In cases of none or low fibre dosages (0 and 20 kg/m³) the specimens failed in shear or compression. With higher fibre contents from 40 to 60 kg/m³ only compression failure was observed.

As it was expected in the most cases, the average number of complete fibres in the failed cross-section almost proportionally increased with rising fibre content (Table 4). Extreme low values relating to the average were only observed for the mixtures F3[20] without bar reinforcement and F2[20], F3[20] with a longitudinal reinforcement of 2 Ø 6 mm as well as F1[40], F3[60] with reinforced steel of 2 Ø 12 mm.

In respect to the use of no bar reinforcement the load loss between $F_{0.5,m}$ and $F_{max,m}$ reduced under increasing fibre content. In contrast, the load loss between $F_{3.5,m}$ and $F_{0.5,m}$ increased under growing fibre amount especially relating to fibre F1 and F3. For the high strength fibre F2 the percentage of load loss between $F_{3.5,m}$ and $F_{0.5,m}$ roughly raised under growing fibre amount. In contrast to the normal strength fibres F1 and F3 the percentage load reduction nearly stagnated between $F_{3.5,m}$ and $F_{0.5,m}$ with increasing fibre concentration. By the use of reinforcing steel the loading rate heightened between the respective load points. A significant difference of the evolution of working capacity between fibre type F1 and F2 in combination with bar reinforcement could not be observed. With respect to the corrugated fibre F3 it could be seen that with 2 Ø 6 mm the loading rate at a deflection of 3.5 mm was clearly lower compared to the straight fibre types. In connection with a steel bar reinforcement of 2 Ø 12 mm a significant difference

Load bearing capacity of fibres for a fibre content of 20 kg/m³

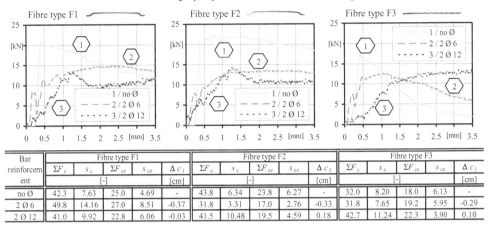

Bar reinforcement	Fibre type F1					Fibre type F2					Fibre type F3				
	ΣF_c	s_c	ΣF_{cc}	s_{cc}	Δc_1	ΣF_c	s_c	ΣF_{cc}	s_{cc}	Δc_1	ΣF_c	s_c	ΣF_{cc}	s_{cc}	Δc_1
		[-]			[cm]		[-]			[cm]		[-]			[cm]
no Ø	42.3	7.63	25.0	4.69	-	43.8	6.34	23.8	6.27	-	32.0	8.20	18.0	6.13	-
2 Ø 6	49.8	14.16	27.0	8.51	-0.37	31.8	3.31	17.0	2.76	-0.33	31.8	7.65	19.2	5.95	-0.29
2 Ø 12	41.0	9.92	22.8	6.06	-0.03	41.5	10.48	19.5	4.59	0.18	42.7	11.24	22.3	3.90	0.10

Figure 8. Comparison of load bearing capacity of fibre types F1, F2 and F3 for different bar reinforcements (f. content 20 kg/m³).

Load bearing capacity of fibres for a fibre content of 40 kg/m³

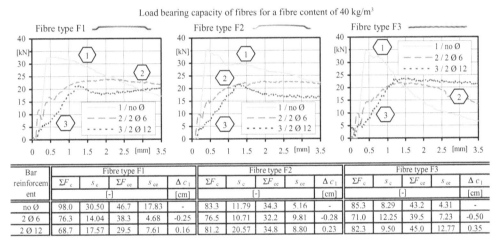

Bar reinforcement	Fibre type F1					Fibre type F2					Fibre type F3				
	ΣF_c	s_c	ΣF_{cc}	s_{cc}	Δc_1	ΣF_c	s_c	ΣF_{cc}	s_{cc}	Δc_1	ΣF_c	s_c	ΣF_{cc}	s_{cc}	Δc_1
		[-]			[cm]		[-]			[cm]		[-]			[cm]
no Ø	98.0	30.50	46.7	17.83	-	83.3	11.79	34.3	5.16	-	85.3	8.29	43.2	4.31	-
2 Ø 6	76.3	14.04	38.3	4.68	-0.25	76.5	10.71	32.2	9.81	-0.28	71.0	12.25	39.5	7.23	-0.50
2 Ø 12	68.7	17.57	29.5	7.61	0.16	81.2	20.57	34.8	8.80	0.23	82.3	9.50	45.0	12.77	0.35

Figure 9. Comparison of load bearing capacity of fibre types F1, F2 and F3 for different bar reinforcements (f. content 40 kg/m³).

between the load bearing capacity at 0.5 and 3.5 mm deformation comparing fibre types F1 to F3 was not observed. The initial crack load ($F_{max,m}$) is normally independent of the counted number of fibres, confirming that this parameter is dominated by the matrix behaviour. But in test series with a fibre amount of 60 kg/m³ a higher initial crack load was noticed compared to the lower dosed fibre concretes especially for the test series without rebars.

On average a proportional increase of fibre concentration does not cause a relative increase of post cracking loads up to deflections of 3.5 mm or higher. Among other things this resulted from different beginnings of fibre fracture during the loading. In this process early breaking caused rapid load losses, late or no breaking indicated ductile material behaviour. Furthermore, the effectiveness of fibres plays a great role in reaching a definite load bearing capacity after the matrix fracture, depending on allocation, orientation and embedded length. All these points were especially influenced by concrete composition (aggregate size and shape, concrete content, w/c-ratio, admixtures), fibre type, rheological properties, casting method, consolidation and others. As it is shown in Table 4, there is a dependence of the post crack load, on the number of effective fibres.

Figure 8 to 10 present parts of load-deflection curves, contributed by the fibres. These curves were obtained for beams with different fibre contents.

113

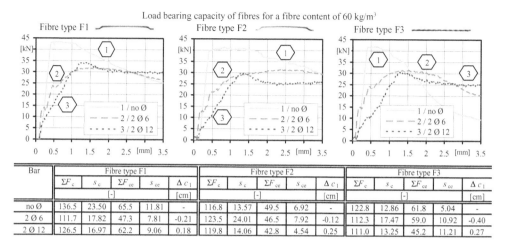

Figure 10. Comparison of load bearing capacity of fibre types F1, F2 and F3 for different bar reinforcements (f. content 60 kg/m³).

Bar	Fibre type F1					Fibre type F2					Fibre type F3				
	ΣF_c	s_c	ΣF_{ce}	s_{ce}	Δc_1	ΣF_c	s_c	ΣF_{ce}	s_{ce}	Δc_1	ΣF_c	s_c	ΣF_{ce}	s_{ce}	Δc_1
		[-]			[cm]		[-]			[cm]		[-]			[cm]
no Ø	136.5	23.50	65.5	11.81	-	116.8	13.57	49.5	6.92	-	122.8	12.86	61.8	5.04	-
2 Ø 6	111.7	17.82	47.3	7.81	-0.21	123.5	24.01	46.5	7.92	-0.12	112.3	17.47	59.0	10.92	-0.40
2 Ø 12	126.5	16.97	62.2	9.06	0.18	119.8	14.06	42.8	4.54	0.25	111.0	13.25	45.2	11.21	0.27

The solid lines (curve 1) in the figures demonstrate the contribution of fibres in specimens without bar reinforcement. In this case the load-deflection curves, obtained for HSC beams (assumed according to [2]), were subtracted from those obtained for HSSFRC specimens without bar reinforcement. The dashed and dotted lines (curve 2 and 3) represent the contribution of fibres to the load bearing capacity of the specimens with bar reinforcement of 2 Ø 6 mm and 2 Ø 12 mm, respectively. These load-deflection curves were generated by subtraction of loads, determined for beams with bar reinforcement without fibres (reference beams R⁰) from beams with fibres and longitudinal reinforcement. In addition to the fibre number the average differences of concrete cover of the bar reinforcements (Δc_1) are given. Δc_1 was calculated by the concrete cover of HSSFRC subtracted by the concrete cover of HSC. Negative values of Δc_1 signify a smaller internal lever arm of the reference beams, compared to the beams with fibres and bar reinforcement. The opposite apply to positive values of Δc_1. Regarding F1, it is shown that independently of the fibre content the load bearing capacity of fibres in pure steel fibre beams was always higher up to deflections of about 2.3 mm, compared to specimens with fibres and bar reinforcement. It can be explained by the generation of multiple cracks in the case of HSSFRC beams with reinforcing bars. The bar reinforcement initially absorbed a big portion of the first crack energy as a result of better bonding between the bars and the matrix, compared to that between the fibres and the matrix. With increasing deflection multiple cracks were noticed. In this connection smaller crack widths in respect to the same deformation were measured,

compared to beams with a single crack. Slower crack opening caused a slower but continuous activation of fibres. Regarding curves 2 and 3 it could be seen that independent of fibre type and content for the smaller bar reinforcement ratio a stronger rise of the curves is observed. Referring to this up to deformations between 1.0 mm and 1.5 mm a higher working capacity of fibres is observed for specimens with 2 Ø 6 mm, compared to those with 2 Ø 12 mm. It can be explained by a higher activation of the fibres during the load transfer after repeated matrix failure on the basis of lower bond forces of the smaller longitudinal reinforcement. F2 demonstrated independent of the fibre content higher load levels up to a deformation of 3.5 mm without bar reinforcement in comparison with reinforcing steel. Resulted from a higher number of broken fibres during the loading for the corrugated fibre F3 a distinct load decrease after the first crack without bar reinforcement was noticed. This caused a clearly lower load bearing performance for beams without reinforcing bars after deflections between 1.0 and 1.5 mm compared to specimens with 2 Ø 6 mm and 2 Ø 12 mm.

4 CONCLUSIONS

Strength and geometry of fibres have a direct influence on the load bearing capacity of HSS-FRC beams without bar reinforcement. Regarding to specimens with straight hooked fibre types F1 and F2 for all fibre amounts that were used in this study high strength fibres resulted in a clearly better ductile behaviour and higher load levels in the post cracking range, compared to normal strength fibres. Specimens with high strength fibre had lower number of broken fibres in the failed cross-sections, compared to those with normal strength ones. It is because a predominantly continuous and slow pull-out of the fibres from the concrete matrix. Specimens with the corrugated fibre type F3 were characterised by a distinct worse ductility, compared to the straight hooked ones, F1, due to higher number of broken fibres.

The load bearing capacity of specimens with normal and high-strength fibres, combined with bar reinforcement, gradually increased up to deflections of approximately 1.5 mm but could not reach the load level of HSSFRC without bar reinforcement. In case of higher deflections, specimens with normal-strength fibres, combined with bar reinforcement, showed higher energy absorption capacity, compared to that of HSC beams with normal-strength fibres and without reinforcing bars. It is because the number of broken fibres at the same deformation in the post-cracking range became lower and the crack widths were smaller due to multiple cracks, compared to beams with reinforcing bars.

The application of high-strength fibres in steel bar reinforced HSC seems to be unnecessary because an increase of the load bearing capacity compared to normal-strength fibres was not observed.

A reduced number of broken fibres in the failed cross-section were obtained in fibre reinforced specimens with bar reinforcement. Ductility of

beams with the corrugated fibre type without bar reinforcement could be improved by increasing the bars reinforcement ratio.

In cases of none or low fibre contents (0 and 20 kg/m³) the specimens failed in shear or compression. With a higher fibre content of 40 kg/m³ primary compression failure was observed. For fibre a volume of 60 kg/m³ a failing in shear of the high-strength concrete beams with a longitudinal reinforcement of 2 Ø 12 mm was not obtained.

For the design of steel bar reinforced HSSFRC constructions, the load bearing capacity at ultimate and serviceability stages may be overestimated if the residual loads based on the test results of the HSSFRC beams without bar reinforcement will be accepted.

REFERENCES

S-H. Cho & Y-I. Kim, "Effects of Steel Fibers on Short Beams Loaded in Shear," ACI Structural Journal, V. 100, No. 6, November-December 2003, pp. 765–774.

Deutscher Beton-und Bautechnik-Verein e.V. (Hrsg.), DBV-Merkblatt "Stahlfaserbeton," Wiesbaden, October 2001.

T. Richter, "Hochfester Beton—Hochleistungs-beton," Schriftenreihe Spezialbetone Band 3, Verlag Bau+Technik GmbH, 1999.

U. Gossla, "Tragverhalten und Sicherheit betonstahlbewehrter Stahlfaserbetonbauteile," Deutscher Ausschuss für Stahlbeton Heft 501, Beuth Verlag, 2000.

Advances in Cement-Based Materials – van Zijl & Boshoff (eds)
© 2010 Taylor & Francis Group, London, ISBN 978-0-415-87637-7

Influence of fibre content and concrete composition on properties of Self-Compacting Steel Fibre Reinforced Concrete

T. Mueller & K. Holschemacher
Leipzig University of Applied Sciences (HTWK Leipzig), Leipzig, Germany

ABSTRACT: A Self-compacting steel fibre reinforced concrete (SCSFRC) is a high-performance building material that combines positive aspects of fresh properties of self-compacting concrete (SCC) with improved characteristics of hardened concrete as a result of fibre addition. Considering these properties the application ranges of steel fibre reinforced concrete (SFRC) and SCC can be covered. This paper presents a description of two different mix compositions of SCSFRC based on a powder and a combination type. In the beginning of the investigations selected fresh and hardened concrete properties under variation of the fibre content (0, 20, 30 and 40 kg/m³) were determined. In total 9 concrete series were produced to investigate the influence of fibre content and concrete composition on flexural tensile strength, fracture behaviour and workability of SCSFRC. The bases for evaluating the efficiency of the different fibre contents are force-displacement curves of SFRC beams, gained in four point bending tests. By summarising the test results the paper gives a support for the choice of the suitable fibre contents and mix compositions in concordance to the requirements of SCSFRC.

1 INTRODUCTION

In 1988 Okamura [6] established the basic description of SCC. Approximately ten years later first in situ applications were observed in Germany. However, over the past years the use of SCC in housing, tunnel and bridge constructions has increased as a result of the excellent rheological properties and the published German regulations [1]. In most cases SCC is chosen because of the good workability, the easy placement and pumping, a congested reinforcement and a complex formwork or the high quality of surface finish. To achieve these properties SCC must satisfy some important workability criteria including very advantageous rheological and self-venting properties as well as adequate resistance against segregation and bleeding.

For over 30 years steel fibres in normal concrete have been successfully used in construction practice in Germany. In general, SFRC is applied in the fabrication of industrial floors but is also utilized in housing, tunnel and prefabricated constructions.

SCC reinforced with steel fibres raises its field of application because the mechanical performance of the concrete in the hardened state is improved. Compared to conventional SCC the most important benefits are the hindrance of the development of macrocracks, the delay of the propagation of microcracks to macroscopic cracks and the

improved ductility after microcracks. SCSFRC is also tougher and demonstrates higher residual strengths.

The type (geometry, shape, surface, allocation) and content of fibres as well as the SCC matrix directly influence the workability of SCSFRC. The main goal of this study was to develop two different types of SCSFRC based on a powder and a combination type in terms of workability and hardened properties suited to casting structural members subjected to bending. For the design of SCFRC members subjected to bending, the flexural tensile strength, which characterizes the material behaviour at a selected deflection or crack mouth opening displacement, is of particular importance. In dependence of the applied fibre content and fibre type the load bearing capacity and the flexural tensile strength are affected decisively. For several years design guidelines for SFRC members have been available, which were published by various organisations like RILEM, the German Concrete Association (DBV) and others. These guidelines are based on standard bending tests which are suitable for the description of the structural behaviour after cracking. However, these guidelines provide no service for evaluating the efficiency of a certain fibre type and amount. In this paper various tests of fresh and hardened properties of selected SCFRC mixtures are presented.

2 TEST PROGRAM

The test program was divided into separate parts in which each of the following was studied in detail: influences of geometry, aspect-ratio and tensile strength of fibres as well as the effects of different concrete properties. The applied fibre contents were 20, 30 and 40 kg/m³. The experimental investigations were carried out according to [3].

2.1 Materials and mixture proportions

The properties of one selected steel fibre type used for the different SCSFRC mixtures are shown in Table 1. The presented fibre type has a length (l_f) of 50 mm, a diameter (d_f) of 1.0 mm and an aspect-ratio ($\lambda = l_f/d_f$) of 50. The fibres have a tensile strength (f_t) of approximately 1100 N/mm². The value n_f represents the number of fibres per kg.

Table 1. Overview of investigated steel fibre.

Geometry		Variables		
		$l_f =$	**50**	mm
		$\Delta l_f =$	**–3/+2**	mm
		$d_f =$	**1**	mm
		$\Delta d_f =$	**±0.04**	mm
		$\lambda =$	**50**	–
		$l_H =$	**4–7**	mm
Form:	straight	$h_H =$	**1.8**	mm
Surface:	plane	$\alpha_H =$	**30–45**	°
Anchorage:	hooked ends	$f_t =$	**1100**	N/mm²
Allocation:	separate	$n_f =$	**3150**	kg⁻¹

Table 2 shows the different concrete compositions of various SCC and SCSFRC mixtures.

The following materials were used:

- Cement CEM II/B-S 32,5R with conformity with ENV 197 from CEMEX OstZement GmbH
- Fly ash EFA-Füller from BauMineral GmbH
- Superplasticizer ISOLA BV/FM10 based on polycarboxylic ether (PCE) from CEMEX Admixtures GmbH
- Stabilizer ST 5 from CEMEX Admixtures GmbH
- 0/2 mm sand, 2/8 mm and 8/16 mm gravel as well as 8/16 mm crushed gravel (chips).

The SCC mixture design of Type 1 (powder type) based on the procedure presented by Okamura. For the mixture design of Type 2 (combination type) a normal-slump traditional concrete was used as template. A slightly higher proportion of fly ash was used and the addition of a stabilizer was required to maintain stability. The three SCSFRC mixtures (fibre dosage of 20, 30 and 40 kg/m³) of Type 1 and 2, respectively were essentially the same apart from the usage of different amounts of steel fibres in comparison with the SCC compositions. The concrete composition Type 3 was only tested with a fibre volume of 40 kg/m³ according to Table 2. For Type 3 the mineral size fraction of 8 to 16 mm gravel was replaced through 8 to 16 mm crushed gravel compared to the SCSFRC mixture Type 2.

All concrete mixtures were mixed in a 0.5 m³ single-shaft compulsory mixer. The yield for each mixture was 0.235 m³. In order to reduce the influence of the moist aggregates for all concrete compositions dried aggregates were used.

Table 2. Overview of investigated concrete compositions.

Composition		Type 1	Type 2	Type 3
Components		Content (kg/m³)		
Cement	c	300.0	350.0	350.0
Total Water	w	169.8	178.3	178.3
w/c ration		0.566	0.509	0.509
Additives	Steel Fibres	0/20/30/40	0/20/30/40	40
	Fly Ash	300.0	175.0	175.0
w/(c+FA) ratio		0.283	0.340	0.340
Admixtures	PCE	3.7/3.8/3.6/4.1	2.6/3.2/2.9/3.2	3.2
	Stabilizer	–	1.0	1.0
Aggregates	0–2 sand	484.5	634.7	634.7
	2–8 gravel	445.7	379.7	379.7
	8–16 gravel	576.1	565.2	–
	8–16 chips	–	–	565.2

2.2 Fresh properties tests

Various workability tests, as mentioned, were carried out for SCC and SCSFRC mixtures in accordance with the German Association of Structural Concrete (DAfStb) guideline [2]. The tests were carried out in the lapse of 10 minutes after emptying the concrete mixer.

The *slump flow test* consists of the determination of the mean diameter of the horizontal concrete spread on a base plate after lifting the slump cone without any compaction. In this test the flowability of the SCC and SCSFRC mixtures were investigated, as shown in Figure 1. The target minimum slump flow for the investigations was 70 cm. During the slump flow test the viscosity of mixtures was estimated by measuring the time taken to reach a spread diameter of 50 cm from the moment the slump cone is lifted up. There was no restriction offered to the freely flowing SCC and SCSFRC mixtures.

An attachment to the slump flow test is the *J-ring* presented in Figure 1. This test is used to determine the influence of reinforcement already during the slump test. This way, passing ability and rheological behaviour under influence of reinforcement were tested. The J-ring has a diameter of 30 cm. Thereby 16 plane steel bars with a diameter of 18 mm are situated evenly along the perimeter. Slump flow and T-50 cm time were measured during the J-ring test, which indicated the restricted slump flow and T-50 cm time.

In the *V-Funnel test* the filling ability and viscosity of the mixtures were evaluated by measuring the time taken for the concrete to completely empty out through the V-funnel (Figure 1), which has a rectangular opening of 75 mm × 65 mm.

Another method for determining the stability of SCC/SCSFRC on the fresh concrete is the *wash test*. The mixtures were filled into a 450 mm high cylinder with a diameter of 150 mm that has slots to take separating sheets that divide the cylinder into three equal sections (Figure 1). The dividing plates were inserted after the concrete sample had been set. The resulting three batches of SCC/SCSFRC were then washed separately over an 8 mm sieve (for 16 mm maximum aggregate size), and the sieve residues were weighed after drying. The difference between the percentages of coarse particles in the individual segments from the percentage of coarse particles in the overall sample was calculated. An SCC is considered as stable to sedimentation if the difference of the measured masses of the 8/16 aggregate fractions of the three individual batches from the target mass was not more than 15 wt.% [5].

2.3 Hardened properties tests

The concrete compressive strength was measured on cubes with an edge length of 150 mm. Furthermore four-point bending tests were carried out for determining the postcracking behaviour. For this purpose SFRC beams with a cross section of 150 × 150 mm and a length of 700 mm were used as specimens (Figure 2). The casting of the specimens, the curing procedure and the test set-up were chosen according to the German regulations [3] which are similar to those of RILEM.

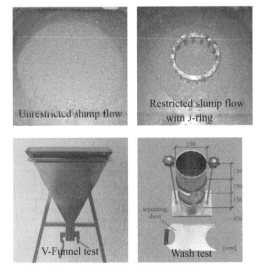

Figure 1. Various fresh properties tests for SCC and SCSFRC.

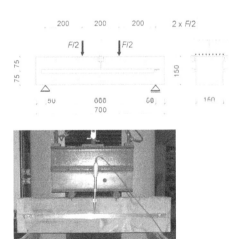

Figure 2. Experimental set-up of the four point bending test.

The beams were loaded orthogonal to the casting direction. The load was applied under displacement control at a rate of 0.2 mm/min. The deflection was recorded by using one LVDT on each side of the beam. The hardened properties of the concretes were tested 28 days after casting.

3 EXPERIMENTAL RESULTS

3.1 Properties of fresh concrete

During and after the tests a uniform distribution and a random orientation of fibres without any signs of balling or clustering was observed. Table 3 presents the workability test results of different SCC and SCSFRC mixtures.

To assess the workability of the SCC and SCS-FRC compositions the flow time or V-funnel time, as a measure of the viscosity, and the slump flow, as a criterion of the yield value have to be compared. The characteristic value pairs of concrete Type 1, 2 and 3 for the investigated fibre contents are shown in Figure 3. On average the slightly higher slump flow in combination with the higher V-funnel time of mixture Type 1 compared to Type 2 could be attributed to the slightly finer material. Regarding Type 1, the higher addition of superplasticizer is primarily responsible for the bigger slump flow versus the other mixture Types 2 and 3. A large slump flow generally results in a low funnel flow time. Unfortunately too high slump flows as observed in M1[40] and M2[20] caused poor results in the wash test (indicating segregation) and endangered the stability of the concrete mixture. Under variation of the content of superplasticizer the fresh concrete properties (V-funnel time and slump flow) of Type 1 compared to Type 2 fluctuated more.

By taking the results of the wash test and microstructural examinations carried out on the hardened concrete samples (cutting surface or surfaces) of the corresponding SCC and SCSFRC it is possible to describe a range in the diagram in which there is adequate flowability with low-segregation workability. In Figures 4 and 5 the cutting surfaces of cylinders with a height of 50 cm and a diameter of 15 cm for the concrete compositions Type 1 to 3 are shown. The cutting surfaces show a homogeneous coarse particle distribution. Larger air voids can be seen only in the microstructure of the cylinder M1[30], M2[30] and M3[40]. These mixtures were stable, but did not exhibit adequate self-de-aeration. In these cases the viscosity was a bit too high. The number and size of air voids in the rest of the sawed specimens were acceptable. In the upper zone of the cylinders an enrichment of paste and only a few coarse aggregate grains were not observed. This behaviour would characterize the start of the unstable range of the concrete mixture.

In secondary investigations two different SCS-FRC mixtures (Type 1 and 2) with a fibre dosage of 40 kg/m³ were produced in a 0.075 m³ pan mixer to determine the influence of water addition and reduction as well as the change with time of the workability, shown in Figure 6. The yield for each mixture was 0.06 m³.

These tests indicate that Type 2 could better compensate the changing of water content (±4%) compared to Type 1. With reference to the water addition and reduction the range between mixtures Type 2 in the diagram was significantly lower compared to Type 1. This fact can be explained by the application of stabilizer in the combination type (Type 2) to prevent segregation and bleeding, which results in a better robustness and workability in practice. Another difference between Type 1 and 2 was the longer thixotropic behaviour of

Table 3. Workability properties of various mixtures.

Workability test parameter	Type 1				Type 2				Type 3
	M1[0]	M1[20]	M1[30]	M1[40]	M2[0]	M2[20]	M2[30]	M2[40]	M3[40]
Slump flow (cm)									
Unrestricted	76.5	77.3	73.3	83.0	73.5	75.5	71.0	71.0	70.5
Restricted	66.0	68.0	56.3	75.5	71.5	62.0	55.5	68.5	68.0
T-50 cm time (s)									
Unrestricted	4.5	5.0	7.0	3.5	3.5	3.3	4.0	4.0	4.0
Restricted	14.0	10.0	25.0	8.0	8.0	8.0	11.5	6.5	7.0
V-funnel time (s)	22.0	19.5	26.0	12.0	11.5	10.0	15.0	12.0	15.0
Wash test (wt.%)									
Top	−1.6	−1.8	−2.8	−10.2	−4.2	−9.1	−2.7	−2.0	3.2
Center	−0.3	−0.2	−2.8	−6.7	−1.4	−1.6	4.2	−0.2	−3.5
Buttom	1.8	2.0	5.5	17.0	5.7	10.8	−1.6	2.2	0.3

0), 20), 30), 40) fibre content in kg/m³.

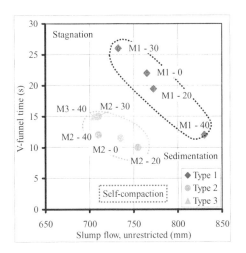

Figure 3. Workability windows for investigated SCC and SCSFRC mixtures as a function of slump flow (unrestricted) and V-funnel time.

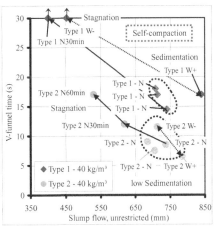

- N: Normal mixture
- N30min: Determination of the fresh concrete properties after 30 min
- N60min: Determination of the fresh concrete properties after 60 min
- W+: Mixture with an addition of 4% water
- W-: Mixture with a reduction of 4% water

Figure 6. Rheological properties of two SCSFRC (40 kg/m³ fibre content) as a function of the slump flow and V-funnel flow time with workability window.

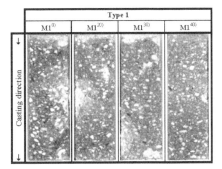

Figure 4. Cutting surface of cylinders (Type 1).

Figure 7. Discretization in cells of specimen's fracture surface (left); Casting of specimen (right).

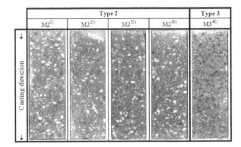

Figure 5. Cutting surface of cylinders (Type 2 and 3).

Type 2 based on the smaller content of fine grain particles in the concrete composition. This results in a clear better workability over the time for the combination type, as it can be seen in Figure 6. The investigations of the fresh concrete proper-

ties occurred without an extra dosage of super-plasticizer after 30 and 60 minutes. Regarding the powder type (Type 1) after 30 min the V-funnel time could no more be measured. However for Type 2 a slower reduction of the self-compacting characteristics was obtained.

3.2 Properties of hardened concrete

To evaluate the degree of the fibre segregation, the specimen's fracture surface was discretized in four rows and four columns of cells as shown in Figure 7.

In each case for every SCSFRC mixture the average number of complete fibres of nine test beams in the casting and loading direction were counted (visible on the fracture surface of the beam), shown for the casting direction in Figure 8. An increase of fibre percentage in the casting

121

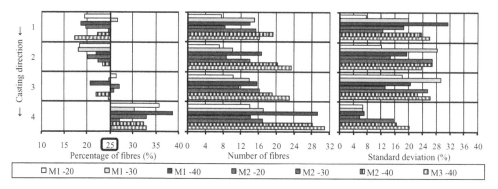

Figure 8. Average Distribution of complete fibres in casting direction.

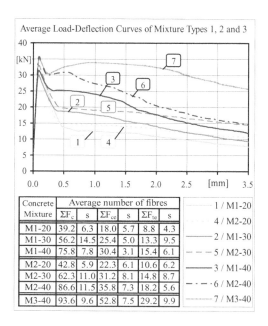

Figure 9. Fracture pattern of beams tested in 4-point bending.

Average Load-Deflection Curves of Mixture Types 1, 2 and 3

Concrete Mixture	Average number of fibres					
	ΣF_c	s	ΣF_{ce}	s	ΣF_{te}	s
M1-20	39.2	6.3	18.0	5.7	8.8	4.3
M1-30	56.2	14.5	25.4	5.0	13.3	9.5
M1-40	75.8	7.8	30.4	3.1	15.4	6.1
M2-20	42.8	5.9	22.3	6.1	10.6	6.2
M2-30	62.3	11.0	31.2	8.1	14.8	8.7
M2-40	86.6	11.5	35.8	7.3	18.2	5.6
M3-40	93.6	9.6	52.8	7.5	29.2	9.9

1 / M1-20
4 / M2-20
2 / M1-30
5 / M2-30
3 / M1-40
6 / M2-40
7 / M3-40

Figure 10. Average load deflection curves of the investigated SCSFRC mixtures and number of fibres in the cross section (fibre contents of 20, 30 and 40 kg/m³).

direction related to the complete fibre distribution was observed. This fibre segregation was more pronounced on specimens produced with mixture Type 1. In dependence of the fibre volume the average number of fibres in the fracture surface for Type 1 was also lower compared with Type 2 and 3. In most cases the highest variance of fibres in percentage was noticed in the middle part of the beam in the casting direction. In all cases the failed surfaces of SCC and SCSFRC mixtures revealed uniform distribution of aggregates and normal air void developments, confirming segregation resistance and stability of the mixtures produced, shown in Figure 9.

In Figure 10 the influence of mixture type and fibre amount on, the pre- and postcracking behaviour of SCSFRC is shown in terms of load-deflection curves. Each curve is the average relationship of nine test beams. Additionally the average numbers of complete and effective fibres as well as the effective fibres in the tension

zone of the failed cross section are given. The abbreviations ΣF_c and ΣF_t describe the number of fibres in the complete cross section (c) as well as in the tension zone (t) of the fractured surface. Thereby the tension zone was defined simplified as the half of the cross-section. The number of effective fibres is characterized with an additional index e (ΣF_{ce}, ΣF_{te}). Effective fibres are those that are not located directly at an edge, are not oriented almost parallel to the failed cross-section and stick out more than 5 mm. These values also apply for Table 4. Additionally the standard deviations of the number of fibres is given.

122

Table 4. Material properties and average number of fibres for SCC and SCSFRC with different fibre contents (20, 30, 40 kg/m³).

CC	Fibre content V_f (kg/m³)	Compressive strength $f_{ck,cube}$ (N/mm²)	Initial crack load $F_{max,m}$ (kN)	Load corresponding to 0.5 mm deflection $F_{0.5,m}$ (kN)	Load corresponding to 3.5 mm deflection $F_{3.5,m}$ (kN)	Load loss between $F_{max,m}-F_{0.5,m}$ ΔF_1 (kN)	(%)	Load loss between $F_{0.5,m}-F_{3.5,m}$ ΔF_2 (kN)	(%)	Complete ΣF_c	ΣF_{ce}	Tension zone ΣF_t	ΣF_{te}
1	0	59.44	31.35	–	–	–	–	–	–	–	–	–	–
	20	58.44	31.79	14.79	7.75	17.01	53.5	7.03	47.6	39.2	18.0	19.2	8.8
	30	58.73	30.30	18.62	9.32	11.68	38.5	9.30	50.0	56.2	25.4	28.6	13.3
	40	61.43	31.73	25.15	11.94	6.57	20.7	13.21	52.5	75.8	30.4	39.3	15.4
2	0	61.04	32.80	–	–	–	–	–	–	–	–	–	–
	20	59.11	30.55	14.76	10.73	15.78	51.7	4.03	27.3	42.8	22.3	21.1	10.6
	30	57.29	33.25	19.65	14.51	13.60	40.9	5.14	26.1	62.3	31.2	30.8	14.8
	40	64.65	36.35	30.87	14.60	5.48	15.1	16.28	52.7	86.6	35.8	42.8	18.2
3	40	62.72	35.62	32.48	25.68	3.14	8.8	6.80	20.9	93.6	52.8	52.0	29.2

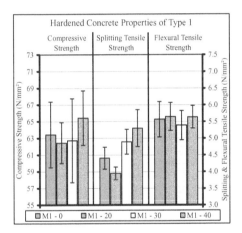

Figure 11. Selected hardened concrete properties of mixture Type 1.

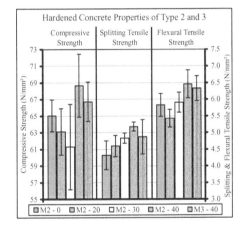

Figure 12. Selected hardened concrete properties of mixture Type 2 and 3.

Regarding mixture Type 1 and 2, there was a steady increase of the load level under increased fibre content in the post cracking range, as depicted in Figure 10. It can be seen that under increasing fibre amount an increase of the brittleness was obtained, caused by an earlier failure of the fibres. For a fibre content of 40 kg/m³ the load level of Type 3 was significantly higher in comparison to Type 1 and 2. In that case 74% (compared to type 1) and 48% (compared to Type 2) more effective fibres in the failed cross-section were counted. In summary it can be said that the use of concrete composition Type 1 demonstrated independent from the fibre concentration lower loads

in the post cracking range compared to mixture Type 2. This fact could be explained by the higher value of cementitious material and the lower w/c-ratio as well as the higher average number of fibres resulted in better bonding characteristics and higher energy absorption capacity (area under the curve).

An overview of selected material properties and number of fibres in the failed cross-section for different concrete compositions and fibre contents (0, 20, 30 and 40 kg/m³) is given in Table 4. For all beams of one test series the fibre number and distribution were determined. Additionally, the strength values based on test series ($f_{c,cube,m}$; $F_{max,m}$; $F_{0.5,m}$;

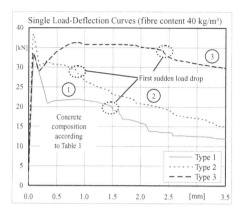

Figure 13. Selected single load-deflection curves of mixture Type 1, 2 and 3 (fibre content 40 kg/m³).

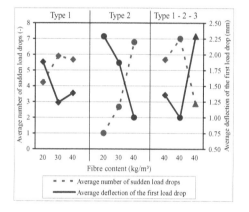

Figure 14. Comparison of average number and deflection of sudden load drops (Type 1, 2 and 3).

$F_{3.5,m}$) and the load loss between $F_{max,m}$ and $F_{0.5,m}$ as well as between $F_{0.5,m}$ and $F_{3.5,m}$ were stated.

As it was expected, the average number of fibres in the failed cross-section increased nearly proportionally with rising fibre content. The load loss between $F_{max,m}$ and $F_{0.5,m}$ reduced under increasing fibre content. On the contrary the load loss between $F_{0.5,m}$ and $F_{3.5,m}$ raised under growing fibre amount. On average a proportional increase of fibre concentration does not cause a proportional increase of post cracking loads up to deflections of 3.5 mm or higher. Among other things, this resulted from different beginnings of fibre fracture during the loading. In this process early breaking caused rapid load losses, late or no breaking caused ductile material behaviour. Furthermore the effectiveness of fibres plays a great role reaching a definite load bearing capacity after the matrix fracture, depending on allocation, orientation and embedded length. All these points were especially influenced by concrete composition (aggregate size and shape, concrete content, w/c-ratio, admixtures), fibre type, rheological properties, casting method, consolidation and others. The initial crack load (F_{max}) was independent of the counted number of effective fibres, confirming that this parameter is dominated by the matrix behaviour. On this account with increasing compressive strength a higher initial crack load may be expected. More importantly, there is a dependence of the post crack load on the number of effective fibres.

Figures 11 and 12 present different hardened concrete properties (average compressive, splitting tensile and flexural tensile strength) of the investigated concrete compositions. With respect to the compressive strength only with a fibre amount

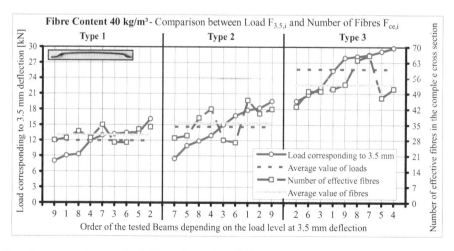

Figure 15. Comparison between load $F_{3.5,i}$ and number of fibres $F_{ce,i}$ for different mixture types (fibre content 40 kg/m³).

of 40 kg/m³ higher strength values were observed compared to the fibreless mixtures Type 1 and 2. It can be seen that under increasing fibre content a raise of the splitting tensile strength was obtained independent of the mixture type. However for Type 1 higher splitting tensile strengths of SCS-FRC compositions in comparison to the SCC were obtained only for mixtures equal or greater than 30 kg/m³ fibre content. With regard to the flexural tensile strength for mixture Type 1 relative homogenous strength values were achieved independent of the fibre amount. By contrast under growing fibre dosage concrete variant 2 demonstrated increased flexural tensile strengths, but only from a fibre content of 30 kg/m³ the strength value of the SCC mixture was exceeded.

Figure 13 shows selected single load-deflection curves of different mixture types with a fibre dosage of 40 kg/m³. In this connection the breaking of fibres was illustrated, which resulted in sudden load drop in the load-deflection curve.

A comparison of the average number of sudden load drops and the average deflection of the first load drop is shown in Figure 14. In respect of mixture Type 1 the average number of load drops varied in a small range between 4.2 and 5.9. Under growing fibre amount from 30 to 40 kg/m³ the number of load drops even decreased. This similar behaviour could be explained by the constant level of the flexural tensile strengths of mixture Type 1. However for Type 2 the average number of load drops clearly increased under growing fibre content. One reason could be the rising flexural tensile strength under increasing fibre concentration resulted in higher bond behaviour between matrix and fibres. Regarding mixture Type 3 in spite of the relative high flexural tensile strength significantly lower number of sudden load drops was observed compared to Type 1 and 2. In this connection a better ductile post cracking behaviour can be noticed for Type 3. It was shown that independent of the mixture type under increasing average number of sudden load drops the average deflection of the first load drops decreased.

Figure 15 represents the comparison between the post cracking load corresponding to 3.5 mm deflection ($F_{3,5,i}$) and the number of effective fibres ($F_{ce,i}$) counted in the fracture surface of the tested beams. In this connection three different concrete compositions with a fibre dosage of 40 kg/m³ were regarded. On average Type 3 indicated clearly higher post cracking loads that resulted from the highest number of effective fibres counted in the fracture surface. For Type 1 the lowest average values were detected. Furthermore it was observed, that higher number of effective fibres did not necessarily cause an increase of the residual loads. As mentioned before, the effectiveness of crack bridging fibres can vary, depending on their distribution, orientation and embedded length.

4 CONCLUSIONS

For SCSFRC the J-ring test is conditionally significant to evaluate the rheological behaviour. To strengthen the expressiveness the space between the bars should be adapted to the fibre geometry and content. Up to fibre contents of 40 kg/m³ the workability of the presented SCSFRC mixtures was not affected by addition of the fibres. The combination type (Type 2) demonstrated a better workability, stability and rheological performance under variation of water content and workability time compared to the powder type (Type 1). In consideration of the two concrete mixtures (Type 1 and 2) it was observed that the load bearing capacity was improved by increasing fibre amount. Also the ductile material behaviour increased but the brittleness decreased under raising fibre concentration. In general for the mixture Type 1 and 2 single cracks were obtained during the four point bending test. Only for the concrete composition Type 3 multiple cracks were observed in some cases. In this connection the load-deflection curves achieved a supercritical load bearing behaviour. In reference to these results it is obvious that the addition of crushed gravels (Type 3) has a positive effect on the fibre orientation and bond behaviour and resulted in a better load bearing capacity after the peak load.

REFERENCES

[1] W. Brameshuber, "Selbstverdichtender Beton," Schriftenreihe Spezialbetone Band 5, Verlag Bau+Technik GmbH, 2004.
[2] Deutscher Ausschuss für Stahlbeton (Hrsg.) DAfStb-Richtlinie "Selbstverdichtender Beton," (SVB-Richtlinie), Ausgabe November 2003.
[3] Deutscher Beton- und Bautechnik-Verein e.V. (Hrsg.), DBV-Merkblatt "Stahlfaserbeton," Wiesbaden, October 2001.
[4] H.B. Dhonde, Y.L. Mo, T.C.C. Hsu & J. Vogel, "Fresh and Hardened Properties of Self-Consolidating Fiber-Reinforced Concrete," ACI Materials Journal, V. 104, No. 5, September–October 2007, pp. 491–500.
[5] S. Kordts & W. Breit, "Assessment of the fresh concrete properties of self compacting concrete," Beton 53 (2003) Heft 11, pp. 565–571.
[6] H. Okamura & K. Ozawa, "Mix-design for Self-Compacting Concrete," Concrete Library, JSCE, No. 25, pp. 107–120, June 1995.

Advances in Cement-Based Materials – van Zijl & Boshoff (eds)
© 2010 Taylor & Francis Group, London, ISBN 978-0-415-87637-7

Application of Textile-Reinforced Concrete (TRC) for structural strengthening and in prefabrication

M. Butler, M. Lieboldt & V. Mechtcherine
Institute of Construction Materials, TU Dresden, Germany

ABSTRACT: Textile-Reinforced Concrete (TRC) is a composite material consisting of high-performance filament yarns made of alkali-resistant glass or carbon and a fine-grained concrete matrix. The main features of TRC are its high tensile strength and ductile behaviour accompanied by high deformations. This material can be used both for new structures and for the strengthening or repair of existing structural elements made of reinforced concrete or other traditional materials. The paper presents some TRC basics as well as examples of TRC use in structural repair and strengthening and also in prefabrication of structural elements, demonstrating the wide range of possible TRC applications.

As examples for strengthening applications, the repairs of a steel-reinforced concrete hypar-shell and of a barrel-shaped roof are presented. In the case of the hypar-shell a sufficient strengthening could be attained by applying three layers of textile sheets on the top side of this shell, with the total thickness of the TRC layer of only 15 mm. In the case of the barrel-shaped roof a more complicated arrangement of TRC layers both on the top side and on the bottom side was necessary for the structural retrofitting.

As prefabrication applications, two examples are presented: a façade panel and a hybrid pipe system with an inner plastic layer. In the case of the façade panel, a decrease in weight by a factor greater than 4 could be achieved via utilisation of TRC instead of the traditional steel-reinforced concrete. The hybrid pipe system combines the advantages of polymer pipes with the advantages of TRC (high stiffness, high ductility etc.). Complete technology has been developed for the discontinuous and continuous production of straight pipes as well as of elbow elements using different wrapping techniques.

1 INTRODUCTION

Textile-Reinforced Concrete (TRC) is a composite material consisting of fine-grained concrete and textile reinforcement of high-performance filament yarns made of polymer, glass or carbon. Shown in Figure 1 are samples of biaxial and multi-axial reinforcement fabrics made of alkali-resistant glass. The filament yarns comprise several hundred to several thousand individual filaments approximately 5 to 25 μm in diameter. The multi-filament yarns in the textile reinforcement structures can be placed according to the acting stresses by varying the amount and orientation of the multi-filament yarns (density and distance of yarns as well as the angle of wefts). In comparison to conventional glass-fibre-reinforced concrete (GFRC), this leads to a reduction of the fibre reinforcement up to 80%, and thus to an essential increase in efficiency of the fibre utilisation (Brameshuber 2006).

The fine-grained concrete most often used by the authors for TRC production has a w/b-ratio of 0.3 with a binder content of 40 to 50% by volume. The high content of binder is essential for sufficient bonding between the fine concrete and the filaments of the textile reinforcement as well as for the workability of the fresh concrete (Schorn et al. 2006). The maximum grain size of the aggregates ranges between 1 and 2 mm, depending on the distance between the yarns of the textile reinforcement, the spacing of the textile reinforcement layers and on the dimensions of the structural element. Two types of textile are shown in Figure 1.

Several fabrication techniques can be used: lamination, low-pressure spraying, casting or injecting,

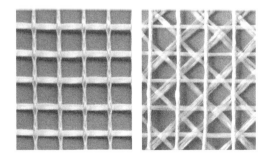

Figure 1. Biaxial and mulitaxial reinforcement fabric made of AR-Glass.

as well as combinations of these techniques (Brameshuber 2006). The addition of short glass fibres to the fine-grained matrix in tandem with textile reinforcement leads to an enhancement in the first crack strength and to a finer crack distribution. Shown in Figure 3 is a pronounced increase in the mechanical performance of the textile reinforced concrete in a tensile test (cf. Figure 2) due to the addition of 0.6% by volume of short glass fibres.

The performance and the durability of the textile-reinforced concrete are influenced by both the interaction between the textile reinforcement and the surrounding fine-grained concrete matrix and the durability of the textile reinforcement in an alkaline environment (Orlowsky & Raupach 2008, Butler et al. 2009). The individual filaments, depending on their position in the cross-section of the yarn, are to various degrees bonded with the matrix. While the outer filaments are completely enveloped by the matrix, the inner filaments (core zone) have often no direct contact with the binder paste and later to the hardened matrix. This means that the filaments in the core zone interact only through frictional forces, which are rather weak. A higher utilisation of all filaments can be achieved by polymer impregnation of the yarns (Gao et al. 2004). In this case, the filaments of the core zone are involved in bearing the loads to nearly the same degree as the outer filaments (Butler et al. 2006).

A few examples of the utilisation of TRC for strengthening existing structures as well as in prefabrication will be presented in the following sections. Primary attention will be directed to the production aspects and will be accompanied by some details on the utilised materials.

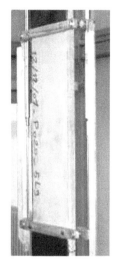

Figure 2.   Set-up for TRC tensile testing.

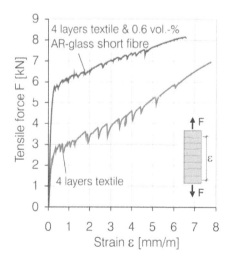

Figure 3.   Effect of adding short glass fibres on the mechanical performance of TRC (representative results).

2   STRENGTHENING OF RC-HYPAR SHELL

The first practical application of the innovative strengthening method using textile-reinforced concrete was carried out in 2006 in the retrofit of a steel-reinforced concrete roof shell at the University of Applied Sciences in Schweinfurt, Germany (Weiland et al. 2007, Ortlepp et al. 2008). The 80 mm thick steel reinforced concrete hypar-shell has a side length of approximately 27 m and a maximum span of approximately 39 m. Due to large deformations (up to 200 mm drop) in the shell's cantilevered areas the designed tensile stresses in the upper steel reinforcement layer of the shell were exceeded significantly (cf. Figures 4 and 5). The strengthening of these particular areas was recommended in response to snowfall levels and wind factors.

Strengthening with shotcrete required layer thickness of approx. 60 to 80 mm resulting in a rather high dead weight which led to a high additional load on the shell. Strengthening with glued carbon fibre composite (CFC) lamellae is characterized by a low dead load but a uni-axial load-carrying effect only, resulting in a very concentrated load transfer.

Strengthening with textile-reinforced concrete couples the advantages of a two-dimensional reinforcement layer with a low dead weight. Because the textile reinforcement adapts easily to the curved surface of the shell, strengthening with textile-reinforced concrete represents an excellent, new

Figure 4. Steel reinforced concrete hypar-shell of FH Schweinfurt (Germany).

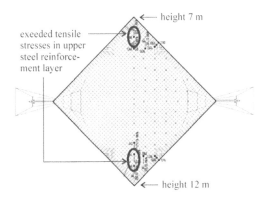

Figure 5. Overstressed areas in cantilevered areas of the shell.

Figure 6. Spraying of a thin layer of fine-grained concrete.

Figure 7. Placing of textile reinforcement.

Figure 8. Embedding the textile reinforcement.

alternative to the classical strengthening methods for such curved concrete structures.

To establish an adequate interaction between the existing concrete and the textile-reinforced strengthening layer a good adhesive tensile bond is required between them. This can be achieved by sandblasting the old concrete surface (Brückner et al. 2006). After sandblasting and generous wetting of the old concrete surface, the first layer of fine-grained concrete was sprayed in a wet spraying process (Figure 6). Afterwards, the first layer of textile reinforcement, made of carbon multi-filament yarns, was placed (Figure 7) and embedded in the freshly sprayed concrete matrix (Figure 8).

In this manner two further alternating layers of fine-grained concrete and textile reinforcement were applied subsequently. A final layer of concrete was then sprayed and smoothed on top. Altogether, the reinforcing layer, consisting of 5 layers fine-grained concrete and 3 layers carbon-yarn textile, shows a total thickness of 15–18 mm. The composition of sprayed matrix is given in Table 1.

Table 1. Material composition of fine-grained concrete.

Material	Amount [kg/m³]
CEM III/B 32.5 LH/HS/NA	628.0
Fly ash	265.6
Microsilica suspension (solid fraction 50%)	100.5
Sand 0–1 mm	942.0
Water	214.6
Super plasticizer (FM 30, BASF)	10.5

Because of the thinness of the reinforcing concrete layer, proper curing was essential. The drying or setting of the thin, sprayed concrete layers during the processing of the individual strengthening layers also had to be avoided, which required a well adapted installation process.

Since textile-reinforced concrete is not yet a standardized construction material, it was necessary to apply for a "special case" technical approval for this strengthening measure from the appropriate authorities. The design of the strengthening was performed by ARGE SCHALENBAU (Rostock, Germany) in cooperation with the Institute of Concrete Structures at the Technische Universität Dresden (TU Dresden). TORKRET AG performed the workmanship on construction site. A detailed description of this strengthening measure is given in Weiland et al. 2007 and Ortlepp et al. 2008. These publications also contain explanations of how to calculate the amount of necessary textile reinforcement for the required load bearing of the TRC layer.

3 STRENGTHENING OF A BARREL-SHAPED CONCRETE ROOF

Schladitz et al. (2009) give a second example for structural strengthening. This case concerned a historical school of engineering in Zwickau, Germany, which was built in 1903 and rehabilitated during last two years to house in the future the tax city office. One part of this school has a large barrel-shaped concrete roof which covers a room measuring 16 m × 7 m. The roof consists of curved steel-reinforced concrete slabs of 80 mm thickness, which are supported by eleven beams with a width of 200 mm and a height of 250 mm. Nine of the ten slabs include rectangular light openings (Figure 9, 10).

The recalculation of static load-bearing behaviour of the roof indicated insufficient load-carrying capacity regarding German concrete code requirements (DIN 1045-1). The construction of a new structure or a truss-support of the roof was precluded by the German monument conservation authorities. Therefore the strengthening of the structure was essential.

The application of the traditional shotcreting method would lead to a considerable dead-weight increase, resulting in additional loading to the structure and its subordinated supports. Furthermore, the thickness of the existing slender beam-slab construction would increase, which was irreconcilable with the demands of the monument conservation authorities.

The use of externally glued lamellae would lead to additional measures in order to secure the bond between the lamellae and the concave underside of the beam and slab. In addition, a complex

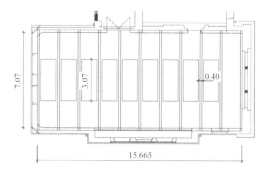

Figure 9.   Top view of the barrel-shaped roof.

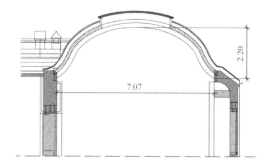

Figure 10.   Section drawing of the barrel-shaped roof.

load distribution within the slabs with rectangular openings would require complicated lamella arrangements to counter stress concentrations. Also, CFC lamellae would be strongly affected in case of fire.

Finally, the TRC strengthening was selected as the most adequate method of rehabilitation since it best complies with the requirements of the monument preservation, fire resistance and static-constructive considerations. In this special case the existing plaster at the inner surface of the beam-slab structure could be replaced by the TRC strengthening layer.

The determination of the internal forces as well as the modelling of the non-linear behaviour of the roof in the ultimate limit state was carried out at the Institute of Statics and Dynamics of Structures at the TU Dresden by means of a layered FE Model (Möller et al. 2005). The dimensioning of textile reinforcement resulting from numerical calculation of forces was performed at the Institute of Concrete Structures at the TU Dresden (Schladitz et al. 2009). In the ultimate limit state a carbon textile reinforcement of 62 mm^2 is required at the lower end of the T-beam cross-section. This corresponds to 5 layers of TRC when using the carbon textile as shown in Figure 11 and has a width of 200 mm.

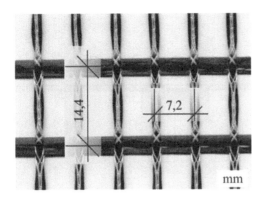

Figure 11. Textile fabric made of carbon mulitfilament yarns.

Figure 12. Placing of the textile reinforcement.

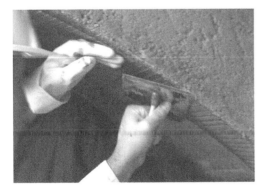

Figure 13. Embedding of the textile reinforcement.

On both the upper and lower sides of the curved slab structure, two layers of textile reinforcement was necessary to ensure the designed ultimate load carrying capacity.

Before applying the strengthening TRC layers, the existing plaster was removed and the old concrete surface sandblasted until the granular structure of the concrete was exposed. Defects within the old concrete surface were rehabilitated and re-profiled. The surface-bond strength of the old concrete subsurface was proven by adhesive tensile tests to be higher than 1.5 MPa.

After pre-wetting the substrate surface, the individual strengthening layers were applied using a wet spraying method (cf. Section 2) until the required number of layers were applied (cf. Figures 12 and 13). The composition of the spayed fine-grained concrete matrix is given in Table 1.

All reinforcing layers were applied fresh-in-fresh in order to ensure a good bond between each individual fine-grained concrete layer. This process was assured by exact cutting and laying plans for reinforcement as well as by accurate construction work and good supervision. Immediately after the application of TRC layers, the concrete was covered by wet fleece and additionally by foil to prevent moisture losses. The concrete was kept moist by periodic water spraying for the first 7 days after application.

4 THIN FAÇADE PANEL

4.1 *Construction and design aspects*

A thin-walled, large façade panel was developed for parking garages. This panel combines low weight and thickness with excellent surface quality and high load-bearing capacity. Since steel reinforcement was replaced by textile reinforcement, no protection against corrosion by concrete cover was necessary. The textile could be positioned directly under the parapet surface, which provided a very efficient utilization of the reinforcement when loaded in bending. As a result, a reduction in weight was achieved to just 1/3 that of a 100 mm thick parapet panel made of conventional steel reinforced concrete. MoreMoreover, a further reduction in weight by 20 percent could be achieved by using lightlightweight concrete as matrix material (Franzke et al. 2003).

The dimensions of the façade plate are 249 cm × 151 cm. It consists of a rectangular lateral frame (100 mm width, 50 mm thickness) and a thin slab of 20 mm thickness within the frame (Figure 15). The frame guarantees the stiffness of the element and provides space for the arrangement of metal elements to fasten the panel to the building. The panel should be fastened only at the corners of the frame, so that stresses due to wind uplift and dead load are distributed to these four points.

Fine-grained concrete made of white cement and mineral pigments was used for the outer layer to achieve the desired decorative appearance of

Table 2. Composition of concrete matrices of panel in kg/m³.

| Material | Inner layer | | Outer layer |
	Ordinary	Light weight	Sandstone
CEM I 42.5 R	625	600	–
CEM I 42.5 R/DW	–	–	600
Fly ash	181	–	–
Microsilica susp.	235	235	–
Sand 0/1	936	–	500
Inflated glass 0/4	–	205	–
Sandstone 0/1*	–	–	100
Water	194	100	210
Plasticiser	7	15	8

* Grounded material.

the front surface, while the inner part of the panels consisted of fine-grained concrete made of ordinary cement without any pigments. Additionally, different techniques were alternatively tested for finishing the front surface. In the case of washed-out concrete, Granodiorite with grain size of 0/2 mm and crushed sandstone were alternatively used as aggregates (Table 2).

4.2 *Production technology*

A tilting table was used for the production of the prototypes. Such a table allows simulation of production technologies used in practice. In the case of washed-out concrete, a hydration retarder was applied to the surface of the formwork before casting. Then a 5 mm thick layer of coloured concrete was spread over it in a wet spraying process. Shortly afterward the fitted pieces of biaxial textile reinforcement were embedded in this concrete layer. This procedure was repeated once, so that two layers of textile reinforcement were placed near the surface. The fastening elements were subsequently placed and a stainless steel bar was positioned along the circumference of the panel (cf. Figure 14). Afterwards, the form was filled with the fine-grained concrete to 10 mm under the top edge of the formwork.

In the next step, another two layers of textile reinforcement embedded in two layers of coloured fine-grained concrete were placed one after the other. The finish surface (the future back surface of parapet) was finally smoothed and covered with polyethylene sheets for curing. The fresh panel was cured for 14 hours and then lifted from the formwork. The front surface, which had been in contact with hydration retarder, was washed off using a high-pressure water stream. Figure 15 shows a finished panel with a sand stone imitating surface.

Figure 14. Placement of fixing elements and steel bar.

Figure 15. Sand stone imitating panel surface.

5 COMPOSITE PIPES

5.1 *Concept of the hybrid pipe system*

The newly designed hybrid pipe consists of an inner polymer tube and a number of outer textile-reinforced concrete layers (cf. Figures 16 and 17). In this way, the advantageous properties of each material type are used most efficiently.

The thin inner polymer pipe is highly resistant against chemical attack and incrustations, and it is impermeable to fluids and gases. Furthermore, the high hydraulic smoothness of the inner surface of the plastic tube minimizes hydraulic resistance.

The outer layer made of textile-reinforced concrete bears the tensile stresses from internal pressure as well as external loads (traffic loads, earth loads, etc.). The utilisation of textile reinforcement provides concrete high tensile strength and ductility, and it allows significant minimisation of the thickness of the concrete layer. The lightweight pipes keep the transport and assembly costs low and reduce the consumption of raw materials. The thin, high-strength and ductile pipes can be efficiently placed e.g. by pressing them through the

Figure 16. Demonstration model of the TRC-polymer hybrid pipe (3 layers reinforcement).

Figure 17. Cross-section of the pipe wall (6 layers reinforcement).

ground. Due to the high tensile strength of TRC, the system can sustain high internal pressures during pumping of fluids and gases.

5.2 Fine-grained concrete matrix and textile reinforcement

The matrix composition was developed to meet specific requirements resulting from the production technique and the quality of the fibre-matrix bond. The formation of the bond between multi-filament yarn and matrix, which mainly occurs in the so-called fill-in-zone, requires a very high content of fine-grained materials (cement and pozzolanic additives). The flour-fine materials can penetrate especially well into the open space between the filaments of the yarn and thus provide the basics for a good bond. Furthermore, the durability of the glass filament yarns is guaranteed by low alkalinity in comparison to conventional concrete. Blast furnace cement or composite cements should be preferably used if AR glass is applied as reinforcement material (Butler et al. 2009). The reduction of alkalinity of the pore solution in

the fine-grained concrete mixture is additionally achieved by a high content of pozzolanic additives. The well tried composition of the fine-grained concrete mixture consists of 3 MP (mass parts) of fine sand with a maximum grain size of 2 mm, 3 MP binding agent and 1 MP water. The binding agent consists of 2 MP cement CEM III/B 32.5 N-NW/HS/NA and 1 MP pozzolans. The pozzolanic part is composed of 90% fly ash and 10% microsilica (solid material). This exemplary composition can be modified by inclusion of different additives. The determining criteria for the optimisation were the adhesion to the plastic pipe, good workability and high stability of the fresh concrete (Hempel et al. 2006).

The fibre content (fineness of the reinforcement filament yarns) and orientation can be varied according to the expected tensile and flexural stresses. The internal pressure induces the hoop tension in the pipe wall, which stress level is approximately twofold higher than the stresses acting in longitudinal direction. Therefore several types of bi-axial (cf. Figure 1a) textiles made of alkali-resistant glass were examined for their strength, but also with regard to the processing-relevant properties. Knitted fabrics with 2,400 tex (yarn count) in the circumference direction and 1,280 tex in longitudinal direction proved to be most useful (Hausding et al. 2006).

5.3 Production technology of straight pipes

A combination of pultrusion and multiple-turn coiling was chosen as the production technique of the pipes. During the production process, the textile fabric is guided through a funnel filled with fine-grained concrete and then this textile "soaked" with fresh concrete is wound around a spinning plastic pipe (cf. Figure 18). The textile must be kept under uniform tensile stress during the winding process, which is assured by the friction resistance when the textile fabric passes through the funnel filled with concrete. The freshly applied layers are compacted by a pressuring device. Each layer of the textile reinforcement and concrete matrix is typically approximately 3 mm thick, however, the thickness of the layers can be easily changed by modification of the used materials and by adaptation of the production technique. The overall thickness of the textile-reinforced concrete "jacket" depends on the number of layers required. In the last work step, the surface is smooth-finished (Figure 19, Helbig et al. 2006, Lieboldt et al. 2007).

5.4 Production technology of elbow pipes

A special production procedure was developed for manufacturing elbow pipes. However,

Figure 18. Application of the textile reinforced concrete.

Figure 19. Smoothing of the pipe surface.

Figure 20. Demonstration model for the arrangement of textile reinforcement.

Figure 21. Fabricated elbow pipe, textile reinforcement embedded in concrete.

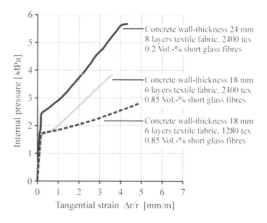

Figure 22. Influence of the percentage of the textile reinforcement on the mechanical performance in the internal pressure tests.

this procedure is also suited for manufacturing straight segments (continuous pipe strings) with elliptical or circular cross sections. In the developed prototype of a production facility, a rotary tray bears the application units for the fine-grained concrete and the reinforcing textile. The non-rotary pipe is arranged in the middle of the tray. During the winding process the rotation velocity of the tray with the application units is coupled with the continuous, axial pipe feed. The necessary reinforcement content is generated by a determined setting of the overlapping of the reinforcement structure (Figure 20, for the demonstration purposes, no concrete matrix was applied here). The total thickness of the TRC layer depends on the overlapping of the textile layers, the thickness of the textile and on the maximum aggregate size. The production of elbow pipes requires an appropriate reinforcement in the form of a band-like textile lattice structure. The shape of the textile corresponds to the lateral surface of a truncated cone. The bend of the lattice-like textile lying in a plane depends on the respective cross-section of the pipe. In the winding process, the cross section of the circular pipe is to be wound in the form of a thread flank.

5.5 *Mechanical performance of the hybrid pipes*

Internal pressure tests and crushing tests based on DIN EN 512 were performed to examine the mechanical performance of the new pipes. The internal pressure tests represent the loading of the pipes during the transportation of fluids and gases, which induces tensile stresses in the pipe walls. The crushingcrushing tests reproduce the loading of pipes, resulting e.g. from earth pressure or earth movements. Different types of glass fibre textile reinforcement with regard to the fineness of the filament yarns were tested. Another varying parameter was the content of short glass fibre. Even with a short glass fibre content of less than 1% by volume, a significant increase in the Limit of Proportionality (LOP, loading at which a transition of the linear-elastic to strain-hardening regime occurs) was noted.

Illustrated in Figure 22 is the influence of the percentage of textile reinforcement and the number of reinforcement layers as observed in the internal pressure tests. By nearly doubling the fineness of the yarn in the circumferential direction from 1280 tex to 2400 tex, a higher stiffness in the strain-hardening regime (after the formation of fine first cracks in the TRC layer) could be achieved. At the same time the ultimate resistance to the internal pressure increased from 2.8 MPa to 3.8 MPa. However, the limit of the linear-elastic range (LOP, uncracked state) was not affected.

In the case of increasing in the number of layers from six to eight, the LOP value increased by approximately 40% from 1.8 MPa to 2.5 MPa, while the ultimate resistance to the internal pressure increased from 3.8 MPa to nearly 5.8 MPa. The accompanying crushing strength increased from 126 kN/m to 155 kN/m (Lieboldt et al. 2007).

6 SUMMARY

The application of textile-reinforced concrete opens new possibilities in the strengthening of existing concrete structures and the prefabrication of thin-walled concrete elements.

In the case of strengthening, a significant increase in the ultimate load of existing structures can be achieved with the help of TRC as it was shown in two examples in this paper. The application of the strengthening TRC layers is simple and can be adapted to complex geometric forms. Furthermore, TRC is a flexible strengthening system enabling augmentation of maximum structural loads upon current demands.

The use of TRC in prefabrication of new components was also shown in two examples. The weight and thickness of the façade panel was reduced to 30% of its original weight and to 2 cm thickness, respectively. The geometry of the panel was adjusted to suit typical load-bearing characteristics of TRC. Also, special cementitious matrices were developed, which fit requirements concerning manufacturing, design and durability of the panel.

In the case of the hybrid pipes, the positive properties of the applied materials, polymer and TRC, were combined ideally and compensated for respective disadvantages. In this way it was possible to reduce the amount of material necessary for both the inner polymer tube and the concrete layer while the load-bearing capacity of hybrid pipes improved greatly. Very high axial rigidity of the pipe was achieved even with textile reinforcements between 2 and 6% by volume in combination with the addition of short fibres of up to 0.85% by volume. The test specimens were absolutely watertight up to the failure pressure.

ACKNOWLEDGEMENTS

The authors would like to acknowledge the financial support from Deutsche Forschungsgemeinschaft (DFG), from Arbeitsgemeinschaft industrieller Forschungsvereinigungen „Otto von Guericke" e. V. (AiF) and from Bundesministerium fuer Wirtschaft und Technologie (BMWi).

The authors would like also to thank their partners at the TU Dresden for provision of data concerning the strengthening applications of TRC as well as the governmental institutions and private companies involved in these projects.

REFERENCES

Brameshuber, W. (ed.) 2006. *Textile Reinforced Concrete. State-of-the-Art Report RILEM Technical Commitee 201-TRC.* RILEM Report 36, RILEM Publications S.A.R.L.

Brückner, A., Ortlepp, R. & Curbach, M. 2006. Textile Reinforced Concrete for Strengthening in Bending and Shear. *Materials and Structures* 39(8): 741–748.

Butler, M., Hempel, R. & Schorn, H. 2006. Bond behaviour of polymer impregnated AR-glass textile reinforcement in concrete. In *Proc. International Symposium Polymers in Concrete (ISPIC), Guimaraes,* April 2006: 173–183.

Butler, M., Mechtcherine, V. & Hempel, S. 2009. Experimental investigations on the durability of fibre–matrix interfaces in textile-reinforced concrete. *Cement & Concrete Composites* 31: 221–231.

Franzke, G. et al. 2003. Development of large-size, thin-walled textile-reinforced balustrade panels. In *Proc. 12th International Techtextil-Symposium, Frankfurt/ M.*, Frankfurt/M.: paper 4.24.

Gao, S.L., Mäder, E. & Plonka, R. 2004. Coatings for glass fibers in a cementitious matrix. *Acta Materialia* 52(16): 4745–4755.

Hausding, J., Lieboldt, M., Franzke, G., Helbig, U. & Cherif, Ch. 2006. Compression-loaded multi-layer composite tubes. *Composite Structures* 76: 47–51.

Helbig, U., Horlacher, H.-B., Lieboldt, M. & Franzke, G. 2006. Development of textile reinforced concrete multilayer composite pipes. *3R international* 45(8): 424–429.

Hempel, R., Lieboldt, M., Franzke, G. & Helbig, U. 2006. New pipelines made of plastic and textile reinforced concrete. *SAMPE Symposium and Exhibition 2006*, Long Beach, April 30–May 4, USA.

Lieboldt, M., Helbig, U. & Engler, T. 2007. Pressure pipes made of textile-reinforced concrete and plastics. *BFT Betonwerk + Fertigteil-Technik* 73(5): 24–33.

Möller, B., Graf W., Hoffmann, A. & Steinigen, F. 2005. Numerical simulation of structures with textile reinforcement. *Computers Structures* 83(19–20): 1659–1688.

Orlowsky, J. & Raupach, M. 2008. Durability model for AR-glass fibres in textile reinforced concrete. *Materials and Structures* 41(7): 1225–1233.

Ortlepp, R., Weiland, S. & Curbach, M. 2008. Restoration of a hypar concrete shell using carbon-fibre textile reinforcement concrete. In Limbachiya, M.C. & Kew, H.Y. (eds), *Proceedings of the International Conference Excellence in Concrete Construction through Innovation, London, 9–10 September 2008*. London: Taylor & Francis Group: 357–364.

Schladitz, F., Lorenz, E., Jesse, F. & Curbach, M. 2009. Strengthening of a barrel-shaped roof with Textile Reinforced Concrete. In *Proceedings 33rd IABSE Symposium, Bangkok, 9–11 September 2009* (in preparation).

Schorn, H., Hempel, R. & Butler, M. 2006. The influence of short glass fibres on the working capacity of textile reinforced concrete. In Hegger, J., Brameshuber, W. & Will, N. (eds), *Proceedings. 1st International Conference on Textile Reinforced Concrete, Aachen*: 45–54, RILEM Proceedings PRO 50, RILEM Publications S.A.R.L.

Weiland, S., Ortlepp, R., Hauptenbucher, B. & Curbach, M. 2007. Textile Reinforced Concrete for Flexural Strengthening of RC-Structures—Part 2: Application on a Concrete Shell. In: Aldea, C.-M. (ed.), *Design & Applications of Textile-Reinforced Concrete. Proceedings of the ACI Fall Convention, Puerto Rico*, SP-251CD-3, 2008—CD-Rom.

Advances in Cement-Based Materials – van Zijl & Boshoff (eds)
© 2010 Taylor & Francis Group, London, ISBN 978-0-415-87637-7

Evaluation of mechanical behaviour of weathered textile concrete

S.W. Mumenya
University of Nairobi, Kenya

M.G. Alexander & R.B. Tait
University of Cape Town, South Africa

ABSTRACT: High Performance Fibre Reinforced Cementitious Composites (HPFRCC) are characterized by a stress–strain response in tension that exhibits strain-hardening behaviour accompanied by propagation of multiple cracks. This process is often referred to as pseudo-ductility due to multiple cracking with relatively large energy absorption capacity. This study analyses the multiple cracking patterns formed in weathered Textile Concrete samples due to direct tensile testing, and links the cracking patterns to the tensile behaviour. The specimens used for the study were thin laminates which were produced by casting six layers of specially made polypropylene textile in fine-grained mortar. The samples were cured under controlled laboratory conditions for 28 days, and thereafter exposed to different weathering regimes for different periods. The weathered samples were tested in direct tension over a range of stresses. For all the samples tested, it was observed that the tensile behaviour was characterised by strain hardening and multiple cracking, which gave high tensile strains in excess of 20 percent at final failure. It was further found that the cracking patterns varied mainly with age, weathering history and stress levels. A general trend of increasing crack widths and crack spacings with ageing was observed which was accredited to increased hydration accompanied by an increase in fibre/matrix bond strength.

1 BACKGROUND

Within the last decade, a new class of civil engineering materials referred to as High Performance Fibre Reinforced Cementitious Composites (HPFRCC) has been under development. Textile concrete (TC) and Engineered Cementitious Composites (ECC) are sub-classes of this new group of materials [1]. The fundamental difference between HPFRCC and conventional Fibre Reinforced Concrete (FRC) is its behaviour in tension. In FRC, strain localization occurs immediately after the first crack is formed, whereas in TC, the process of fibre bridging causes further multiple matrix cracks to develop thus preventing propagation of the initial crack. This process results in non-localized multiple cracking at increasing tensile stress level beyond the first cracking stress, commonly known as macroscopic pseudo strain-hardening-behaviour [2]. In Textile Concrete, this phenomenon is accompanied by high ductility, hence TC has presented a new range of thin element applications in the construction industry.

A theory of crack quantification was developed by Aveston et al. [3], based on a purely frictional fibre/matrix bond for a matrix with a well defined single-valued breaking stress. The theory by Aveston et al. [3] was comprehensively reviewed and modified by Hannant [4] and applied firstly in prediction of cracking patterns and secondly, in describing the pseudo-ductility of HPFRCC. It is this latter application of cracking theory that is the focus of this study.

The tensile behaviour of "pseudo-ductile" cementitious materials reinforced with textiles has attracted the attention of researchers in recent years due to the numerous potential applications of these materials especially as thin elements [5, 6]. Textile Reinforced Concrete is a material that has been found to offer viable solutions for rehabilitation of structures, and developments along these lines continue [7, 8, 9]. By studying the cracking patterns on weathered TC samples, this study contributes to a better understanding of the effect of environmental exposure on the mechanism of multiple cracking in HPFRCC.

2 MATERIALS AND EXPERIMENTAL DETAILS

2.1 *Fibres*

The fibres used in this study were "hybrid fibres", specially produced in South Africa and designed

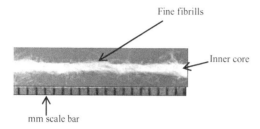

Figure 1. Typical "hybrid fibre".

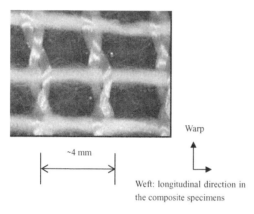

Figure 2. The textile magnified 8 times.

for use with cement matrices. In FRC the term 'hybrid' is used for use of more than one fibre type in a mix, for instance short fibres combined with long fibres, perhaps also of two different base materials. In this study, the term 'hybrid' refers to the use of a fibrillated form of polymeric fibre as the core that consists of two strands of fibrillated polypropylene (PP) fibres of 110 D'Tex each. To this core, an outer fluffy layer composed of fine 3 d'tex PP filaments has been spun to form a sheath as illustrated in Figure 1. The spinning process arranges the fine fibrils helically as well as extending them lengthwise along the core. The sheath is further secured onto the core by providing overlaps of approximately 40 mm.

2.2 Textile

The textile used for production of TC composites was manufactured in conventional mills. The yarn in the weft direction was a "hybrid fibre" illustrated in Figure 1, whereas in the warp direction, two strands of fibrillated polypropylene fibres were used. The textile is woven into a mesh, commercially referred to as "CemForce", illustrated in Figure 2. The longitudinal specimen axis was in the Weft direction to maximize the fibre contribution.

2.3 Cementitious mixture composition

The binder was a thoroughly blended mixture of ordinary Portland cement (CEM I 42.5) and Ultra-Fine Fly Ash (UFFA) locally referred to as "Superpozz", with 90% of particles finer than 11 μm. To achieve an appropriate mix, coarse aggregate was avoided and dune sand with maximum size 600 μm, fineness modulus of 1.71 and controlled moisture content (air dry) was employed. The relative density of the sand was 2.60. A sand/binder ratio of 1.0 was used, with nominal water: binder ratio of 0.5. "Superpozz" was used at a cement replacement level of 10% by mass, primarily to assist in achieving flow.

Mixing was undertaken for two minutes in a Hobart A120 planetary mixer, and the flow characteristics determined using the ASTM flow table test [10]. The mortar mix had a minimum setting time of two hours and had minimum segregation and bleeding.

2.4 Specimen production

The specimens were produced from thin laminates of Textile Concrete. The width of the specimens tapered from the ends to the centre, and the ends of the specimens were symmetrical about the centre line. The radii of all fillets were equal. Six layers of fabric mesh woven from polypropylene fibres, referred to as Cemforce were embedded in the fine-grained cementitious mortar to produce the desired laminates of 8 mm nominal thickness, which was desirable for accelerated ageing exposure. Plastic moulds were specially manufactured to facilitate hand lay-up casting techniques and easy removal after setting of the specimens. The laminates were cast horizontally in the specially made moulds and water cured for 28 days. Figure 3 illustrates the composite specimen used in the study.

2.5 Exposure regimes

Before exposure to different environments, samples were water cured for 28 days. After curing the

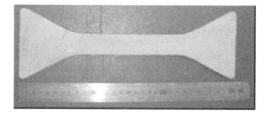

Figure 3. Tensile test sample. Overall length = 300 mm. Centre test length = 125 mm. Gauge width = 25 mm.

samples were subjected to different environmental regimes as follows:

i. 100 hot/cold cycles involving raising temperatures from $23 \pm 1°C$ to $50 \pm 1°C$ in 30 minutes and allowing a cooling time of 45 minutes. The Relative Humidity was 20 ± 2 percent.
ii. 100 Wetting/Drying cycles within temperatures that varied between $35 \pm 1°C$ and $23 \pm 1°C$ over 24 hour period. The Relative Humidity varied between 52 percent and 100 percent.
iii. Continuous carbonation exposure for 6 months at a temperature of approximately $30 \pm 1°C$ and Relative Humidity varying between 50 and 70 percent.
iv. Natural exposure in a moderate climate for 11 months.
v. Natural exposure in a tropical climate for 11 months.
vi. Natural exposure in a tropical climate for 16 months.
vii. Control samples at $21 \pm 1°C$ and Relative Humidity of 53 ± 5 percent for periods ranging between 28 days and 20 months.

After weathering, specimens were visually examined for any signs of damage or surface crazing prior to tensile testing.

In order to compare the tensile behaviour of laboratory-produced Textile Concrete (TC) samples with applications in the South African Industry, specimens were prepared from a disused TC board for tensile testing. The Industrial specimens were of the same width with the laboratory samples but the thickness, mix, and number of layers was different due to a different production process. The Industrial specimens were four-layered textile laminates, and the board had undergone natural weathering for 24 months, thus providing an additional natural exposure regime to the moderate and tropical environments.

Rectangular specimens of size $100 \times 25 \times 6$ mm were cut from the board and the ends were cast in epoxy to form gripping areas similar to the laboratory-produced samples. The Industrial specimens were appropriately labeled and subsequently tested in tension to failure.

2.6 Tensile testing

Aluminium grips were designed to closely fit the tapered edges of the specimens, and had adaptable attachments at the ends compatible with the holders of a ZWICK Universal Testing Machine (UTM) of 100 kN capacity but with special 10 kN load cells. The tensile testing rig, with a mounted specimen is illustrated in Figure 4.

Loads were measured using a 10 kN load cell with an accuracy of 0.5 percent and the cross-

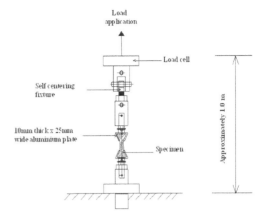

Figure 4. Schematic of tensile test rig.

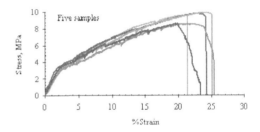

Figure 5. Tensile behaviour of eight month old control specimens.

head travel was used to measure displacements. The specimen was loaded in tension at the rate of 10 mm per minute under displacement control of the crosshead separation. The tensile behaviour of the specimens was characterised by the stress-strain relationship as illustrated by typical traces for eight month old control samples in Figure 5.

2.7 Crack quantification

The cracks that developed after testing were quantified with respect to the age and weathering history. For each batch of samples, the average crack spacing, average crack density and average crack width were measured at five locations chosen at random over 150 mm long and 50 mm wide gauge sections of representative samples. The crack density was a measure of the amount of cracking within the gauge sections of the specimens and crack spacings were obtained from direct linear measurements. Crack widths were measured with the aid of an attached (Vickers hardness type) scale on a light microscope with a resolution of 1 μm.

Table 1. Key parameters of the stress-strain curves.

Sample age and exposure conditions	Conditions at end of linear region		Peak conditions		Strain at failure Strain, %	Area under the curve ($\times 10^4$) J/m^3
	Stress, MPa	Strain, %	Stress, MPa	Strain, %		
8 months control	3.22	2.15	8.95	20.00	23.82	377.50
12 months control	3.25	2.00	9.91	20.34	22.34	413.00
14 months control	3.23	1.88	10.26	21.00	22.00	407.10
16 months control	2.83	1.98	11.22	25.00	30.07	521.40
12 months hot/cold	3.77	2.19	8.73	12.00	20.04	299.68
14 months wet/dry	2.99	1.36	11.43	19.00	21.00	397.53
16 months carbonated	3.95	1.79	12.33	15.95	17.95	320.56
12 months moderate	3.93	1.95	9.68	12.96	17.97	279.63
12 months tropical	3.35	1.51	9.78	18.97	20.96	351.22
24 months industrial	2.91	0.49	12.08	19.96	30.29	695.30

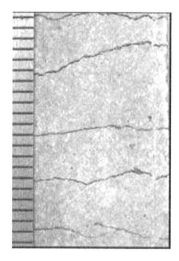

Figure 6. Typical multiple cracking at strain hardening stage.

3 RESULTS

Apart from the naturally weathered (N1) samples on which minor surface crazing was observed prior to mechanical testing, there was no visible damage in other samples. The tensile behaviour of TC was characterized by the stress-strain curves as exemplified in Figure 5. The figure shows that TC behaviour in tension is essentially linear up to stresses of approximately 3 MPa. Below this stress level there were no visible signs of macro cracking but cracks were visible in the gauge areas beyond approximately 3 MPa stress levels. Multiple cracking occurred between stresses of approximately 3 MPa and 8 MPa. Continued loading after the samples had fully undergone multiple cracking caused fibre damage, as opposed to debonding

during multiple cracking, and final failure was by fibre rupture.

A typical multiple cracking pattern exemplified by Figure 6 characterised the cracking patterns in all specimens tested.

Table 2 shows crack quantities for samples aged between two months and 24 months.

4 DISCUSSION

4.1 *The stress-strain curves*

For conventional un-reinforced cementitious matrices, an initial linear stage in stress-strain curves occurs at low strains of less than 0.02 percent [8], a value which was too low in this study to be represented on the overall stress-strain curves exemplified in Figure 5. Therefore, only the microcracking and multiple cracking stages are considered in this discussion.

The translation between microcracking and multiple cracking stages is marked by the end of linear region. The effects of micro cracks on the overall matrix behaviour were not major, thus the microcracking region was approximately linear. Figure 5 illustrates that multiple cracking accounted for the strain hardening behaviour of the composite. The average strains increased from approximately 2 percent (at the end of the linear portion) to values ranging between 15 and 20 percent at the peak conditions (Table 1). This illustrates that although the strength and modulus of PP are not very high, the extended surface of outer fluffy layers on the PP fibre were effective in providing sufficient physical bonding characteristics to the cement matrix which was sustained after weathering. A key feature in the stress-strain curves is the variation in the gradients of the curves both during multiple cracking and also just before final rupture, indicating

Table 2. Crack quantification.

Exposure history	Age months	No of tests	Failure condition		Crack quantities at end of test		
			Load, N	Strain, %	Average crack density, mm	Average crack spacing, mm	Area under the curve ($\times 10^4$), J/m^3
Control*	2	5	901	2.4	0.12	8.6	141
Control	2	6	2900	25.7	0.12	8.5	202
Control	7	5	2670	20.4	0.18	5.5	124
Control	8	5	2670	23.8	0.18	5.8	141
Control	12	5	2550	22.3	0.19	5.7	141
Control*	12	5	901	1.7	0.08	12.2	33
Control	14	5	2921	22.0	0.16	6.7	147
Control	15	5	2969	25.0	0.21	4.9	152
Control	16	6	2848	30.1	0.07	14.6	86
Hot/Cold	12	5	1951	20.0	0.20	5.8	77
Wet/Dry	14	5	2988	21.0	0.18	6.1	139
Wet/Dry	12	5	903	1.0	0.06	15.9	142
Moderate	12	5	2696	18.0	0.20	4.9	198
Tropical	12	5	2475	21.0	0.22	4.8	108
Tropical (before test)	16	6	0	0	Single crack	Single crack	364**
Tropical	16	6	2980	21.0	0.05	20.7**	444**
Industrial	24	5	3020	30.2	0.21	4.8	130

* Samples were loaded up to elastic stage.
** There were several wide cracks interspersed by finer cracks hence the high values.

a change in mechanism from fibre debonding and load transfer (multiple cracking) to interface damage and final fibre rupture.

The high failure strain in the 16 month old sample, which seemed anomalous to other similar-age specimens, was due to development of fine branching-off cracks in the vicinity of the major cracks which contributed to further extension of the composites. A high failure strain was also observed in the Industrial samples, which was attributed to development of several closely spaced cracks that were fairly evenly distributed over the gauge length of the specimens. For the conditions tested, there was no loss in pseudo-ductility but there was a change in cracking patterns.

4.2 *Cracking patterns*

The quantification process was inevitably characterised by high variability as illustrated in Table 2. The results indicated that there was a correlation between crack density and ageing. For example, specimens cured for two months under controlled conditions developed cracks at spacings of approximately 5 mm after tensile testing. Conversely, specimens tested at a later age of approximately 16 months developed relatively wider cracks at wider spacings of between 8 mm and 20 mm.

Although the variation in crack widths was mainly attributed to ageing, the correlation was not linear as other factors such as the moisture state of the specimen and bonding nature could also have played a role. A strong bond and dense matrix (exemplified by carbonated samples) resulted in wide crack spacings compared with samples exposed to a hot/cold environment which had a weaker bond but developed closely spaced cracks. However, a general trend of increasing crack widths and crack spacing with ageing was due to increased hydration which causes an increase in brittleness of the matrix.

5 CONCLUSIONS

The following conclusions were drawn from this study:

i. The pseudo-ductility of Textile Concrete is sustained after exposure to different environments at ambient conditions.
ii. Crack quantification was characterised by high variability due to the complex nature of the mechanism of cracking.
iii. The cracking patterns in Textile Concrete (TC) are a function of loading level, age and the weathering history.

iv. The outer fluffy layers on the special PP fibres that were used in the study provide sufficient physical bonding characteristics to the cement matrix, which is sustained after ageing and exposure to ambient environments.

v. The variation in crack widths was mainly accredited to ageing, however, the correlation was not linear as other factors such as the moisture state of the specimen and bonding nature could also have played a role.

The results of the study reported here were based on a single matrix mix that was prepared under controlled laboratory conditions. Further studies need to be undertaken to investigate the role of the matrix composition on the pseudo-ductility of Textile Concrete, including a matrix mix from Industry.

The crack quantification had high variation. Therefore it is recommended that the method be refined possibly by utilizing photographic techniques in order to capture the crack patterns as a function of time as the test progresses. A refined method would quantify the microcracks at the linear range of the stress-strain curve.

REFERENCES

ASTM C230 1990. International Standard on Specification for Flow Table for Use in Tests of Hydraulic Cement.

Aveston, J., Cooper, G. & Kelly, A. 1971. Single and multiple fracture. In 'The Properties of Fibre Composites', IPC Science and Technology Press, Surrey, U.K., pp. 15–26.

Bruckner, A., Ortlepp, R. & Curbach, M. 2008. Anchoring of shear strengthening for T-beams made of textile reinforced concrete (TRC)' Mater Struct 41: 407–418.

Hannant, D. 1978. Fibre Cements and Fibre Concretes, John Wiley.

Hegger, J., Sherif, A., Brukermann, O. & Konrad, M. 2004. Textile Reinforced Concrete: Investigations at Different Levels, in A. Dubey, ed., 'Thin Reinforced Cement-Based Products and Construction Systems', number SP-224, American Concrete Institute, pp. 33–44.

High Performance Fiber Reinforced Cement Composites (HPFRCC 2)—Proceedings of the Second International RILEM Workshop-RILEM Proceedings 31. Edited by A.E. Naaman and H.W. Reinhardt. Ej FN Spon, an imprint of Chapman and Hall, 1996,506. ISBN O-4 19-2 1180-2, Ann Arbor, USA, June 1995.

Kanda, T., Lin, Z. & Li, V. 2000a. 'Modeling of tensile stress-strain relation of pseudo strain hardening cementitious composites', ASCE Journal of Materials in Civil Engineering 12(2), 147–156.

Kabele, P. 2003a. Analytical model of multiple cracking in fibre reinforced cementritious composites under uni-axial tension, in 'International Conference on Structural Health Monitoring and Intelligent Infrastucture', Tokyo, Japan.

Mechtcherine, V. & Jun, P. 2008. Behaviour of Strain Hardening Cement-Based Composites in Tension and Compression. Proceedings of the Seventh RILEM International Symposium on Fibre Reinforced Concrete: Design and Applications (BEFIB 2008). Edited by Ravindra Gettu, RILEM Publications. ISBN 978-2-35158-064-6, Chennai India, pp. 471–481.

Naaman, A.E. 2008. Development and Evolution of Tensile Strain Hardening FRC Composites. Proceedings of the Seventh RILEM International Symposium on Fibre Reinforced Concrete: Design and Applications (BEFIB 2008). Edited by Ravindra Gettu, RILEM Publications. ISBN 978-2-35158-064-6, Chennai, India, pp. 1–28.

Advances in Cement-Based Materials – van Zijl & Boshoff (eds)
© *2010 Taylor & Francis Group, London, ISBN 978-0-415-87637-7*

Study on mechanical behavior of synthetic macrofiber reinforced concrete for tunneling application

A. Pineaud, J.-P. Bigas & B. Pellerin
CHRYSO, Sermaises, France

A.P.J. Marais
CHRYSO SA Ltd, Jet Park, South Africa

ABSTRACT: Synthetic microfibers have recently been introduced into the market. Such fibers can enhance properties of hardened concrete. As an example, shear strength and ductility of the material can be improved.

An especially interesting application is shotcrete tunneling because it has requirements on ductility. The aim of this study is to investigate the influence of synthetic macro fibers on flexural strength and post peak behavior based on a typical shotcrete mix design.

Three different fibers have been selected for an experimental study. The tested fibers have 50 mm lengths and differ by their shape and surface morphology. All the selected fibers are tested at dosages of 3, 5 and 7 kg/m³.

This study was performed in order to clearly establish the benefits of such type of fibers on shotcrete mix design.

This study aims to determine the influence of the fiber's different nature and respective dosage on:

– rheological behavior at fresh state; effect on superplasticizer dosage requirement at constant workability measured by slump test.
– 28 days compressive and flexural strength.
– flexural post peak behavior in order to evaluate toughness enhancement.

1 INTRODUCTION

Fiber Reinforced Concrete is gaining more and more interest in structural applications, because of enhanced mechanical properties and durability (Prisco, 2004). These enhanced properties are of special interest in the tunneling sprayed fiber reinforced concrete application (Aftes, 1999) (Hanck, 2004) in order to replace wire mesh.

Recently numerous synthetic macrofibers have been introduced into the market providing a wide range of fibers products. Synthetic fibers can reduce early age cracking. Their use should also be to improve resistance to handling stresses for precast applications, to improve cohesiveness for pumped concrete or to reduce rebound and material waste for shotcrete applications.

The effect of fibers on the concrete's behavior concerns both fresh and hardened states. The enhancement of concrete properties is depended on concrete mix design, materials, fiber type, diameter and length, and on the quantity of fibers added. Macrofibers can improve properties of hardened concrete. From a dosage as low as 3 kg/m³, those fibers can enhance flexural and toughness properties of concrete.

Widely used standard test methods for flexural thoughness and first crack strength are able to characterize the performances of synthetic macrofibers.

The aim of this work was to compare different types of synthetic fiber at different dosages, in order to bring support to the selection of fibers for tunneling jobsites.

2 MIX DESIGN AND TESTING

2.1 Mix design

C45/55 class concrete was used for this study. Concrete mix design was based on a typical sprayed concrete mix design. For this specific study and in order to ease experimental work and to reduce experimental variations, specimens were casted without the use of a shotcrete accelerator. The total amount of superplasticizer was adjusted between 0.7% and 0.9% based on binder, in order to achieve the same initial slump of 220 ± 20 mm for every concrete batch.

Macrofibers were introduced from 3 to 7 kg/m³, with the coarse and fine aggregates into the mixer at the beginning of the batching process. As mixing was performed on lab scale, the macrofibers tested were not packaged into water soluble bags. No additional mixing time was needed in order to obtain homogeneous and cohesive concrete. The lab mixing time was 210 sec.

Concrete mix proportioning is displayed in Table 1.

Three different 50 mm synthetic fibers were tested: commercially available CHRYSO® Fibre S50 (referenced as BET 694010 in figures), BET 694011 (with specific surface treatment) and BET 694012 (with special surface morphology). Geometrical and mechanical properties of the studied fibers were presented in Table 2.

BET 694010 and BET 694011 are fibrillated polypropylene fibers produced by extrusion in order to obtain a mesh of interconnected fibers. Added to concrete at mixing stage, the fibrillated fibers create a linked fiber filament network which enhances their anchorage into the cement matrix. Some providers had chemical surface treatment on the polypropylene fibrillated fibers in order to improve the anchorage.

Table 1. Mix design.

Gravel 4–10 mm, kg/m³	644
Sand 0–4 mm, kg/m³	1206
CEM I 52.5R, kg/m³	400
HRWRA CHRYSO®Fluid	0.7 to 0.9% of
Premia 310	cement binder
Effective water, liter	190
Density, kg/m³	2490
Slump, mm	220 ± 20

Table 2. Characteristics of the fibers.

	BET 694010	BET 694011	BET 694012
Length (L_f), mm	50	50	50
Equivalent diameter (ϕ_f), mm	1,0	1,0	1,0
Aspect ratio (L_f/ϕ_f)	50	50	50
Tensile strength, MPa	600	600	400–800
Density, kg/m³	920	920	1360
Melting point, °C	160	160	253
Elastic modulus, GPa	5,0	5,0	11,3
Colour	white	white	grey

Table 3. 28d compressive strengths, Mpa.

	3 kg/m³	5 kg/m³	7 kg/m³
BET 694-010	52	53	51
BET 694-011	49	52	53
BET 694-012	53	53	53

Figure 1. Detail from $70 \times 70 \times 280$ mm³ notched beam after breaking.

BET 694012 is an extruded polyester fiber oriented with special surfaces in order to improve bonding capabilities.

2.2 Testing

Fresh properties of concrete were measured with Abram's cone; targeted slump was 220 ± 20 mm whatever the fiber nature and dosage (3, 5 and 7 kg/m³ except BET 694012 that was only tested at 3 and 5 kg/m³).

The concrete was poured into different testing moulds that and were vibrated on a vibrating table to ensure compaction.

The compressive strength was measured on 150 mm cubic specimens at 28 days. Specimens were tested on a Toni Tecknik 3000 kN maximum load servo-hydraulic testing machine. The average of three cubes were recorded and reported in Table 3.

The flexural test procedure was adapted from ASTM and EN standards. Tests were conducted on $70 \times 70 \times 280$ mm notched beams. Notches were obtained with a diamond saw (Figure 1).

Three point bending tests were performed with an Instron electromechanical press.

Without LVDT sensor, test was controlled using cross bar displacement.

Displacement rate has been selected in order to ensure that the test was stable (no brittle failure) and to make sure that the test duration was not too long (no creep effects). A speed of 5 µm/s for this test was found acceptable; higher levels lead to brittle failure.

Stopping criterion was selected according to ASTM and EN standards. Given the displacement, load level was very low, 5 mm was found sufficient. Using these experimental conditions, tests didn't last more than 17 min.

A typical cracking pattern is presented in Figure 1. The experimental conditions allow for, as expected, a single crack, which starts from the notch.

Three specimens were tested for every different concrete batch.

3 RESULTS

3.1 Fresh state and compressive strength

Eight concrete mixes, (3 kg, 5 kg and 7 kg/m³ dosage for each fiber, except BET 694012) were cast in order to measure the effect of fiber introduction on fresh concrete state and on the superplasticizer dosage requirement at constant water content.

Introducing macrofiber can result in a slump loss and a slight decrease in workability when placing the fiber reinforced concrete. In our case, increasing fiber dosage has a limited effect on initial concrete consistency. The targeted 220 mm was obtained for every concrete designed with the same water content. A light (0.2%) adjustment of HRWRA was necessary for BET 694010 and BET 694012 in order to keep constant water content and initial slump.

No significant differences have been measured between the different concrete batches regarding the effect of fibers content on 28 days compressive strengths. (Table 3).

In the case of this mix design, the introduction of fibers, whatever their dosage, shape or surface morphology, has no influence on compressive strength characteristics.

3.2 Flexural tests

The results presented were focused on the effect of dosage and type of fiber on maximal load. (F max in Newton) Energy absorption (W in Joule) was calculated as the area under the load-displacement curve.

In order to evaluate fiber performances, load displacement curves were used to describe pre-peak behavior, maximal load at peak value and to appreciate the post peak residual strength.

Load displacement curves for BET 694010 series (Figure 2) shows a typical behavior of fiber reinforced concrete.

The mechanical behavior should be described in three classical stages. First part was the pre-peak phase, second part was the peak at maximum load corresponding to the flexural strength and third part was the post-peak residual mechanical strength useful to characterize the toughness level.

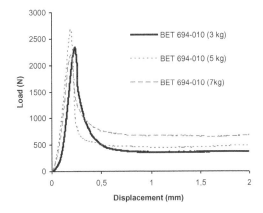

Figure 2. Load-displacement curves vs. dosage of BET 694010.

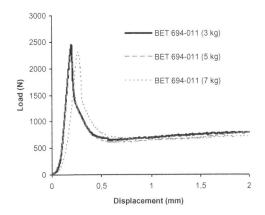

Figure 3. Load-displacement curves vs. dosage of BET 694011.

This method allows for the comparison of the different results. All beams failed at approximately the same load. This can be explained by the same concrete mix design for all specimens tested; differences can only be measured at the post-peak phases.

No important effect of the volume fraction has been detected on flexural test, in the case of BET 694010.

For BET 694011, similar tendency was observed (Figure 3). We observed quite similar maximum load values and post-peak tendencies do not seem to be influenced a lot by fraction volume.

With BET 694012, the curves seem to be slightly influenced by the dosage (Figure 4). The post peak behavior with 5 kg showed enhanced toughness.

According to the curve analysis, results do show evidence that the volume fraction was the first order parameter on this kind of mix design, in order to explain flexural test results. The maximal

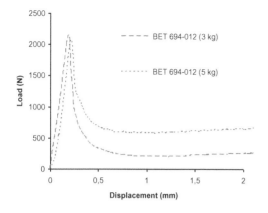

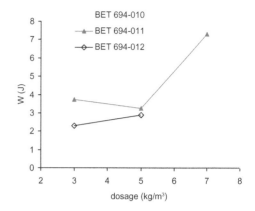

Figure 4. Load-displacement curves vs. dosage of BET 694012.

Figure 6. W vs. dosage.

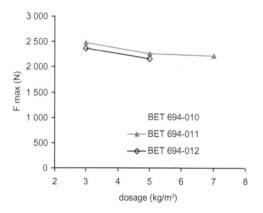

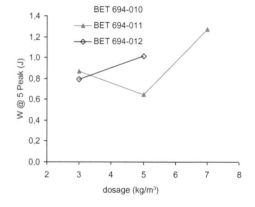

Figure 5. Maximal load vs. fiber dosage.

Figure 7. W @ 5 peak vs. dosage.

load values and the slope of the curves look quite similar, deeper analysis will be necessary in order to have a definite assessment.

3.2.1 *Maximal load*

Maximal load mean values (3 samples) for each concrete mix are displayed in Figure 5.

The results show no significant effect on maximal load of the type of fiber and their dosage.

This can be attributed to the similar length and aspect ratio of each tested fiber. Regarding dosage influence, the similar values tend to show that maximum pre-peak reinforcement has been reached at 3 kg with this type of fibers. Which means that, with tested fiber, tensile strength of concrete cannot be improved by increasing fiber dosage higher than 3 kg/m^3.

3.2.2 *Area under Load-displacement curve*

Energy absorption of concrete has been characterized by the area under load-displacement curve

(W, in J). These results are presented in Figure 6 as a function of dosage for each type of fiber.

This figure shows that W increases with increasing fiber dosage, except for BET 694010 and BET 694011 at 5 kg/m^3 (likely due to experimental artifact).

The enhancement of W shows better toughness and post-peak behaviour at 7 kg/m^3 dosage rate.

W values with BET 694011 were greater than with BET 694010. This shows that specific surface treatment that was conducted on BET 694011 has a positive effect on concrete bending ductility. We can deduce that this treatment may upgrade the slipping properties of the fiber in the concrete after the peak (cracked concrete).

Based on Figure 6, BET 694012 seems to be less effective than the other fibers. This result was not in accordance with the surface aspect and the shape of this fiber is designed to enhance anchorage.

Figure 7 shows the area under load-displacement curve calculated at five times the displacement at peak (W @ 5 peak, in J).

Comparison of fibers on this figure was slightly different from Figure 6: according to Figure 7, BET 694012 gives similar to or higher ductility than the others. As in Figure 7 the area calculation was limited at 5 times displacement at peak. We can conclude that, for short displacements, BET 694012 leads to similar energy absorption when compared with to the others fibers.

Observation of load-displacement curves (Figure 8) shows differences on pre-peak behavior with this fiber.

Figure 8 presents typical load-displacement curves for the three fibers at 3 kg/m³.

The BET 694012 curve shows higher stiffness in the first part. We noticed that the post peak behavior on this curve was more brittle than the others. These observations demonstrated that BET 694–012, despite a better anchorage in concrete matrix, dissipates less energy by slipping during post-peak period. This can be attributed to the damage of the concrete around the fiber caused by its more efficient anchorage.

Differences can be explained by the fibers morphology. First, BET 694-010 and BET 694-011 were fibrillated at each end, one with a special treatment of the surface in order to enhance the anchorage and limit slipping. The fibrils anchor each fiber so their bonding capabilities were superior to the conventional monofilament fiber (Trottier, 2004).

Secondly the BET 694012 plate shows a corrugated surface in order to maximize bonding to the shotcrete matrix.

According to these results, fibrillated fibers enhance post-peak ductility and lead to powerful slippage. Concerning corrugated fibers, their good anchorage leads to higher stiffness before cracking but lower slippage properties.

4 CONCLUSION

Three different 50 mm synthetic fibers were tested in our experimental program. The following conclusions can be drawn from this lab scale evaluation

- Volume fraction was not a very sensitive parameter. Our study showed that with typical concrete mix design being used, a 3 kg/m³ dosage rate represents a sufficient volume fraction in order to improve mechanical properties of concrete mixes.
- Maximal load was slightly influenced by the fiber type and their coating or surface treatment. The tested fibers showed similar maximum load behavior.
- Post-cracking behavior was improved with BET 694011 and BET 694012. The beneficial effect of the chemical surface treatment on the surface (BET 694011) and shape of the fiber (BET 694012) seems to improve concrete properties.
- Chemical coating tends to improve slippage properties, leading to better post-peak toughness.
- Corrugated surface of the fibers leads to better anchorage and so higher pre-peak stiffness but less powerful slippage.

For tunneling sprayed concrete the advantages of adding synthetic fibers mainly relies on mechanical properties enhancement. The results show that different solutions were available. In situ tests will be performed as a continuation of this experimental work in order to finalize the main conclusions of this study.

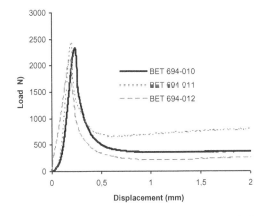

Figure 8. Typical load-displacement curves at 3 kg/m³.

REFERENCES

AFTES–AITES, 1999, Fiber-reinforced sprayed concrete, technology and practice, GT6R1F3.
ASTM C78-02 Standard test method for flexural strength of concrete (using simple bean with third-point loading), West Conshohoclen, PA: American Society for Testing and materials, 2002.
BS EN 12390-5: 2009, Testing hardened concrete. Part 5: Flexural strength of test specimens, London: British Standard Institution, 2009.
Hanck, C. & Al. 2004, Macro-synthetic fiber reinforced shotcrete in a norvegian road tunnel *edition Taylor and Francis group, Shotcrete 2004*. London.
Trottier J.-P. & Mahoney M. 2004, Innovating synthetic fibers, Shotcrete, Fall 2004.

Advances in Cement-Based Materials – van Zijl & Boshoff (eds)
© 2010 Taylor & Francis Group, London, ISBN 978-0-415-87637-7

SHCC precast infill walls for seismic retrofitting

Sun-Woo Kim, Sang-Hyun Nam, Jun-Ho Cha, Esther Jeon & Hyun-Do Yun
Chungnam National University, Daejeon, South Korea

ABSTRACT: It is noted that frame structures incorporating infill walls have shown definite economic and performance advantages over conventional rigid frame structures when the structures are required to resist large lateral loads due to earthquake ground motion. The objectives of this study are to evaluate the effect of notch as a damage fuse element in the infill walls, and to investigate the effect of cementitious composite properties, particularly strain hardening and multiple cracking, on the seismic performance of infill walls when subjected to displacement reversals. The experimental investigation consisted of cyclic loading tests on 1/3-scale models of infill walls. Material ductility and notch were main variables in the test. Compared with RC infill wall, even though SHCC (strain-hardening cement composite) infill wall had less effective section due to notched midsection, SHCC infill wall specimen showed higher strength and energy dissipation capacity than those of RC infill wall specimen without notched midsection.

1 INSTRUCTION

1.1 Background

Structures must resist lateral force as earthquake without collapse and significant damage to the structural members. On the seismic design, the Korean Building Code requires that the structures should obtain satisfactory design load and seismic reinforcement detail so that adequate ductile response and limiting condition for maximum displacement are assured. Especially, the code specifies that the seismic design should be applied for three or higher story buildings to resist earthquake of magnitude 7.0 on the Richter scale. The types of buildings in Korea are largely classified into 54.3% of residential buildings, 24.9% of commercial buildings, 11.1% of industrial buildings, 5.8% of educational buildings and 3.9% miscellaneous buildings. Among them, only 52.3% of buildings have been constructed since 1990 that seismic philosophy was included in the Korean Building Code. If earthquake occurs, the rest of structures which had been constructed before 1990 will be collapsed with large distortion at the base of building that is strongly affected by the seismic force. Therefore, there are so many researches being carried out to improve seismic performance of existing structures instead of the uneconomic method of demolishing existing structures and building new structures with new seismic performance design.

1.2 Objectives

Ahead of the evaluation on the combined action of the infilled frame, this study is to evaluate on the damage process and shear behavior of infill walls. Two main concepts were proposed to improve the seismic performance of infill walls. The one is making notch to induce damage from the corner to midspan of infill wall. Most of infilled frames would fail at the top of the column due to shear force. So, through notched midspan, the damage of column can be transferred from top to midspan of the column. The other is use of SHCCs (strain-hardening cement composites), which represent improved ductility and the dispersion of microcracks by hybrid fiber (poly ethylene and polyvinyl alcohol), to infill walls in order to absorb damage energy.

The objectives of this study are (a) to evaluate the effect of notch as a damage fuse element in the infill walls and (b) to investigate the effect of cementitious composite properties, particularly strain hardening and multiple cracking, on the seismic performance of infill walls when subjected to displacement reversals. The investigation focuses mainly on the comparative behaviors of precast infill walls in terms of cracking mechanism, ultimate strength, and energy dissipation capabilities.

2 MATERIALS

2.1 Used material composition and properties

The SHCC used in this study utilized hybrid 0.375 ultra-high molecular weight polyethylene (PE) fibers and 0.375% polyvinyl alcohol (PVA) fibers, as listed in Table 1, and cement, fine aggregates (grain sizes ranging from 105 to 120 μm), and methyl cellulos-based viscosity modifying admixture

Table 1. Physical properties of reinforcing fibers.

Fiber	Density g/cm³	Length mm	Dia-meter μm	Aspect ratio m/m	Tensile strength MPa	Elastic modulus GPa
PE	0.97	15	12	1250	2500	75
PVA	1.30	12	39	307	1600	40

Table 2. Mixture proportions for cement composites.

Cement composites	W/C %	Fiber %	C kg/m³	W	MC	SP
Concrete	0.45	0.0	700	315	38	892
SHCC*	0.45	0.75	700	315	38	892

* Fibers in SHCC were 0.375% of PE and 0.375% of PVA.

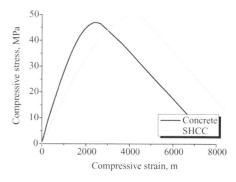

Figure 1. Compressive behavior of cement composites.

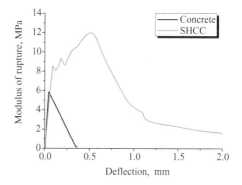

Figure 2. Flexural behavior of cement composites.

(VMA) 0.2% at cement weight fractions. Total fiber content in the SHCC was 0.75%. Concrete contained coarse aggregates (maximum grain size 18 mm), cement, water, and a high range water-reducing admixture to enhance the fresh properties of the mixture. The water-to-cement ratio of all matrices was 0.45, and the specified compressive strength of the cylindrical specimens was 40 MPa. The mixture proportions on each of cement composite used in this study were listed in Table 2.

All specimens were cured at 23°C and 95–100% relative humidity (RH) for about 1 day, whereupon they were demolded. After then, all specimens were cured in water at 23 ± 2°C during 28 days.

Only D6 deformed bars were used as the vertical, horizontal and diagonal reinforcement. Tensile tests were performed on five steel samples of rebar. D6 showed yield strength of 291 MPa at 0.19% strain, and ultimate strength of 375 MPa at 0.58% strain.

2.2 *Mechanical properties of cement composites*

The monotonic compressive tests were carried out on the cylindrical specimens, which had 100 mm diameter and 200 mm height, according to KS F 2405 (which is similar to ASTM C 39). All of the compressive testing was performed at a strain rate of 0.30% per minute. Uniaxial monotonic tensile tests were performed on the dumbbell shape specimens to examine the effect of reinforcing fibers on the tensile capacity of SHCC. The comparison between reinforced SHCC and concrete is based on the material properties of both cement matrices. As shown in Figure 1, concrete and SHCC have similar ranges of compressive strength f_{cu}

(47 to 50 MPa). The concrete exhibited 61% higher elastic modulus than that of SHCC. It may be due to the reason that the SHCC was mixed with silica powder instead of coarse aggregate in order to improve bond strength between reinforcing fiber and cement matrices.

The flexural tests were carried out on the rectangular beam specimens according to KS F 2408. All beams had 100×100 mm² cross-section, 400 mm length and 300 mm net span. As the results of four-point bending tests, flexural strength of SHCC was 12 MPa, which was twice as high as concrete as shown in Figure 2. Especially, deflection capacity of SHCC was 0.51 mm, which was tenfold as large as concrete.

Direct tensile tests were performed on the dumbbell shape specimens in order to characterize the tensile behavior of SHCC used in this study. For concrete, the tensile tests were not performed because the contribution of the tensile stress in the concrete to the flexure capacity of the wall is very small and can be neglected. Although total fiber content of the SHCC matrix is 0.75% (PVA 0.375% and PE 0.375%), all five specimens were

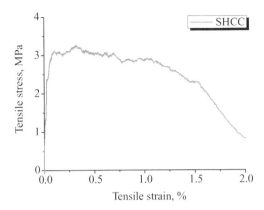

Figure 3. Tensile behavior of SHCC.

Table 3. Strength test results of cement composites.

Cement matrix	Compression		Flexure		Tension	
	f_{cu} MPa	E_c GPa	f_r MPa	δ mm	f_t MPa	ε_t %
Concrete	46.9	28.7	5.9	0.05	–	–
SHCC	49.9	17.8	12.0	0.51	3.3	0.31

tested up to 2% tensile strain as shown in Figure 3. The measured strain was applied to the displacement control parameter of the test.

The results of the compressive, flexural and direct tensile tests are listed in Table 1, and each test results presented is the average of each five specimens.

3 TESTING PROGRAM

3.1 Test specimens

Total 4 specimens have been planned as listed in Table 4. The reinforcing fibers used in the experimental program were listed in Table 1, and the mix proportions were summarized in Table 2. It was noted that several parameters were investigated selectively: two different fibers (polyvinyl alcohol and poly ethylene) and two different matrices (concrete and SHCC), two different reinforcing methods (conventional and diagonal) and two different wall configuration (rectangular type with and without notched midsection). Specimen PIWC-C and PIWS-C are conventionally reinforced walls, and specimen PIWC-ND and PIWS-ND are diagonally reinforced concrete walls. Regular concrete matrix was used in specimen PIWC-C and PIWC-

Table 4. Description of specimens.

Specimen	$\dfrac{l_w \times h_w \times t_w}{\text{mm}}$	Notch	Reinforcing-method	Cement composites
PIWC-C	$1400 \times 800 \times 70$	no	conventional	concrete
PIWC-ND	$1400 \times 800 \times 70$	yes	diagonal	concrete
PIWS-C	$1400 \times 800 \times 70$	no	conventional	SHCC
PIWS-ND	$1400 \times 800 \times 70$	yes	diagonal	SHCC

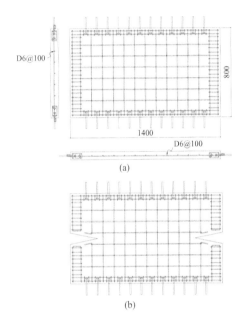

Figure 4. Specimen geometries and reinforcement details (unit: mm): (a) Infill wall specimens; (b) notched infill wall specimens.

ND, and SHCC was used in specimen PIWS-C and PIWS-ND, respectively.

The infill wall specimens were scaled to about 1/3, which was 1,400 mm wide and 800 mm height for specimens with of aspect ratio, $h_w/l_w = 0.57$. A notch depth was 250 mm, which is about 18% of total midsection length. A 70 mm thick element was needed to allow D6 rebar as horizontal, vertical and diagonal reinforcement. Figure 4 shows the reinforcement details of the specimens.

3.2 Experimental setup and instrumentation

The experimental setup was designed to simulate the rectangular frame structure under lateral reversal cyclic loading. It comprised two steel

Figure 5.   Test setup.

frame columns pin-connected with lateral loading beam and reaction floor. The capacity of the loading actuator is 1,000 kN with a maximum displacement of 150 mm in in both pushing and pulling directions. The test specimens were instrumented to monitor the applied loads and displacements at the top of wall specimen. Displacement transducers were used to measure the drift and shear deformation of wall. A displacement routine used for reversed cyclic loading is as shown in Figure 5.

4   TEST RESULTS

4.1   *Failure mode*

Figure 6 shows cracking and failure mode of the infill wall specimens.

In specimen PIWC-C, which was conventionally reinforced concrete wall, an initial crack was observed at the bottom of specimen at 0.05% of drift. At 0.10% of drift, inclined crack propagated from the bottom to end of midsection of specimen. At 0.15% of drift, previously developed cracks propagated to the top of wall. A major shear crack along the specimen occurred at 0.25% of drift. Since then, the width of existing inclined cracks was increasing only, and there was no occurrence of new cracks. At 0.45% of drift, new inclined cracks occurred at the top of specimen and previously developed shear crack width increased rapidly. At 1.0% of drift, shell concrete at the bottom boundary zone of wall fell off and shear capacity of the wall decreased.

In specimen PIWC-ND, which is diagonally reinforced concrete infill wall, an inclined crack that connected notch with the bottom of specimen and horizontal direction crack at the bottom boundary zone of specimen occurred at 0.05% of drift. As drift increased, many cracks were observed around the previously occurred inclined cracks and these cracks propagated from the top to the midsection

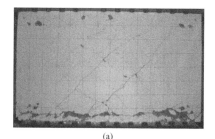

(a)

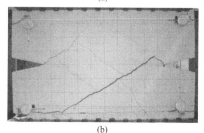

(b)

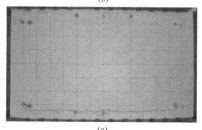

(c)

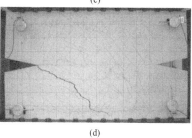

(d)

Figure 6.   Cracking pattern and failure mode of specimens: (a) PIWC-C; (b) PIWC-ND; (c) PIWS-C; (d) PIWS-ND.

of specimen. At 0.10% drift, very large shear crack was occurred in both positive and negative directions. Since then, without more cracks occurring, only width of the existing cracks increased. This growth of the crack width was continued until 0.4% of drift when specimen failed due to drop of capacity and 10 mm increase of crack width.

In PIWS-C specimen, which is conventionally reinforced SHCC infill wall, the first inclined crack was observed at the middle of the wall during the cycles to 0.05% drift and some flexural cracks appeared at the bottom corner of the wall. As drift

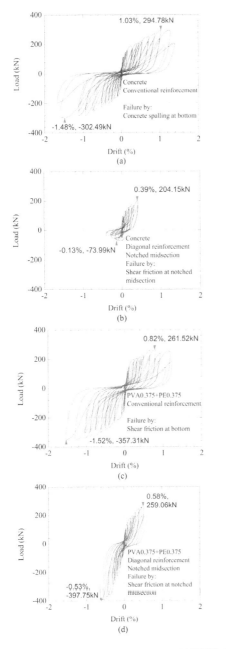

Figure 7. Load vs. displacement response: (a) PIWC-C; (b) PIWC-ND; (c) PIWS-C; (d) PIWS-ND.

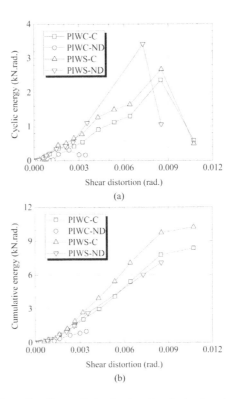

Figure 8. Energy dissipation based on the load vs. shear distortion hysteresis loops: (a) Cyclic energy dissipation; (b) Cumulative energy dissipation.

increased, many cracks appeared near the inclined cracks. However, until 0.25% drift, the shear cracking which appeared at 0.15% drift in the PIWC-C specimen was not observed. At 0.25% drift, horizontal cracking started at the bottom of wall and previously occurred shear cracks (300~400 mm length) extended across nearly the full panel width. However, only the length of cracks increased, widely crack opening was not observed except that crack width was increased at the bottom joint of wall. The formation of cracks propagated to the center of wall as drift increased.

At 1.0% drift, a severe shear crack formed and specimen separated along the shear crack. However shear capacity of wall did not decreased in spite of the shear crack formation. At 1.2% drift, the shear crack opened widely, resulting in complete failure of the infill wall.

In specimen PIWS-ND, which was diagonally reinforced SHCC infill wall, an initial crack was observed at 0.05% drift. As drift was increasing, a lot of new hair cracks were observed near inclined cracks. These cracks propagated to the center of wall. The cracks which started from notch propagated to the bottom of wall at 0.15% drift. However, a great number of individual hair cracks were spread over the specimen as drift increasing. At 0.50% drift, the width of crack which started from notch became larger. At 0.63% drift, specimen PIWS-ND failed due to widely opening crack at the bottom of wall.

153

4.2 Load vs. displacement response

The load vs. displacement response is one of the most important sources of information for overall behavior of the specimen. Figure 7 shows the load vs. drift response. The drift is defined as the ratio of the relative displacement from the data recorded by the displacement transducers mounted along the middle of infill wall specimen.

The conventionally reinforced concrete infill wall, specimen PIWC-C, shows stable hysteretic behavior up to lateral drift of 1.0%. It shows considerable loss of stiffness at 1.0% drift and unstable, degrading hysteresis beyond 1.0% drift. In the specimen PIWC-C, the capacities attained in the positive (pushing) and negative (pulling) directions are 294.78 kN and 302.49 kN, respectively.

The diagonally reinforced concrete infill wall with notch, specimen PIWC-ND, exhibits unstable response with poor energy dissipation. The maximum load of specimen PIWC-ND is 204.15 kN at 0.375% drift in the positive direction, which is 31% lower than that of specimen PIWC-C.

To evaluate the effectiveness of SHCC to improve the energy dissipation and drift capacity in an infill wall, the SHCC with hybrid PVA and PE fibers in a 0.75% volume fraction was used. The maximum load of PIWS-C specimen is 357.31 kN at 1.5% drift in negative direction, which is 18% higher than that of the PIWC-C specimen.

The maximum load of PIWS-ND specimen is 397.75 kN at 0.5% drift in the negative direction, which is twice higher than that of the PIWC-ND. In case of same reinforcement details, SHCC provided better behavior to infill walls than concrete. Therefore, it is noted that the SHCC material can prevent infill wall with notched midsection from rapid degradation of shear force.

4.3 Energy dissipation capacity

The energy dissipated by each specimen during full cycle was calculated as the area enclosed by the load vs. shear distortion loop. The cumulative energy was calculated by summing up the energy dissipated in consecutive cycles throughout the test. The cyclic energy dissipated vs. shear distortion and the corresponding cumulative energy for each specimen is plotted in Figure 8. Using the cumulative energy based on the load vs. shear distortion hysteresis loops, the energy dissipated by specimen PIWS-C was 610% that of specimen PIWC-C, and the cumulative energy of specimen PIWS-ND was 250% that of specimen PIWC-ND, respectively. It can be observed that the total cyclic energy dissipation capacities for SHCC specimens were generally higher than that of RC specimens. Conventionally

reinforced concrete infill wall, specimen PIWC-ND, showed the least energy dissipation capacity due to notched midsection. And the use of fibers in the infill wall with notch had a significant effect on the cumulative energy dissipation showing increase of 250% over that of RC infill wall.

5 CONCLUSIONS

In case of same reinforcing details, SHCC infill walls exhibited higher strength and deformation capacity than that of RC infill walls. Also this improvement on the behavior affected energy dissipation capacity of the infill wall. But shape or depth of the notch on the infill wall will have to be carefully considered because deformation capacities of notched infill walls, in spite of diagonal reinforcing, were less than those of conventional infill walls.

ACKNOWLEDGEMENT

This work was supported by the Korea Research Foundation Grant funded by the Korean Government (MOEHRD) (KRF-2006–311-D00916), and some researches were supported by the 2nd Brain Korea.

REFERENCES

Architectural Institute of Korea. 2005. *Korean Building Code*. Ministry of land, transport and maritime affairs. *http://www.mltm.go.kr*. Republic of Korea.

Fukuyama, H. et al. Test on High-Performance Wall Elements with HPFRCC. *Proc. International Workshop on High Performance Fiber Reinforced Cementitious Composites in Structural Applications*: 23–26.

Nagai, S. et al. 2002. Shear capacity of ductile wall with high performance fiber reinforced cement composite. *Proc. of the 1st fib Congress*: 767–774.

Sugano, S. 1981. Guidelines for Seismic Retrofitting (Strengthening, Toughening, and/or Stiffening), Design of Existing Reinforced Concrete Buildings. *Proceeding of the Second Seminar on Repair and Retrofit of Structures*. Ann Arbor. Michigan. pp. 189–246.

Yun, H.D. et al. 2008. Effect of Reinforcing Fiber Types on the Behavior Characteristics of SHCCs. *Journal of the Architectural Institute of Korea* 24(5): 141–148.

Yun, H.D. et al. 2006. Seismic Performance of Lightly Reinforced Concrete Frames with High Performance Fiber-Reinforced Cement Composite Infill Walls. *Journal of the Architectural Institute of Korea* 22(5): 31–38.

Yun, H.D. et al. 2007. Mechanical properties of high-performance hybrid-fibre-reinforced cementitious composites (HPHFRCCs). *Magazine of Concrete Research* 59(4): 257–271.

Improving infrastructure sustainability using nanoparticle engineered cementitious composites

Michael D. Lepech
Stanford University, Stanford, CA, USA

ABSTRACT: Contributing 5% of global anthropogenic greenhouse emissions through cement production alone, the concrete industry is a major contributor to climate change. Vehicles that use concrete transportation infrastructure release another 30% of anthropogenic greenhouse emissions. Within this research, carbon nanoparticles are used to tailor the fiber-matrix interface of ECC materials for improved mechanical performance. Using life cycle assessment, the high energy content of carbon nanoparticles (2 to 100 times more energy intensive than aluminum) is balanced against their ability to improve ECC performance, increase infrastructure durability, and reduce traffic emissions, thereby improving transportation system sustainability. Specifically, carbon nanoparticles are used to tailor the crack bridging stress versus opening relationship within ECC to improve tensile ductility and crack width control. Results show that depending on the application, carbon nanoparticles can improve transportation infrastructure sustainability by reducing a number of environmental indicators including primary energy (MJ) and global warming potential (kg CO_2-eq).

1 INTRODUCTION

1.1 *Environmental impact of infrastructure*

The volume of concrete used in infrastructure around the world comprises large flows of material between natural and human systems. Worldwide use of concrete for construction projects exceeds 12 billion tons annually. To support this consumption, cement production in the year 2008 totaled 2.9 billion metric tons (van Oss 2009). Rates, compositions, and spatial distributions of such major flows are major determinants of the degree to which societies are sustainable. While concrete infrastructure continues to grow around the world, evidence suggests that the performance of current infrastructure systems is deficient in terms of the social, environmental, and economic dimensions of sustainability.

Social: 33% of U.S. major roads are in poor and mediocre condition, while 27% of US bridges are structurally deficient or functionally obsolete. Roadway improvements would reduce accidents and save lives. Of the 41,059 US traffic fatalities in 2007, one third have been attributed to inadequate roadway conditions. Furthermore, construction causes significant traffic delays. Total urban traffic congestion costs the nation $78.2 billion in wasted fuel and lost time each year (TRIP 2009).

Environmental: Cement production, an energy intensive process, is responsible for 5% of global greenhouse gas emissions (WBCSD 2002), and significant amounts of NO_x, particulate matter (PM) and other pollutants such as SO_2. The production of 1 ton of cement clinker requires approximately 1.7 tons of non-fuel raw materials and results in the release of 1 ton of CO_2 (van Oss 2002).

Economic: Roads represent one infrastructure system with significant economic impacts. For example, driving on roads in need of repair or improvement costs motorists in the US an average of $335 per driver in extra vehicle operating costs each year, or $67 billion total (TRIP 2009). Also, delayed shipments of freight can lead to productivity losses directly impacting business and industry.

The unsustainability of current infrastructure, coupled with the anticipated rapid growth of new ones, lead to an obvious need for improvement. Recent research suggests that substituting concrete with advanced cementitious composites that have superior mechanical performance can have a profound impact on each of the three dimensions of sustainability. These impacts occur throughout an infrastructure system's life cycle, which encompasses resource extraction, materials processing, construction, use, maintenance and repair, and end-of-life management. In this research the nature of these impacts is characterized and materials are engineered for overall performance improvements throughout the material, structure, and system lifecycle.

Due to the long life of most transportation infrastructure systems, and the large amounts of energy and emissions associated with their use (e.g. vehicles traveling over the infrastructure) a major component of the sustainability of these systems is material and structural durability. Along these lines, the limited durability of reinforced concrete is responsible for significant amounts of infrastructure repair, rehabilitation, and replacement. The combination of corrosion-prone steel reinforcement with brittle concrete materials is the root of many concrete infrastructure failures. Kütner (2009) estimated that 90% of concrete failures are related to reinforcement corrosion and associated brittle spalling of concrete cover. Alternative technologies that suppress the brittle nature of concrete promise to improve the mechanical performance (and sustainability) of infrastructure. One such technology is engineered cementitious composites (ECC), a ductile HPFRCC material. Given the demand for infrastructure systems worldwide, the potential application of ductile ECC for sustainable infrastructure construction could be large.

ECC is a unique group of fiber reinforced cementitious composite materials with high ductility. As a successful sample of materials engineering using the paradigm of the relationships between material microstructure, processing, material properties, and performance (Li 1992), the fiber, matrix and interface of ECC are carefully tailored under the guidance of micromechanical models that link composite ductility to individual phase properties (Li 1998). ECC strain-hardens in tension, accompanied by sequential development of multiple cracking after first cracking. Tensile strain capacity exceeding 5% has been demonstrated in ECC reinforced with polyethylene (PE) and polyvinyl alcohol (PVA) fibers (Li et al. 2002). An example of the uniaxial tension stress-strain relationship is shown in Figure 1.

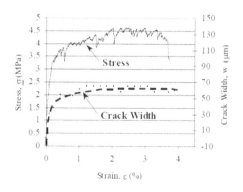

Closely associated with the strain-hardening and multiple cracking behavior is the small steady state crack width. Even at a strain of 4–5%, crack widths of ECC remain below 100 μm. Such small crack widths imply a significant improvement in structural durability (Lepech 2009). This development of crack widths is also shown in Figure 1. Through careful material design, the fiber volume fraction in ECC remains moderate, typically below 2%. As a result, unlike many other high performance FRCs, ECC can be prepared in standard concrete mixers.

1.3 *Sustainable ECC pavement overlay*
applications

Numerous researchers have studied the life cycle costs and impacts of pavement systems (i.e. asphalt versus concrete) (Horvath & Hendrickson 1998; Zapata & Gambatese 2005). Results from these studies have shown that Portland cement concrete pavements exhibit lower energy consumption and related impacts during the first three life cycle stages (extraction of raw material, manufacture, and placement) as compared to asphalt pavements. But energy consumption and emissions throughout service life is heavily dependent upon local parameters, such as traffic volume, truck loads, and climate.

Zhang et al. (2008) has looked at the overall sustainability of unbonded concrete overlays, HMA overlays, and ECC overlays using economic, environmental, and social metrics. Zhang et al. found that ECC overlays were able to improve the sustainability of rigid pavement overlays by using ECC materials to extend overlay service life and reduce surface roughness thereby improving vehicle fuel economy. Most important among these findings was the large impact that service life and the suppression of reflective cracking failure mechanisms have on pavement overlay system sustainability.

The motivation behind this research is the development of new ECC composites that leverage the unique properties of carbon nanoparticles to tailor ECC performance for use in rigid pavement overlay applications. These new versions of ECC are controlled to maintain high ductility, high fatigue resistance, and the "kinking-and-trapping" mechanism for suppression of reflective overlay cracking (Zhang and Li, 2002).

2 TAILORING ECC MATERIALS USING CARBON NANOPARTICLES

2.1 *Micromechanical tailoring*

As mentioned above, the main goal of this effort is to maintain high ductility, fatigue resistance, and suppression of reflective cracking mechanisms. As discussed previously by Lepech et al. (2008) for pavement applications this can be accomplished by improving ECC tensile ductility and suppressing fracture localization as long as possible.

The fundamental micromechanics that govern this ductile behavior through the formation of multiple microcracks within ECC provide the basic tools for nanoparticle material composite design. The basis of multiple cracking and strain hardening within ECC is the propagation of steady state cracks which were first characterized by Marshall and Cox (1988) for continuous aligned fiber reinforced ceramics, and extended to fiber reinforced cementitious composites by Li and Leung (1992). The formation of multiple steady-state cracking is governed by the bridging stress versus crack width opening relation along with the cracking toughness of the mortar matrix. To achieve this strain-hardening phenomenon, the inequality shown in Equation 1 must be satisfied.

$$J'_b = \sigma_0\delta_0 - \int_0^{\delta_0} \sigma(\delta)d\delta \geq J_{tip} \approx \frac{K_m^2}{E_m} \quad (1)$$

where J'_b is the complimentary energy, σ_0 and δ_0 are the maximum crack bridging stress and corresponding crack opening, J_{tip} is the fracture energy of the mortar matrix crack tip, K_m is the fracture toughness of the mortar matrix, and E_m is the elastic modulus of the mortar matrix.

In addition to the fracture energy criterion, a strength criterion expressed in Equation 2 must also be satisfied.

$$\sigma_0 > \sigma_{cs} \quad (2)$$

where σ_0 is the maximum crack bridging stress and σ_{cs} is the cracking strength of the mortar matrix.

From Equation 1, the complimentary energy can be see graphically in Figure 2 as the area to the left of the crack-bridging stress versus crack opening relation (σ–δ curve).

For the design of robustly ductile composites, Kanda et al. (2003) found that the ratio between complimentary energy, J'_b, and matrix fracture energy, J_{tip}, (commonly referred to at the strain hardening potential) should be on the order of 3. However, this value must theoretically only be greater than unity to achieve strain-hardening. Therefore, to increase the robust ductile and fracture suppression performance of ECC materials (and thereby improve their infrastructure performance) for pavement applications, three material tailoring routes can be used.

1. maximizing complimentary, J'_b
2. minimizing matrix fracture energy, J_{tip}
3. maximizing J'_b and minimizing J_{tip} simultaneously.

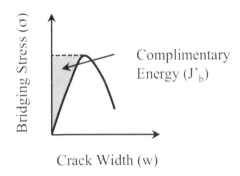

Figure 2. Characterization of crack bridging stress versus crack width opening (σ–δ) curve and complimentary energy (J'_b).

In this work we will focus exclusively on using carbon nanoparticles to maximize the value of complimentary energy, J'_b, by tailoring the interfacial properties between the fiber and cementitious matrix.

Tailoring of the interface between fibers and matrix within ECC to achieve high ducility was explored exensively by Li et al. (2002) using poly-vinyl-alcohol (PVA) fibers. As a hydrophilic fiber, PVA bonds very strongly to the surrounding cementitious matrix in ECC. As noted by Li et al., this resulted in higher fiber rupture tendencies in PVA-ECC materials and a reduction (or complete loss) of strain capacity. To combat this phenomenon, a proprietary oiling agent was applied to the surface of the PVA fiber by the manufacturer to decrease bonding behavior and promote greater fiber pullout. The success of this approach was seen in a larger complimentary energy associated with oiled PVA σ–δ curves, and the increased tensile ductility of the oiled PVA-ECC material.

However, the use of surface oiling to tailor the fiber-matrix interface has some disadvantages.

1. The PVA-fiber oiling agent is proprietary and therefore receives little additional research focus and ongoing improvement,
2. The oiling agent is tailored for PVA fiber and therefore not immediately transferrable to other fiber types,
3. The manufacturing of the oiling agent is a trade secret and therefore no information on ecological impacts resulting from oiling production is available.

Therefore, rather than oiling the surface of the fibers prior to mixing in ECC materials to control the interfacial properties, carbon nanoparticles are used in this work to control the interface by forming a engineered layer around the fiber surface to act as a controlled sleeve around the fiber for tailoring purposes. Using this approach, fiber tailoring is done on-site (laboratory or field) to correspond more closely to changing matrix constituents or environmental conditions.

2.2 Processing and testing of nanoparticle ECC

A number of mix designs using carbon nanoparticles were mixed, changing the dosage of carbon nanoparticles as percent weight of fiber. Prior to mixing in the ECC matrix, virgin PVA fibers were placed in a bath of C_{60} carbon nanoparticles to facilitate coating of the fibers by a structure of the fullerene particles. The structure of individual C_{60} particles and the structure of these fullerene particles on the surface of the PVA fiber are shown in Figure 3.

Following the coating of the fibers, they are mixed within the ECC matrix using normal mixing equipment.

The amount of cabon nanoparticles of the surface of the fibers was controlled using weight measurements of the bulk fiber after coating and additional bath coating or hydro-mechanical removal of the particle structures. The nanoparticles mix designs used in this research are shown in Table 1.

These mix designs were evaluated using single fiber pullout testing procedures (detailed by Redon et al. 2001) to construct the crack bridging stress versus crack width opening behavior. From single fiber pullout behavior, the σ–δ relationship is constructed using Equation 3 (Li 1992).

$$\sigma(\delta) = \frac{V_f}{A_f} \int_{\phi_0}^{\phi_1} \int_{z=0}^{(L_f/2)\cos\phi} P(\delta, L_e) g(\phi) p(\phi) p(z) dz d\phi$$

(3)

where, σ is the crack bridging stress, δ is the crack opening, V_f is the fiber volume fraction, A_f the cross sectional area of the fiber, L_f is the fiber length, ϕ

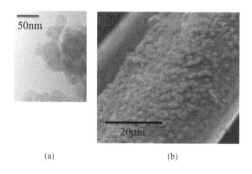

Figure 3. (a) Nanoparticle structure (courtesy of UD-NEST) (b) nanoparticle structures on PVA fiber surface after deposition.

Table 1. Mix proportions (kg/L) for ECC (M45) and nanoparticle ECC (NECC) materials.

	M45	NECC-1	NECC-2	NECC-3
C	0.578	0.578	0.578	0.578
S	0.462	0.462	0.462	0.462
W	0.319	0.319	0.319	0.319
FA	0.693	0.693	0.693	0.693
PVA	0.026	0.026	0.026	0.026
HRWR	0.00751	0.00751	0.00751	0.00751
CN	-	0.0021	0.0034	0.0055

C—Cement; S—Sand; W—Water; FA—Fly Ash; PVA—poly-vinyl-alcohol fiber; HRWR—High Range Water Reducer; CN—carbon nanoparticles.

Table 2. Strain hardening potential for M45ECC and nanoparticle ECC mix designs.

Mix	J'_b/J_{tip}	$\varepsilon(\%)$
M45	1.1	3.6
NECC-1	1.3	3.8
NECC-2	0.8	1.2
NECC-3	0.4	1.1

is the fiber orientation angle, P is the fiber pullout force, L_e is the fiber embedment length, and $g(\phi)$ is the fiber snubbing as a function of fiber orientation angle. Using this 3-D integration of all possible fiber orientations and embedment lengths, the σ–δ curve can be constructed from single fiber pullout (P–δ) test data.

Using Equation 1 to compute J'_b and Equation 3 to construct the σ–δ relationship, the values of J'_b/J_{tip} (also known as strain hardening potential) can be computed for the various mix designs in Table 1. J_{tip} was measured using notched beam fracture tests of NECC mortar. Strain hardening potential values are given in Table 2. Also shown in Table 2 are the average tensile ductility measurements (uniaxial tension tests) for 10 uniaxial tension specimens for each mix.

The uniaxial tension performance of a representative nanoparticle ECC material is shown in Figure 4.

As shown in Table 2 and Figure 4, the use of carbon nanoparticles can be used to tailor the interface of PVA fibers within ECC material and result in an improved performance over previous ECC M45 mix designs that are used for large-scale infrastructure applications. Using carbon nanoparticle interface tailoring, the critical strain hardening potential ratio (J'_b/J_{tip}) of nanoparticle ECC was increased by 18% over the M45 value. An accompanying increase in average tensile strain capacity was also observed.

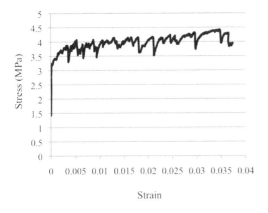

Figure 4. Uniaxial tension stress-strain curve of nano-particle ECC material.

3 PRODUCTION OF CARBON NANOPARTICLES (CNP)

While the use of carbon nanoparticles for fiber-matrix interface tailoring can lead to improved performance of ECC materials, this in itself is of little value. The objective of this research was to leverage the improved performance of nanoparticle ECC materials in the construction of sustainable infrastructure systems. To determine this, the production and use intensity of carbon nanoparticles is quantified.

There are number of methods that are currently in use to produce carbon nanoparticles. (In this discussion we will be looking at both fullerene and nanotube synthesis). As noted by Kushnir and Sandén (2008) the application of carbon nanoparticles has resulted in a number of new technologies for energy storage, energy conversion, and use. These are considered central to large-scale climate mitigation strategies. Supporting these efforts, carbon nanoparticle production has reached 400 metric tons annually, with rapid growth continuing.

Meeting this worldwide demand, CNP producers are using a number of production methods. These include:

1. Fluidized Bed CVD
2. Floating Cataylst CVD
3. HiPco
4. Pyrolysis
5. Electric Arc
6. Laser Ablation
7. Solar Furnace

Fluidized bed chemical vapor deposition (CVD), floating catalyst CVD, and HiPco methods all use gas fuel as a feedstock for carbon (methane, benzene, and carbon monoxide, respectively). Pyrolysis tech-

niques use toluene as a feedstock. Electric arc, laser ablation, solar furnace processes use graphite as a feedstock. Currently, these methods are also producing a varied set of carbon nanoparticle and nanotube products of varied structural quality. Currently, only pyrolysis is used to produce carbon nanoparticles. The other six processes are used to produce primarily single walled or multi-walled carbon nanotubes. For a complete review of these processes and the associated impacts see Kushnir and Sandén (2008).

The production of CNPs using most of these methods requires large amounts of primary energy (both feedstock and process). The comparative energy requirements for production of CNPs for each of these processes is shown in Table 3.

Currently, pyrolysis is the production method most commonly used for the large-scale production of carbon nanoparticles. Due to the large volume of carbon nanoparticles needed for the infrastructure application considered in this research, pyrolysis is indentified as the only commercially viable supply process for infrastructure CNPs at this time. The energy requirements and associated greenhouse gas emissions associated with CNP production through the pyrolysis process are shown in Table 4.

Even though only a small amount of CNPs are used in the production of nanoparticle ECC mate-

Table 3. Production energy intensity for CNPs using various production techniques (from Kushnir and Sandén, 2008).

Production	Thermal Energy (MJ/kg)	Eletrical Energy (MJ/kg)
Fluidized Bed CVD	328	626
Floating Catalyst CVD	295	187
HiPco	47	5769
Pyrolysis	6341	749
Electric Arc	300	2178
Laser Ablation	211	9424
Solar Furnace	300	142

Table 4. Production energy and GHG intensity for CNPs using pyrolysis production techniques (energy values from Kushnir and Sandén, 2008).

Process Phase	Energy (MJ/kg)	Global Warming Potential (kg CO_2-eq/kg)
Net Feedstock	6344	152
Gas Purification	506	98
Compression	72	14
Purification Processes	168	15
Total	7090	279

Table 5. Primary energy and global warming potential emissions per liter of ECC and NECC material.

Mix	Energy (MJ/L)	Global Warming Potential (kg CO$_2$-eq/L)
M45	6.7	0.67
NECC-1	21.6	1.256
NECC-2	30.7	1.618
NECC-3	45.7	2.204
Concrete*	2.5	0.373

* Concrete Mix Proportions (kg/L): Cement—0.39; Sand & Aggregate—1.717; Water—0.166.

rials, their impact on primary energy consumption and global warming potential emissions for the overall composite material is large. This is due to the very high energy requirements and greenhouse gas emissions per kilogram associated with pyrolytic CNP production. The total primary energy requirements and global warming potential emissions per liter of ECC nanoparticle material for the four mix designs shown in Table 1 are shown below in Table 5.

As demonstrated by Table 5, the use of nanoparticles for the production of more sustainable ECC *materials* is not possible. However, as noted in Section 1.1, economic, social and environmental impacts occur throughout an *infrastructure system's* life cycle, which encompasses resource extraction, materials processing, construction, use, maintenance and repair, and end-of-life management. Therefore, the effect of nanoparticle ECC materials on overall infrastructure sustainability should be evaluated using a comprehensive life cycle assessment framework.

4 LIFE CYCLE MODELING OF NANOPARTICLE ECC PAVEMENT OVERLAYS

As noted previously, Zhang et al. (2008) investigated the overall sustainability of unbonded concrete overlays, HMA overlays, and ECC overlays using economic, environmental, and social metrics. The comprehensive life cycle assessment carried out herein follows a similar framework and approach.

4.1 *Life cycle model construction and service life prediction*

As detailed by Zhang et al. (2008) and Qian (2007), the suppression of reflective cracking in a rigid pavement overlay application is central to prolonging overlay service life and improving life cycle overlay sustainability metrics. The ability to suppress localized reflective cracking under repetitive

tire loads is characterized in the laboratory by the fatigue performance of ECC material in bending. This connection between overlay performance and flexural fatigue is discussed further in Qian (2007) and Lepech et al. (2008).

The overall life cycle assessment framework for this research (from Zhang et al., 2008) is shown in Figure 5. This assessment includes material acquisition, material processing, construction, use (e.g. regular vehicle energy consumption and emissions, construction congestion emissions, maintenance procedure impacts), and end-of-life phases. For each phase all energy and material inputs are accounted, along with all emissions, wastes, products, and co-products. This analysis is ISO 14040 series compliant.

As mentioned, within this life cycle analysis the ability to determine the structural performance of nanoparticle ECC overlays is essential. Fatigue

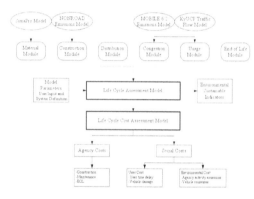

Figure 5. Integrated life cycle assessment model for ECC pavement overlays (from Zhang, 2008).

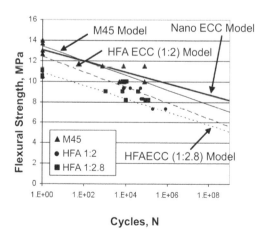

Figure 6. Fatigue curves for ECC M45, nanoparticle ECC, and high volume fly ash (HFA) ECC materials.

modeling was done on nanoparticle ECC materials for comparison with existing ECCs that have undergone extensive fatigue testing. A fatigue prediction model proposed by Lepech et al. (2008) was used to predict the end of service life for ECC nanoparticle pavement overlays. The stress versus load cycle (S-N) curves for a number of ECC materials (including nanparticle ECC) is shown in Figure 6.

Based on Figure 6, nanoparticle ECC shows a distinctly improved behavior in high cycle fatigue (above 1 million cycles). Over the expected 20-year service life of the ECC overlay approximately 5×10^7 equivalent single axel loads (ESALs) are expected. At this number of cycles, nanoparticle ECC materials are capable of withstanding higher stress levels, and thereby can be constructed thinner than conventional concrete, asphalt, or ECC overlays. The designed thickness of overlays using concrete, ECC M45, and nanoparticle ECC materials is shown in Table 6.

The overlay designs analyzed in this study are constructed upon an existing reinforced concrete pavement. The annual average daily traffic (AADT) is approximately 70,000 vehicles with 8% heavy duty trucks (MDOT 1997). In the baseline scenario, the annual traffic growth rate is 0%. The three overlay systems are modeled as 10 km long freeway sections in two directions. Each direction has two 3.6 m wide lanes, a 1.2 m wide inside shoulder, and a 2.7 m wide outside shoulder. The concrete overlay is 180 mm thick and unbonded from the existing pavement by a 25 mm asphalt separating layer. The HMA overlay is 190 mm thick. The nanoparticle ECC is 42 mm thick and constructed on the existing pavement.

4.2 Overlay life cycle modeling results

Total life cycle results represent the environmental impacts from the material module, construction module, distribution module, traffic congestion module, usage module and end of life (EOL) module over a 40-year timeline. The environmental indicators in this study include energy consumption and

greenhouse gas emissions. Overall results are shown in Figure 7 (0% traffic growth), Figure 8 (1% traffic growth), and 9 (variable traffic growth).

In Figures 7 and 8, life cycle emissions are broken down into life cycle phase. As seen, the majority of emissions come from materials production, usage (increased emissions from vehicles riding on rougher pavements), and congestion (increased emissions from vehicles sitting in construction congestion). From Figures 7a and 7b, even though NECC overlays require less construction material and provide better fatigue life, they are more energy or greenhouse gas intensive than a traditional concrete overlay due to the higher energy and greenhouse gas intensity of material production.

Figures 8a and 8b begin to show the importance of traffic growth. In these figures, 1% annual traffic growth is considered. It is here that the improved fatigue performance of the NECC materials is leveraged due to higher traffic levels at the end of service life. As compared to the 0% growth baseline model, NECC is improved when compared with concrete and asphalt. As seen in Figures 9a and 9b, the improved fatigue properties of NECC are further leveraged with 2% traffic growth.

These results demonstrate that for the high traffic volume roadway considered (70,000 vehicles daily) both concrete and NECC materials perform more

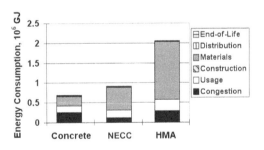

Figure 7a.   Life cycle distribution of energy consumption for 0% traffic growth.

Table 6.   Stress in pavement overlay at 5×10^7 ESALs and required overlay thickness.

Material	Stress at 5x10⁷ Cycles (MPa)	Overlay Thickness (mm)
M45	8.0	52
NECC-1	8.8	42
Concrete	2.3	180*

* Concrete thickness taken from Department of Transportation Design Guidelines.

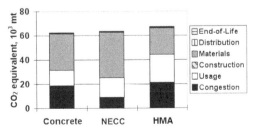

Figure 7b.   Life cycle distribution of greenhouse gas emissions for 0% traffic growth.

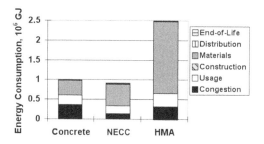

Figure 8a. Life cycle distribution of energy consumption for 1% traffic growth.

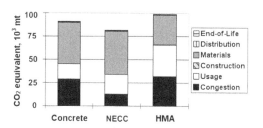

Figure 8b. Life cycle distribution of greenhouse gas emissions for 1% traffic growth.

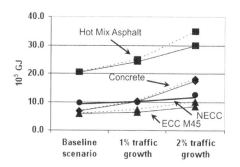

Figure 9a. Pavement overlay life cycle energy consumption as a function of traffic growth rate.

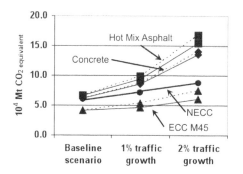

Figure 9b. Pavement overlay life cycle greenhouse gas emissions as a function of traffic growth rate.

sustainably than asphalt. This finding is application specific and would change with lower daily traffic counts. At 0% traffic growth, the use of higher performing NECC materials does not outweigh the increased energy and greenhouse gas intensity of these materials in comparison to traditional concrete. At 1% traffic growth the comparison between concrete and NECC is essentially even. At higher growth rates, NECC materials do become a more sustainable overlay pavement solution when considering the complete pavement overlay life cycle under increasing larger traffic counts.

5 CONCLUSION

The construction and use of transportation infrastructure poses significant sustainability challenges including primary energy consumption and greenhouse gas emissions. To mitigate long-term energy depletion and climate change impacts a new version of ECC incorporating carbon nanoparticles was developed. While these new materials have significantly higher energy and greenhouse gas intensities per liter, when viewed over the full life cycle of a pavement overlay their improved performance can be leveraged to reduce total energy consumption and greenhouse gas emissions for large volume traffic growth scenarios. While significant challenges exist in using these materials in field applications, for countries with double-digit traffic growth patterns (i.e. India or China) the use of advanced materials could serve an important role in long-term energy conservation and climate change mitigation.

REFERENCES

Horvath, A. & Hendrickson, C. 1998. Comparison of Environmental Implications of Asphalt and Steel-Reinforced Concrete Pavements. *Trans. Research Record.* 1626(1): 105.

Kanda, T., Hiraishi, H. & N. Sakata, 2003. "Tensile Properties of ECC in Full Scale Production," *FRAMCOS-5*, Vail, Colorado, pp. 1013–1020.

Kushnir, D. & Sandén, B. 2008. "Energy Requirements of Carbon Nanoparticle Production" *J of Ind. Ecology* 12(3): 360.

Küter, A. 2009. "Management of Reinforcement Corrosion" PhD Thesis. Technical University of Denmark. Lyngby.

Lepech, M., Li, V.C. & Keoleian, G. 2005. "Sustainable Infrastructure Material Design", *4th Int Workshop on LCCA and Design of Civil Infrast Sys*, Cocoa Beach, FL, May 11–17.

Lepech, M.D. & Li, V.C. 2009. "Permeability of Cracked Engineered Cementitious Composites" *Cement and Concrete Composites* (accepted) Como, Italy. June 10–14, 2008.

Lepech, M.D., Keoleian, G.A., Qian, S., Li, V.C. 2008. "Design of Green Engineered Cementitious Composites for Pavement Overlay Applications" *IALCCE08.* Varenna, Lake Como, Italy. June 10–14, 2008.

Li, V.C. & Wu, H.C. 1992. "Conditions for Pseudo Strain-Hardening in Fiber Reinforced Brittle Matrix Composites", J. App. Mech. Review, 45(8), pp. 390–398.

Li, V.C., Wu, C., Wang, S., Ogawa, A. & Saito, T. 2002. "Interface Tailoring for Strain-hardening PVA-ECC," ACI Materials Journal, 99(5), pp. 463–472

Li, V.C. & Leung, C.K.Y. "Theory of Steady State and Multiple Cracking of Random Discontinuous Fiber Reinforced Brittle Matrix Composites" *ASCE Journal of Engineering Mechanics*, Vol. 118, No. 11, 1992. pp. 2246–2264.

Marshall, D.B. & Cox, B.N. 1988. "A J-integral Method for Calculating Steady-state Matrix Cracking Stresses in Composites" *Mechanics of Materials*, No. 8 pp. 127–133.

MDOT. *1997 Noninterstate Freeway Segments: Deficient Segments-URBAN.* MDOT: Ann Arbor.

Qian, S. 2007. *Influence of Concrete Material Ductility on the Behavior of High Stress Concentration Zones.* PhD Thesis. University of Michigan, Ann Arbor.

Redon, C., Li, V.C., Wu, C., Hoshiro, H., Saito, T. & Ogawa, A. "Measuring and Modifying Interface Properties of PVA Fibers in ECC Matrix," *ASCE Journal of Mat. in Civil Engineering*, Vol. 13, No. 6, Nov./Dec., 2001, pp. 399–406.

TRIP. 2009. Key Facts about America's Road and Bridge Conditions & Federal Funding May 2002 http://www.tripnet.org/NationalFactSheetMay2009.pdf

van Oss, H.G. 2009. Mineral Commodity Summaries; Cement U.S. Geological Survey. January 2009. Washington, D.C.

van Oss, H.G. & Padovani, A.C. 2002. Cement Manufacture and the Environment. Part I: Chemistry and Technology. Journal of Industrial Ecology. 6(1): 89–105.

WBCSD. 2002. Toward a Sustainable Cement Industry. Draft report for World Business Council on Sustainable Development. Battelle Memorial Institute. http://www.wbcsd.org/newscenter/reports/2002/cement.pdf

Zapata, P. and Gambatese, J. 2005. Energy consumption of asphalt and reinforced concrete pavement materials and construction. *Journal of Infrastructure Systems*, 11(1): 9–20.

Zhang, H., Keoleian, G.A. & Lepech, M., 2008. "An Integrated Life Cycle Assessment and Life Cycle Analysis Model for Pavement Overlay Systems." *IALCCE08.* Varenna, Lake Como, Italy. June 10–14, 2008.

Zhang, J. & Li, V.C. 2002. "Monotonic & Fatigue Performance in Bending of Fiber Reinforced ECC in Overlay System," J. of Cement and Concrete Research, 32(3): 415–423.

Advances in Cement-Based Materials – van Zijl & Boshoff (eds)
© *2010 Taylor & Francis Group, London, ISBN 978-0-415-87637-7*

Influence of pumping on the fresh properties of self-compacting concrete

D. Feys
Magnel Laboratory for Concrete Research, Department of Structural Engineering,
Ghent University, Gent, Belgium
Hydraulics Laboratory, Department of Civil Engineering, Ghent University, Gent, Belgium

G. De Schutter
Magnel Laboratory for Concrete Research, Department of Structural Engineering,
Ghent University, Gent, Belgium

R. Verhoeven
Hydraulics Laboratory, Department of Civil Engineering, Ghent University, Gent, Belgium

ABSTRACT: Pumping of concrete is a frequently applied casting process. For traditional concrete, slump losses have been reported in literature, but the real cause is still unknown. In case of self-compacting concrete, it is not known at all how the fresh properties evolve due to pumping. This paper will describe the evolution of the fresh properties of SCC due to pumping operations, in which the velocity is increased step wise. Two different effects modify the fresh properties: structural breakdown and an increase in air content. Both effects cause a decrease in viscosity, which is translated in a lower V-funnel flow time and lower pressure losses during pumping. On the other hand, structural breakdown and the increase in air content have an opposite influence on the yield stress. If structural breakdown dominates, the yield stress decreases, if the effects of the increase in air content dominate, yield stress increases. In the first case, as both yield stress and viscosity decrease, segregation can be provoked. In the second case, due to the increase in yield stress, the filling ability of the SCC is reduced, which can lead to improper filling of the formwork. The results show a trend that the more fluid SCCs tend to segregate and the less fluid SCCs tend to loose even more fluidity. Furthermore, the magnitude of these effects appears to increase with increasing velocity in the pipes.

1 INTRODUCTION

On site, concrete can be placed in the formwork in two different ways: by means of a bucket, inducing a discontinuous casting process, or by means of pumping. In case of pumping, the casting rate can become high and savings in time and labor costs can be achieved. On the other hand, very few studies on the pumping of concrete exist (Kaplan, 2001; Chapdelaine, 2007) and the fundamental understanding of this practical process is not completely achieved yet. Especially in the case of Self-Compacting Concrete (SCC), the research field is still completely open as this type of concrete is more fluid compared to the ordinary vibrated concrete types. As a result, the flow behaviour of SCC in pipes is reported to be different (Feys, 2009).

Although from a scientific point of view, the phenomena occurring during pumping of SCC are not completely understood yet, this casting process is applied daily. Sometimes, it is reported that the fresh properties are significantly influenced by the pumping operation, but also in case of ordinary concrete. This paper will describe the test method used and the results of the influence of pumping on the fresh properties of SCC. Before the description of the pumping tests, a short introduction will be given dealing with the rheological properties of fresh SCC.

2 RHEOLOGY

2.1 *Steady state*

In steady state conditions, during which no time-dependent effects influence the results, fresh concrete in general can be described as a Bingham material, showing a linear relationship between the shear stress (related to the pressure loss) and the shear rate (related to the velocity gradient), according to equation 1 (Tattersall & Banfill, 1983).

$$\tau = \tau_0 + \mu_p \cdot \dot{\gamma} \qquad (1)$$

where: τ = shear stress (Pa)
  τ_0 = yield stress (Pa)
  μ_p = plastic viscosity (Pa s)
  $\dot{\gamma}$ = shear rate (s⁻¹)

As can be seen, at least two parameters are needed to describe the fresh behaviour of concrete: the yield stress, which is the resistance to the initiation of flow and the plastic viscosity, which is the resistance to a further acceleration of the flow.

When comparing ordinary concrete and SCC, it is observed that the yield stress of SCC is much lower in order to achieve the self-compactability and that the viscosity of the SCC is generally higher to assure the segregation resistance of the SCC-mixture (Wallevik, 2003a).

Note that under some circumstances the rheological behaviour of SCC is non-linear, but this is beyond the scope of this paper (Feys et al., 2009).

2.2 Time dependent properties

In time, the obtained rheological properties vary in time, of which the cause can be theoretically divided into three main parts: thixotropy, structural breakdown and loss of workability (Wallevik, 2003b; Wallevik, 2009).

Thixotropy is defined as the reversible breakdown and build-up of connections between small particles in the concrete. The "structuration state (λ)" represents the amount of connections. The lower λ, the less connections, the more fluid the concrete (Roussel, 2006).

Structural breakdown is the known as the disruption of chemical connections under the influence of shear. In contrast to thixotropy, structural breakdown does not show, strictly speaking, any rebuild over time (Tattersall & Banfill, 1983).

Loss of workability represents the increase in number of connections of any type in the concrete, which can no longer be broken by the acting shearing forces. As a result, the structuration state permanently increases and the concrete becomes more stiff. Finally, the chemical bonds become very strong transforming concrete from the liquid to the solid state.

Under influence of increasing shear, the distinction between thixotropy and structural breakdown is very difficult to make and as a result, these effects will be examined together in the discussion in this paper. The general effect of this structural breakdown "in its broad sense", as it is considered in this paper, is that λ shows a certain equilibrium

value for each applied shear rate (except for the very low shear rates), which will be achieved after a certain time. The higher the applied shear rate, the lower the equilibrium value of λ and consequently, the more fluid the concrete. For example, due to a sudden increase in shear rate, the stress shows a (mostly) exponential decrease with time, until equilibrium is reached.

2.3 Air content

In the previous sections, concrete is regarded as a homogeneous suspension. In case the sample of concrete on which the measurements are performed is sufficiently large, this assumption can be justified. But concrete does not only contain solid particles and liquid, it also contains a gas phase: air. The exact influence of air on the rheological properties of fresh concrete is currently still under investigation, but at this moment, some qualitative conclusions can be drawn.

The influence of air in a liquid material (or a suspension) is governed by the capillary-number (Ca), which is the ratio of the shearing forces to the surface tension forces (eq. 2) (Rust & Manga, 2002).

$$Ca = \frac{d \cdot \mu_a \cdot \dot{\gamma}}{\Gamma} \qquad (2)$$

where: Ca = capillary-number (-)
  d = bubble diameter (m)
  μ_a = apparent viscosity (Pa s)
  Γ = surface tension (N/m)

In case the Ca-number is low (<1), the shearing forces are not sufficiently high to overcome the surface tension and the bubble remains spherical. As a result, the flow resistance increases with increasing air content. In the other case where Ca > 1, the bubbles deform due to the shearing forces and they align in the flow direction. Consequently, the flow resistance decreases with increasing bubble content.

3 PUMPING TESTS

3.1 Concrete pump

The concrete pump used is a standard available truck-mounted piston pump, depicted in figure 1. Inside the pump, two cylinders alternately push concrete in the pipes or pull concrete from the reservoir. Once the first cylinder is empty and the second is full, a powerful valve changes the connection between the pipes and the cylinders. The operator of the pump can vary the discharge in 10 discrete steps from 4–5 l/s (step 1) to 40 l/s (step 10). For safety reasons, step 5 has never been exceeded in the tests.

Figure 1.   Concrete piston pump.

Figure 2.   105 m loop circuit.

Figure 3.   Pressure sensor and strain gauges.

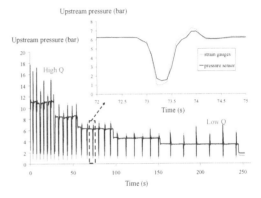

Figure 4.   Determination of discharge by measuring the time needed for a certain amount of strokes. The time between two vertical spikes represents the emptying of one cylinder. The evolution of the pressure during the change of the valve is shown in the inset.

3.2   Circuit

Behind the pump, a 81 m or 105 m (figure 2) loop circuit has been installed by means of steel pipes with an inner diameter of 106 mm. The circuit consisted of five horizontal sections, of which three were instrumented, and an inclined part. At the end of the circuit, the concrete falls inside a reservoir, which can be closed for sampling and discharge calibration, but which is mostly open. In case the reservoir is open, the concrete falls back inside the reservoir of the pump and is ready to be re-used.

3.3   Instrumentation

In the last horizontal section of the circuit, two pressure sensors were installed in order to measure the pressure loss (figure 3). In three of five horizontal sections, including the section with the pressure sensors, strain gauges were attached to the outer pipe wall, recording the expansion and contraction, which is related to the local pressure

(Kaplan, 2001). Only the results of the last horizontal section will be discussed in this paper.

Discharge is somewhat more complicated to measure as there is no direct tool available. On the other hand, due to the pumping mechanism, a pressure drop is observed each time the valve of the pump changes the connection (figure 4). Between two pressure drops, the total volume of one cylinder is pushed inside the pipes (which is called a stroke) and by measuring the time needed for a certain amount of strokes, discharge can be easily determined.

3.4   Concrete

During this part of the research program, four SCC mixtures were pumped in the circuits described. The total amount of concrete needed per test was 3.25 m³ and consequently, all concretes were

167

produced by a ready-mix company and delivered to the lab in a time span of one hour. Mixtures 15 and 17 were commercial products of the mixing plant. Mixtures 14 and 16 were based on laboratory compositions, containing 697 kg of coarse aggregates (up to 16 mm), 853 kg of sand, 360 kg of CEM I 52.5 N (OPC) and 240 kg of limestone filler. Mixture 14 contained 160 l of water, while mixture 16 contained 165 l of water per m³ of concrete. The amount of SP was adopted in order to achieve a target slump flow of 650 mm in case of mixtures 14 and 15 and 700 mm in case of mixtures 16 and 17. The SP were PCE-based, showing a long workability retention. For the commercial mixtures, the same sand, limestone filler and SP were used.

3.5 *Testing procedure*

In order to study the influence of pumping on the fresh properties of SCC, a special testing procedure was developed. It consists of three sub-cycles, repeated five times, each time increasing the maximum discharge. The sub-cycles are divided in three parts:

– Maintaining discharge constant until an equilibrium in pressure loss is observed. This can take more than 10 minutes, especially at the low discharges.
– Discharge calibration and sampling. The concrete samples were used to determine the rheological properties and to execute the standard tests on SCC, like slump flow, V-funnel, sieve stability, air content and density measurements.
– A stepwise decreasing discharge curve, maintaining each discharge for five full strokes, starting from the discharge in the first part of the sub-cycle. This procedure takes in average 2 minutes. At discharge step 1, which is the lowest discharge, no decreasing discharge curve was determined. After the decreasing discharge curve, the maximal discharge is increased by one step and the sub-cycle is repeated, as shown in figure 5.

This procedure was executed once for mixtures 14, 16 and 17 and twice for mixture 15. For mixtures 14 and 16 and the first test on mixture 15, the discharge did not exceed step 4. For mixture 16, a small variation was applied, by repeating the step at discharge 3 for three times (step 3, step 3bis and step 3ter).

4 RESULTS

4.1 *Pressure loss—discharge curves*

Plotting the results of pressure loss as a function of discharge, for each equilibrium point at each discharge, and all downward curves reveals that the pressure loss at a certain discharge decreases when a discharge is applied before, which can be seen in figure 6 for mixture 14. As a result, if a higher discharge is applied, the flow resistance of the SCC in the pipes decreases. The results of the other SCC mixtures are very similar.

4.2 *Rheological measurements*

The results of the rheometer tests executed on the sampled concrete indicate similar results. Although the test results are not always reliable, the general trend shows a decrease in plastic viscosity and in some cases an increase in yield stress. This confirms the pumping results as in case of SCC, the pressure loss is mainly dependent on the viscosity of the concrete (Feys, 2009).

4.3 *Tests on fresh SCC*

The tests on fresh SCC indicate for all mixtures a decrease in V-funnel flow time. For all mixtures, except for mixture 14, an increase in air content and

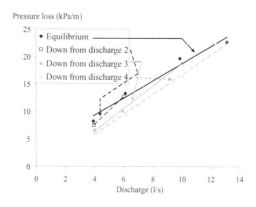

Figure 6. Pressure loss as a function of discharge, showing a lower pressure loss at a certain discharge if a higher discharge was applied before. Results from mixture 14.

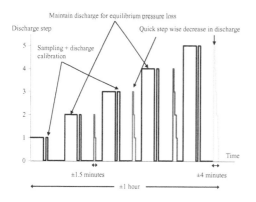

Figure 5. Testing procedure.

Table 1. Overview of the tests on fresh concrete for all mixtures.

	Mix 14			
	Q1	Q2	Q3	Q4
Age (hour)	2:30	2:45	3:00	3:30
Tests on fresh SCC				
Slump flow (mm)	818	758	745	658
V-Funnel (s)	5.23	6.1	3.65	5.82
Sieve Stability (%)	10.9	11.0	14.2	7.5
Air content (%)	1.6	1.8	1.6	1.5

	Mix 15 - test 1			
	Q1-1	Q1-2	Q1-3	Q1-4
Age (hour)	1:30	1:45	2:00	2:10
Tests on fresh SCC				
Slump flow (mm)	645	625	660	570
V-Funnel (s)	5.43	4.18	3.77	3.42
Sieve Stability (%)	4.2	7.0	6.6	4.0
Air content (%)	2.1	2.4	3.2	4.2

	Mix 15 - test 2				
	Q2-1	Q2-2	Q2-3	Q2-4	Q2-5
Age (hour)	2:50	3:00	3:10	3:20	3:30
Tests on fresh SCC					
Slump flow (mm)	525	543	505	498	445
V-Funnel (s)	3.54	3.06	3.29	3.46	3.74
Sieve Stability (%)	3.4	4.5	1.9	0.8	0.3
Air content (%)	3.7	3.9	4.6	5.0	6.2

	Mix 16					
	Q1	Q2	Q3	Q3 bis	Q3 ter	Q4
Age (hour)	2:35	2:45	3:00	3:10	3:20	3:30
Tests on fresh SCC						
Slump flow (mm)	670	675	655	585	620	535
V-Funnel (s)	5.24	4.02	4.78	3.72	3.76	3.89
Sieve Stability (%)	8.7	12.7	6.9	6.8	7.8	5.7
Air content (%)	1.1	1	1.4	1.3	2.2	3.9

	Mix 17				
	Q1	Q2	Q3	Q4	Q5
Age (hour)	1:20	1:30	1:40	1:50	2:00
Tests on fresh SCC					
Slump flow (mm)	785	780	750	765	750
V-Funnel (s)	3.39	3.08	2.66	2.35	2.22
Sieve Stability (%)	10.5	-	11.7	15.6	18.5
Air content (%)	1.4	1.9	3.1	3.9	4.9

a decrease in density is measured. Mixtures 15 and 16 show a decreasing slump flow, while for mixture 17, slump flow remains constant. On the other hand, the sieve-(un)stability value appears to increase for mixture 17. The detailed results of all tests on fresh SCC can be found in table 1. All results are qualitatively in accordance with the rheometer results and the results of the mimninp tests.

5 DISCUSSION

5.1 Structural breakdown

As mentioned in section 2.2, the equilibrium internal structure of the concrete decreases with increasing shear rate. As a result, the concrete becomes more fluid with increasing maximal discharge applied. This result is confirmed by the decreasing pressure losses, decreasing viscosity and decreasing V-funnel flow time. For mixture 17, the structural breakdown theory provides the ability to explain the observed segregation. Due to the decrease in viscosity and the constant yield stress, SCC becomes more sensitive to segregation. On the other hand, the structural breakdown theory is not capable of explaining the effect of increase in yield stress for mixtures 15 and 16.

5.2 Air content

For all mixtures, except mixture 14, the air content increases with increasing discharge up to values of around 5–6%. For these values, the importance of the air bubbles on the rheological properties is no longer negligible, and as a result, the theory presented in section 2.3 should be applied.

Analysis has shown that the viscosity is determined at high Ca-numbers and as a result, it should decrease with increasing air content, as can be seen in figure 7 (Feys et al., 2009b). The yield stress in case of SCC on the other hand is determined at low Ca-numbers for the air bubbles sizes measured (on the hardened concrete). Consequently, the yield stress should increase with increasing air content, which is also visible in figure 7 (Feys et al., 2009b).

As a result, the air content theory is capable of explaining both the decrease in viscosity and increase in yield stress. On the other hand, it is not applicable to mixture 17, as it does not predict any increasing sensitivity for segregation.

From a practical point of view, an increase in yield stress can lead to a decrease of filling ability of the SCC, resulting in the not perfect filling of a formwork.

5.3 Combination of effects

Both effects of structural breakdown and increasing air content act simultaneously on the concrete. According to both theories, the viscosity decreases, but depending on the initial fresh properties of the concrete, the yield stress after pumping can evolve in two ways, as observed in the experiments. In case of SCC with a rather high slump flow, the

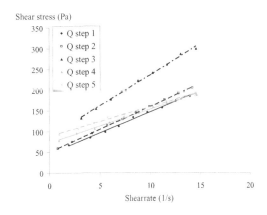

Figure 7. Rheological curves for mixture 15—test 2, showing a clear decrease in viscosity (inclination) and increase in yield stress (intersection with shear stress axis), with increasing discharge step. Note that the results for step 3 are not reliable.

structural breakdown theory is more dominant and an increased danger for segregation is noticed. In case the initial slump flow is low, the air content theory is more dominant and the yield stress increases.

From the restricted amount of results, it is also observed that the magnitude of the effects increases with increasing pumping velocity. As a result, in order to minimize the effects, pumping should be performed at low velocities.

5.4 Mixture 14

Mixture 14 shows different results compared to the other mixtures because no increase in air content is observed. The specific reason for this behaviour can be the very high amount of SP applied. As this mixture contained less water than mixture 16, and approximately the same slump flow was targeted, it contained more than the double of the amount of SP applied in mixture 16. Possibly, this amount of SP increases the surface tension of the air bubbles in the concrete, making it more difficult to deform (as Γ increases in equation 2).

On the other hand, due to large problems during insertion of the concrete, the concrete age was quite elevated (150 min) at the beginning of the test. As a result, some effects of loss of workability can affect the measured results. When omitting these possibly affected results (Q4 in table 1), the evolution of the concrete obeys the structural breakdown theory.

6 CONCLUSIONS

The fresh properties of SCC are described by its rheological values: yield stress and plastic viscosity. The yield stress is the resistance to the initiation of flow, while the plastic viscosity is the resistance to a further acceleration of flow.

By means of full-scale pumping tests in quite long circuits, it is shown that the fresh properties are affected by pumping. A special testing procedure was developed in order to investigate these effects.

From the full-scale pumping tests, it is observed that the pressure loss at a certain discharge decreases when a higher discharge is applied before. The rheometer results and the tests on fresh concrete confirm a decrease in plastic viscosity, but the yield stress can evolve in two different ways.

Two causes are found to influence the rheological properties of SCC during pumping: structural breakdown and an increase in air content. Due to structural breakdown, both yield stress and viscosity should decrease and the danger for segregation increases. According to the increase in air content, viscosity should decrease, but the yield stress must increase, which can lead to an improper filling of the formwork. Both effects appear to become more important with increasing pumping velocities.

ACKNOWLEDGEMENTS

The authors would like to thank the Research Foundation in Flanders (FWO) for the financial support of the project and the technical staff of both the Magnel and Hydraulics Laboratory for the preparation and execution of the full-scale pumping tests.

REFERENCES

Chapdelaine F., 2007, *Fundamental and practical study on the pumping of concrete*, Ph-D-thesis (in French), Universite Laval.

Feys D., 2009, *Interactions between rheological properties and pumping of self-compacting concrete*, Ph-D-thesis, Ghent University.

Feys D., Verhoeven R., De Schutter G., 2009a, Why is fresh self-compacting concrete shear thickening?, *Cem. Conc. Res. 59*, pp. 510–523.

Feys D., Roussel N., Verhoeven R., De Schutter G., 2009b, Influence of the air content on the steady state rheological properties of fresh SCC, without air-entraining agents, Proc. of the 2nd intern, symp. on design, performance and use of SCC, Beijing, pp. 287–295.

Kaplan D., 2001, *Pumping of concretes*, Ph-D-thesis (in French), Laboratoire Central des Ponts et Chaussées, Paris.

Roussel N., 2006, A thixotropy model for fresh fluid concretes: theory, validation and applications, *Cem. Conc. Res. 36*, pp. 1797–1806.

Rust A.C., Manga M., 2002, Effect of bubble deformation on the viscosity of dilute suspensions, *J. Non-Newt. Fluid Mech. 104*, pp. 53–63.

Tattersall G.H., Banfill P.F.G., 1983, *The rheology of fresh concrete*, Pitman, London.

Wallevik O.H., 2003a, *Rheology—A scientific approach to develop self-compacting concrete*, Proc. of the 3rd intern. symp. on SCC, Rejkjavik, pp. 23–31.

Wallevik J.E., 2003b, *Rheology of particle suspensions Fresh concrete, mortar and cement paste with various types of lignosulfonates*, The Norwegian University of Science and Technology, Trondheim.

Wallevik J.E., 2009, Rheological properties of cement paste Thixotropic behaviour and structural breakdown, *Cem. Conc. Res. 39*, pp. 14–29.

Durability mechanics of advanced cement-based materials

Advances in Cement-Based Materials – van Zijl & Boshoff (eds)
© *2010 Taylor & Francis Group, London, ISBN 978-0-415-87637-7*

Aspects of durability of strain hardening cement-based composites under imposed strain

F.H. Wittmann
Aedificat Institute Freiburg, Germany
Centre for Durability and Sustainability Studies, Qingdao Technological University, Qingdao, China

T. Zhao, L. Tian, F. Wang & L. Wang
Centre for Durability and Sustainability Studies, Qingdao Technological University, Qingdao, China

ABSTRACT: The designed service life of reinforced concrete structures is not reached in practice too often. One major reason for this well-known economical and ecological problem is the limited strain capacity of cement-based materials under tensile stress. By adding PVA fibers to cement-based composites their strain capacity can be improved by a factor of up to several hundred. Multi crack formation is at the origin of the pseudo-ductility of strain hardening cement-based composites (SHCC). However, the high strain capacity is beneficial with respect to durability only if the cracks formed under imposed strain do not lead to significantly increased ingress of aggressive compounds such as chlorides or sulfates. In this contribution experimental results of measurement of capillary suction and penetration of chloride ions into SHCC under different imposed strain will be presented and discussed. The influence of hydrophobic treatment has been studied in particular.

1 INTRODUCTION

Limited durability of concrete structures has become a serious economical and ecological problem in many countries by now. Service life of reinforced concrete structures can be limited by diverse mechanisms such as carbonation, penetration of aggressive compounds and leaching or frost action. In practice, however, several deteriorating mechanisms act simultaneously and damage is accelerated by synergetic effects (see e.g. Zhao et al. 2005). Most damage mechanisms are based on mass or heat transport. Uptake of liquids and compounds dissolved in liquids is controlled by capillary suction if the surface is in contact with the corresponding liquid. Capillary suction is a comparatively efficient transport process and dissolved ions can be transported deep into the material in a short period. Diffusion of moisture and ions leads to slow but deeper penetration depth at a later stage.

New pore space is formed by crack formation and this new space is available for transport of liquids. Therefore it may be expected that service life of SHCC in aggressive environment may be reduced under imposed strain although no real cracks are being formed. It has been shown by means of neutron radiography recently that the fracture process zone of cement-based materials may absorb more water than the undamaged material (Zhang et al. 2009 a).

In this contribution the influence of imposed strain on capillary suction of water and salt solutions shall be investigated. It is well known that water repellent treatment of concrete can reduce capillary suction significantly (Zhan et al. 2003). For this reason capillary suction of integral water repellent SHCC has also been studied. The influence of addition of a water repellent agent to SHCC on capillary suction has been studied first by Martinola et al. (2002) and Martinola et al. (2004). Results have later essentially been confirmed by Sahmaran & Li (2009).

Characteristic properties of SHCC make this comparatively new material of interest for diverse applications. Because of the high damping capacity applications in seismic regions are obvious and have been a strong driving force for the materials technology. Protective layers on concrete structures are usually exposed to considerable hygral and thermal gradients. The limited strain capacity of ordinary Portland cement mortar cannot absorb the imposed strain without crack formation. Cement-based coatings made with SHCC have been successfully applied on large dams or on irrigation canals as repair or protective coatings. A totally different field of potential applications is the extension of service life in aggressive environment. As crack formation is a dominant mechanism of degradation of normal reinforced concrete structures sheets of SHCC are used in bridge construction. Local repair of damaged concrete structures may also take advantage of the enormous strain capacity of SHCC.

For structural design the ultimate strain capacity of a given type of SHCC is of primary interest. For service life design and improved durability it has to be shown which strain may be tolerable without loss of tightness with respect to water and in water dissolved ions.

One major aim of this contribution is to point out that for durability reasons the ultimate strain capacity may not necessarily be the correct design value. The desired service life may in many cases not be reached if durability design is based on this value.

2 CAPILLARY SUCTION

If the surface of porous materials such as concrete, mortar or burnt clay bricks is in contact with water or any other wetting liquid, the liquid will be absorbed by capillary action. In the simplest case, as for instance one single capillary with radius r, the absorbed amount of water as function of time can be described by means of the following equation:

$$\Delta W = A\sqrt{t} \tag{1}$$

with A being the coefficient of capillary suction $[kg/m^2 \, s^{1/2}]$. It can be shown that A has the following physical meaning:

$$A = \psi_\rho \sqrt{\frac{r_{eff}\,\sigma\cos\Theta}{2\eta}} \tag{2}$$

In equation (2) Ψ stands for the water capacity of the absorbing material $[m^3/m^3]$, that means the volume within the pore space, which can be filled by capillary suction, ρ stands for the density of the absorbed liquid $[kg/m^3]$, r_{eff} designates an effective radius $[m]$, representing the pore size distribution of a given material, σ is the surface tension $[Nm/m^2]$ and Θ the wetting angle of the liquid, and finally η is the temperature dependent viscosity of the penetrating liquid $[(N\,s)/m^2]$.

The penetration depth of the absorbed liquid as function of time $x(t)$ can be predicted by means of the following equation:

$$x(t) = B\sqrt{t} \tag{3}$$

In equation (3) B is the coefficient of capillary penetration $[m/s^{1/2}]$, and has the following meaning:

$$B = \frac{A}{\psi\rho} \tag{4}$$

Water uptake of an undamaged sample of SHCC can be characterized by the coefficient of capillary suction A and the water capacity Ψ. If the matrix of the material is damaged by imposed strain, both the water capacity Ψ and the effective radius r_{eff} will increase as new micro cracks are being formed. As a consequence the coefficient of capillary suction A will increase. For this reason determination of capillary suction as function of imposed strain may be considered to be a non-destructive test method to quantify the degree of damage induced by combined loads. In the following this method will be used to follow the damage induced to SHCC by imposed strain under tension and compression.

3 EXPERIMENTAL

SHCC with the following composition has been prepared: 715 kg/m^3 Portland cement type 42.5, 306 kg/m^3 Fly ash, 26 kg/m^3 silica fume, 715 kg/m^3 sand with a maximum grain size of 0.3 mm, and 429 kg/m^3 water. This composition results in a water-cement ratio of 0.6 and a water-binder ratio of 0.41. 2% of PVA fibers have been added to the fresh mix. Six specimens could be made in steel molds from one batch. Part of the specimens has been prepared with an addition of 2% silane emulsion in order to obtain integral water repellent SHCC. Specimens were kept under wet burlap for one day and then they were further stored in a humid room (RH > 95%, T = 20°C) until testing at an age of 14 days.

Dumb bell shaped specimens have been produced from the mix. The geometry is shown in Fig. 1.

A strain of 0.5, 1, and 2% has been imposed on the samples at an age of 14 days in a universal testing machine. The number of cracks and their width have been measured along three parallel lines in the zone of unidirectional tensile stress. From the

Figure 1. Shape of the dumb bell specimens for testing under direct tension. Dimensions: width = 90 mm, length = 330 mm and thickness = 30 mm. The size of the volume under uniaxial stress: 120 × 60 × 30 mm (L × W × T).

174

individual values average crack width and standard deviation have been determined. Capillary suction of water has been measured under load by means of a Karsten tube.

After the direct tension test had been finished the end blocks have been cut off by means of a diamond saw. The blocks obtained in this way have the following dimensions: 90 × 65 × 30 mm. On these blocks capillary suction of water and salt solutions has been determined as function of time as described in Zhan et al. (2005). The blocks have then been loaded by compression up to 50% and 100% of the ultimate load. The applied load has been kept constant for 10 minutes. Damage induced by compressive load has been measured by capillary suction.

In addition chloride profiles have been determined as established in undamaged and damaged SHCC, which has been exposed to 5% NaCl solution for 10.75 hours. To obtain the chloride content as function of distance from the surface layers of 1 mm thickness have been ground successively. The chloride content of the powder obtained by grinding was determined by means of an ion sensitive electrode.

4 RESULTS OF TENSILE TESTS

4.1 Stress-strain diagram

The strain hardening of the two SHCC mixes (with and without silane emulsion) has been determined experimentally at an age of 14 days. Typical results are shown in Fig. 2 and Fig. 3. As observed earlier addition of 2% silane emulsion increases slightly the maximum stress but reduces the strain capacity (see e.g. Martinola et al. 2004).

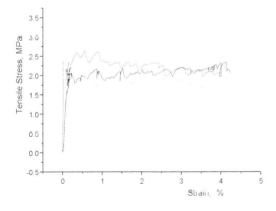

Figure 2. Typical stress-strain diagrams as obtained with neat SHCC produced for this project.

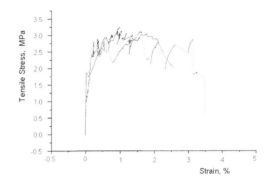

Figure 3. Typical stress-strain diagrams as obtained with SHCC with 2% silane emulsion added.

Table 1. Number of cracks, average crack width, standard deviation, and maximum crack width as function of imposed tensile strain as measured under load und after subsequent unloading on neat SHCC.

Strain %	Under load	Number of cracks	Average crack width μm	Standard deviation μm	Maximum cr. width μm
0.5	yes	12	50.9	17.2	80
0.5	no	12	47.7	15.5	70
1.0	yes	22	55.2	25.2	130
1.0	no	22	49.9	21.5	100
2.0	yes	36	55.6	23.0	130
2.0	no	36	48.7	24.3	120
4.0	yes	42	71.8	26.3	150
4.0	no	42	63.0	27.6	140

4.2 Number and width of cracks

The number of cracks and their width has been measured along three parallel lines in the zone of unidirectional tensile stress. The number of cracks as well as the mean value and the corresponding standard deviation of crack width as observed on neat SHCC as function of imposed strain is shown in Table 1. Results obtained on SHCC to which silane emulsion had been added in the fresh state are shown in Table 2.

The crack width at the same imposed strain has a tendency to be slightly bigger in SHCC to which silane emulsion has been added as compared to neat SHCC. This is also reflected by the lower number of cracks in integral water repellent SHCC.

4.3 Capillary suction as function of imposed strain

First capillary suction of both neat SHCC and integral water repellent SHCC has been measured. Then strain has been imposed by tensile load and

Table 2. Number of cracks, average crack width, standard deviation and maximum crack width as function of imposed tensile strain as measured under load und after subsequent unloading on integral water repellent SHCC.

Strain %	Under load	Number of cracks	Average crack width μm	Standard deviation μm	Max. cr. w. μm
0.5	Yes	13	52.3	28.1	120
0.5	no	13	44.4	22.8	90
1.0	yes	20	61.7	33.7	150
1.0	no	20	52.2	29.0	130
2.0	yes	24	64.4	32.4	170
2.0	no	24	60.7	30.1	150

Table 3. Coefficient of capillary suction A of neat SHCC and A′ of water repellent SHCC under different imposed strain.

Imposed strain %	A g/(m² h^{1/2})	A′ g/(m² h^{1/2})
0	489.1	65.5
0.5	945.4	345.6
1.0	1332.2	406.4
2.0	1469.5	477.0

be determined by fitting with equ. (1). The corresponding values are compiled in Table 3. It can be seen that the amount of water absorbed by neat SHCC after one hour of contact with the liquid is more than seven times higher as compared with the integral water repellent SHCC. In both types of SHCC the absorbed amount of water increases under imposed tensile strain. The related increase is higher in water repellent SHCC but at 2% strain it hardly reaches the value of neat SHCC. Both types of SHCC undergo damage under imposed strain, but the induced micro cracks in integral water repellent SHCC do not contribute to moisture uptake and movement in the same way as they do in neat SHCC.

5 RESULTS OF COMPRESSIVE TESTS

5.1 *Capillary suction*

As mentioned above, after the direct tension test the end blocks of the dumb bell shapes specimens were cut off. Some of these blocks have been loaded under compression up to 50% and 100% of the ultimate load. The ultimate load was found to be about 40 N/mm² for both types of SHCC under investigation. Then capillary suction has been determined in contact with water and with 5% NaCl solution.

Water absorption of neat and integral water repellent SHCC is plotted as function of square root of time in Fig. 6 and Fig. 7 respectively. Application of 50% of the ultimate compressive load leads to a moderate increase of capillary absorption. This is a clear indication of damage induced by the compressive load. Obviously damage of SHCC is much more severe after application of 100% of the ultimate load as shown in Fig. 6. In the case of neat SHCC the water front reached the upper boundary after slightly more than one hour of contact with water. An absorbed amount of water of 5300 g/m² corresponds approximately to the water capacity of the damaged SHCC.

As can be seen from Fig. 7, integral water repellent SHCC absorbs considerably less water than

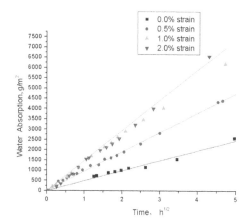

Figure 4. Capillary absorbed water of neat SHCC as function of suction time.

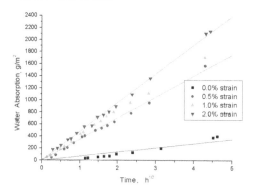

Figure 5. Capillary absorbed water of integral water repellent SHCC as function of suction time.

capillary suction has been measured again. The amount of water absorbed is plotted in Fig. 4 as function of square root of time for neat SHCC and in Fig. 5 for integral water repellent concrete.

From the experimental data shown in Fig. 4 and Fig. 5 the coefficient of capillary suction A can

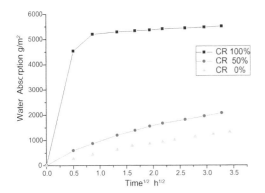

Figure 6. Capillary suction of water by neat SHCC in the unloaded stage and after compressive loading up to 50% and 100% of the ultimate load.

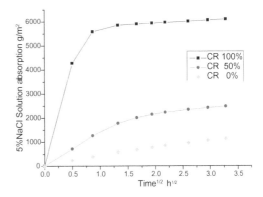

Figure 8. Capillary suction of 5% aqueous NaCl solution by neat SHCC in the unloaded stage and after compressive loading up to 50% and 100% of the ultimate load.

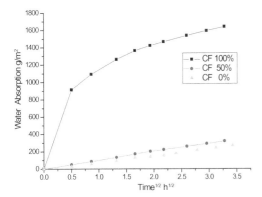

Figure 7. Capillary suction of water by integral water repellent SHCC before loading and after for ten minutes with 50% and 100% of the ultimate load.

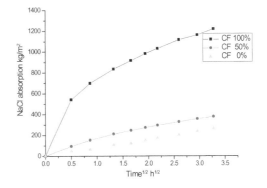

Figure 9. Capillary suction of 5% aqueous NaCl solution by integral water repellent concrete before and after compressive loading.

neat SHCC per unit of time. After application of 100% of the ultimate load for ten minutes capillary water absorption after about 10 hours has reached about one third of the water capacity of the untreated material only. Even at very severe damage of SHCC water repellent treatment can reduce ingress of water by capillary suction significantly.

50% of the ultimate load is rather high. Therefore in the future tests shall be run with lower load levels. In this way the limit load shall be determined which may be applied without significant damage to the material. For structures in an aggressive environment this limit should not be overcome during their life time, otherwise the service life may be shortened considerably. Details will be calculated with an advanced model for durability design.

On specimens similar to those, which were used for water absorption (see Figs. 6 and 7), capillary absorption of 5% NaCl solution has been measured. Results are shown in Fig. 8 and Fig. 9.

Similar to what can be seen in Fig. 6 the volume of liquid, which can be absorbed by capillary suction, is reached just after slightly more than one hour of contact with the salt solution. The higher mass, which is absorbed in contact with 5% NaCl solution, is partly due to the higher density of the salt solution. In this case damage induced by 50% of the ultimate compressive load was slightly more severe than in the tests with water (see Figs. 6 and 7). This is most probably due to normal scatter of experimental results.

Again, absorption of liquid by integral water repellent SHCC is significantly lower as compared with neat SHCC. That means that more salt is transported into the undamaged and damaged neat SHCC and more salt is transported into deeper zones. This will be further elucidated by tests for longer suction time. Furthermore it is planned to expose strain damaged SHCC to natural sea water.

In Table 4 the coefficients of capillary suction as obtained from the data shown in Fig. 8 and Fig. 9

Table 4. Coefficient of capillary suction A of neat SHCC and A′ of water repellent SHCC for water and 5% NaCl solution after loading by compression for ten minutes.

Absorbed liquid	Load %	A g/(m² h^{1/2})	A′ g/(m² h^{1/2})
Water	0	487	65
	50	958	99
	100	9'050	1'840
5% NaCl solution	0	445	78
	50	1'383	166
	100	14'050	1'060

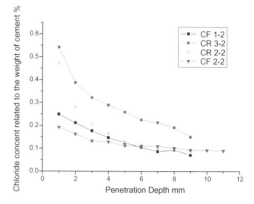

Figure 10. Chloride profiles as measured in neat SHCC (CR 3–2 and CR 2–2) and in integral water repellent SHCC (CF 1–2 and CF 2–2) after contact of the surface with 5% NaCl solution for 10.75 hours.

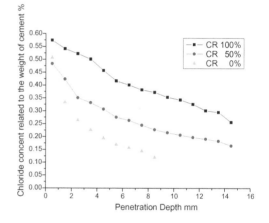

Figure 11. Influence of compressive pre-loading on chloride penetration into neat SHCC after contact with 5% NaCl solution for 10.75 hours.

are compiled. The damage induced by a compressive load of 50% and 100% of the ultimate load can be quantified by these values. Above 50% compressive load the material is seriously damaged and this will have a direct influence on durability. Due to the micro-cracks formed under load water and aggressive ions dissolved in water can penetrate deeper into the material. It will be studied later to which extend strain induced damage can be compensated by water repellent treatment.

5.2 Chloride penetration

Chloride profiles have been determined on neat SHCC and integral water repellent SHCC. The end blocks of the dumb bell shaped specimens have been cut off and one surface has been put in contact with 5% NaCl solution for 10.75 hours. Resulting chloride profiles are shown in Fig. 10.

A small amount of chloride penetrated the integral water repellent SHCC, while within the short period of contact with salt solution a significant amount of chloride has been transported with the water into the pore space of neat SHCC. This result could have been anticipated from the data obtained by measuring capillary suction.

We have observed that capillary suction under compressive load up to 50% and 100% of the ultimate load increases significantly. Hence we may expect that specimens which had been loaded under compression before contact with salt solution would show more chloride and deeper penetration. For neat SHCC this is shown in Fig. 11. Up to a compressive load of 50% a significant increase of the chloride profile can be observed, but at 100% compressive load much more chloride has penetrated. This can be explained by formation of micro-cracks which contribute to the water transport and also increase the water capacity of the damaged material.

This observation is a clear indication of a synergetic effect of mechanical loading and chloride penetration. If the design service life is predicted on the basis of chloride migration values measured in a laboratory and in the unloaded state of SHCC and the real structure or structural element will be in contrast permanently or temporarily under compressive or tensile load the real service life will be significantly shorter than the designed service life.

In Fig. 12 chloride profiles as determined in preloaded integral water repellent specimens of SHCC are shown. For comparison the chloride profile as measured on unloaded specimens is shown in Fig. 12 as well. In this case an applied compressive load of up to 50% of the ultimate load leads to a modest increase of chloride penetration. Water repellent treatment has nearly compensated the strain induced damage. If 100% of the ultimate load has been applied damage of the integral water repellent SHCC becomes so serious that a

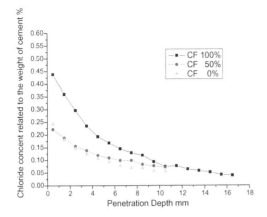

Figure 12. Influence of compressive pre-loading on chloride penetration into integral water repellent SHCC after contact with 5% NaCl solution for 10.75 hours.

significant amount of chloride can penetrate the material in a very short period. From these results we can conclude again that integral water repellent SHCC is not as vulnerable under imposed strain as neat SHCC. Service life of structures under combined mechanical and environmental loads will definitely be extended considerably if integral water repellent SHCC is used.

It has been shown earlier that addition of silane emulsion reduces chloride penetration into concrete but it does not act as a real chloride barrier (Zhang et al. 2009 b). A chloride barrier, however, can be established by means of surface impregnation with silane (Zhan et al. 2005). It has also been shown that capillary suction of cracks in concrete up to a critical crack width can be essentially eliminated by surface impregnation (Guo et al. 2008).

Further research is necessary to quantify the influence of integral water repellent SHCC on service life of structures in aggressive environment. But it is obvious already that there is an interesting potential for more durable and more economic and more ecological construction by using integral water repellent SHCC.

6 CONCLUSIONS

From results presented in this contribution the following conclusions can be drawn:

- Capillary suction is a powerful method to characterize damage due to micro crack formation in cement-based materials.
- Determination of the coefficient of capillary suction also allows us to estimate the penetration of harmful substances into the pore space of SHCC.

- The ultimate strain limit obtained from the stress-strain diagram is not a useful parameter for durability design of structural elements made of SHCC.
- Depending on the environment a reduced strain limit has to be introduced to take the increased ingress of aggressive compounds under imposed strain into the material into consideration adequately.
- Integral water repellent SHCC is an efficient protection of SHCC against uptake of water and ions dissolved in water.
- The high strain capacity of SHCC can be used more extensively if integral water repellent SHCC is produced.

ACKNOWLEDGEMENT

The authors gratefully acknowledge substantial support by Kuraray Co. Ltd., Japan.

In addition financial support by the National Natural Science Foundation of China (50878109) and the National Key Technology R&D Program (2007BAB27B03) is gratefully acknowledged.

REFERENCES

Guo, P., Wittmann F.H., & Zhao T. 2008. On the efficiency of surface impregnation of cracked concrete, *Int. J. Restoration of Buildings and Monuments* 14: 425–434.
Martinola, G., Bäuml, M.F. & Wittmann F.H. 2002. Modified ECC applied as an effective chloride barrier, *Proc. Int. JCI Workshop on Ductile Fiber Reinforced Cementitious Composites (DFRCC)—Application and Evaluation,* Takayama, Japan: 171–180.
Martonola, G., Bäuml, M.F., & Wittmann, F.H. 2004. Modified ECC by means of internal impregnation, *Journal of Advanced Concrete Technology* (ACT) 2: 207–212.
Sahmaran, M. & Li, V.C. 2009. Influence of microcracking on water absorption and soptivitiy of ECC, *Materials and Structures* 42: 593–603.
Zhan, H., Wittmann, F.H. & Zhao, T. 2003. Chloride barier for concrete in saline environment established by water repellent treatment. *Int. J. Restoration of Buildings and Monuments* 9: 535–550.
Zhan, H., Wittmann, F.H. & Zhao, T. 2005. Relation between the silicon resin profiles in water repellent concrete and the effectiveness as a chloride barrier. *Int. J. Restoration of Buildings and Monuments* 11: 35–46.
Zhang, P., Wittmann, F.H., Lehmann, E., Vontobel, P. & Hartmann, S. 2009 a. Observation of water penetration into water repellent and cracked cement-based materials by means of neutron radiography, *Int. J. Restoration of Buildings and Monuments* 15: 91–100.
Zhang, P., Wittmann, F.H. & Zhao, T. 2009 b. Capillary suction of and chloride penetration into integral water repellent concrete, *Int. J. Restoration of Buildings and Monuments* 15: 185–192.
Zhao, T., Wittmann, F.H. & Ueda, T. 2005 (eds.). *Durability of Reinforced Concrete under Combined Mechanical and Climatic Loads,* Aedificatio Publishers, Freiburg, Germany.

Advances in Cement-Based Materials – van Zijl & Boshoff (eds)
© *2010 Taylor & Francis Group, London, ISBN 978-0-415-87637-7*

Driving infrastructure sustainability with Strain Hardening Cementitious Composites (SHCC)

V.C. Li

Advanced Civil Engineering Materials Research Laboratory,
University of Michigan, Ann Arbor, MI, USA

ABSTRACT: Infrastructure sustainability has become increasingly important globally as concerns about infrastructure decay and environmental deterioration rise. This paper poses the thesis that Strain Hardening Cementitious Composites (SHCC) has many characteristics that positively contribute to infrastructure sustainability; recent advances in durability studies, material greening, as well as current developments in smart self-healing and self-sensing functionalities in SHCC are highlighted. The paper also identifies additional research needed to realize the promise of SHCC as an enabling technology for infrastructure sustainability.

1 INTRODUCTION

Many nations, developing or developed, are experiencing increasing pressures resulting from extreme weather patterns, energy and water scarcity, and civil infrastructure decay. While these large societal challenges of the 21st century initially appear unrelated to one another, they may in fact be coupled in one way or another. For example, concrete is the most used engineered material, exceeding 12 billion t/year and amounting to about 2 ton per person per year (van Oss and Padovani, 2002) on average globally for civil infrastructure construction and repair. Accompanied with the large material flow is the high-energy intensity of cement, about ten times that of the averaged economy in the US (WBCSD 2002). Cement production is also responsible for about 5% of anthropogenic carbon dioxide and significant levels of SO_2, NOx, particulate matter and other pollutants (USEPA 2000); it contributes disproportionately to the global warming potential when compared with other human activities It is now the 3rd CO_2 polluter worldwide, after fossil fuels and deforestation. About a third of the drinking water in the US never reaches consumers due to leakage in our aging water infrastructure. And about 40% of the energy in the US is consumed in buildings (ASCE 2009). The intertwining relationship between the built and the natural environment is becoming increasingly evident.

Infrastructure decay has significant impact on the environment. In 2009, the American Society of Civil Engineers prescribed an average grade of D to civil infrastructure in the US (ASCE 2009). In the category of bridges, for example, nearly 20% of bridges are rated as "structurally deficient or

functionally obsolete." It is estimated that $100 billion is needed to repair and rehabilitate the over 100,000 miles of aging levees many of which are over 50 years old. Energy infrastructure is severely lacking behind demand, which has grown 25% since 1990. The concern does not stop at aging infrastructure with substandard performance affecting public safety. The economic cost of returning them to an acceptable level has been quoted at US$ 2.2 trillion. This financial burden on the US economy is obvious. Less known to the public is the fact that as much as 50% of field repair fails and require re-repair for concrete infrastructure (Vaysburd et al, 2004). In each repair, more material is consumed, with attendant energy consumption and pollution emissions. Poor automobile fuel economy and traffic delays on inferior roadways induced energy waste and tailpipe emissions. The impact of infrastructure decay and maintenance needs on the natural environment represent a major concern to governments, industries and general citizens.

It may be argued that greener concrete with enhanced durability for new and repaired infrastructure provides the best solution to the infrastructure decay problem, while assisting in mitigating environmental concerns. Given the well known fact that cracking in brittle concrete is a major cause of infrastructure deterioration, this makes the case for a future generation of concrete that is more damage tolerant, and which suppresses the deterioration mechanisms commonly experienced in infrastructure, such as corrosion of reinforcing steel. Strain-hardening Cementitious Composites (SHCC) is a class of relatively new concrete material that possesses many of the qualities required. Its development in the last decade has been rapid and significant. There is evidence that with further research and development, SHCC may provide a material solution to many problems stressing our built and natural environments.

This paper overviews the characteristics of SHCC that pertains to addressing infrastructure decay and associated environmental concerns. It also describes some on-going and future additional research necessary to fulfill its promise as a preferred concrete of the next generation intelligent infrastructure. It poses the thesis that infrastructure sustainability can be advanced through deliberate materials engineering of SHCC.

2 INFRASTRUCTURE SUSTAINABILITY

2.1 Defining sustainable infrastructure engineering

The World Commission on Environment and Development (WCED 1987) defined sustainable development as development that "meets the needs of the present generation without compromising the ability of future generations to meet their needs." Infrastructure sustainability needs to incorporate the concepts of life-cycle analysis, carbon and energy footprints, new and renewed infrastructures, and material selection for sustainability. Clearly a range of civil engineering disciplines contributes to infrastructure sustainability, including materials engineering, structural design, and construction and maintenance management. Each discipline may dominate over others at different phases of an infrastructure system's life. From a sustainability viewpoint, however, it is necessary to consider all phases of a structure's life cycle holistically; the economic, social and environmental impacts of each phase are typically dependent on each other. To illustrate, a green construction material that has low environmental impact in the material production phase may end up contributing to a large life cycle carbon and energy footprint if repeated repairs are required during the use phase of the built infrastructure. In this example, it is clear that considering sustainability from a green materials engineering viewpoint only is inadequate in addressing infrastructure sustainability. Thus, a working definition of sustainable infrastructure engineering is the integrated material development, structural design and construction, and infrastructure management that are consistent with the principles of sustainable development.

2.2 The SIMSS design approach to infrastructure sustainability

A useful approach for deliberate driving of infrastructure sustainability was offered by Lepech

(2006, 2009). The Sustainable Infrastructure Materials, Structures, and Systems (SIMSS) design approach integrates the materials development, structural design and infrastructure system operations stages with life cycle evaluation (figure 1). This integration emphasizes the interdependencies of the different phases of the life cycle of a structure in their contributions to sustainability indicators, and encourages a more holistic approach in attaining infrastructure sustainability. In SIMSS, scale linkage occurs naturally. For example, the nanometer scale coating on fibres used in an SHCC can be linked to scheduling of repair maintenance for a fleet of kilometer scale bridges in assessing the life cycle carbon and energy footprints of an SHCC bridge system.

At the "System" scale, maintenance for serviceability of an infrastructure dominates the resource input and emissions output. At the "Structure" scale, durability under combined mechanical and environmental loads dictates the time scale of deterioration and repair needs. At the "Material" scale, composite properties, often beyond the elastic stage, determine local mechanisms of physical and chemical response to load and fluid transport. The three scales are connected through sharing of the materials and structural properties apex (figure 1). It is exactly because of these connectivities that materials with properly designed microstructure and production methodology exerts its influence on infrastructure sustainability throughout the three materials, structure and system scales.

The "Evaluation" module in figure 1 embodies the tools of life cycle analysis (LCA) and economic modeling of a given infrastructure system. This module accounts for energy and raw material input and emissions output at each phase of the life cycle, including material production and transport, construction, operation and maintenance, and end-of-life.

Implementation of SIMSS has been performed on bridge, pavement and pipe infrastructure,

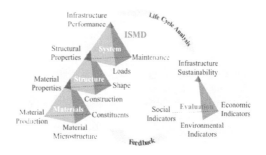

Figure 1. Sustainable Infrastructure Materials, Structures, and Systems (SIMSS) Design Approach (Lepech, 2006; 2009).

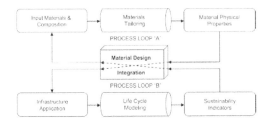

Figure 2. Interactive design framework for sustainable infrastructure systems (Keoleian et al, 2005).

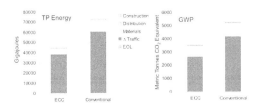

Figure 3. Life cycle analysis results for a R/C bridge deck with ECC link slab versus one with conventional expansion joint (Keoleian et al, 2005).

adopting the iterative methodology of Keoleian et al (2005). Figure 2 shows the iterative looping between materials tailoring (Process Loop "A") and life cycle analyses (Process Loop "B"). In Loop "A", micromechanics is used as the analytic tool to guide the tailoring of material ingredients in the deliberate selection of chemical type, the geometric dimensions, and weight proportions (Li et al, 2001) in order to attain target composite properties. In the case of SHCC, the most important target property would be the tensile strain capacity. This composite property is then translated into infrastructure performance. For sustainability consideration, the structural durability gain as a result of the unique property characteristics of SHCC becomes the focus in linking the materials tailoring loop to the life cycle analyses loop. An infrastructure deterioration model is necessary to make contact between structural durability and system maintenance scheduling (Lepech, 2006). A life cycle model is then used as the analytic tool in Loop "B" to translate alterations in system maintenance scheduling for the use phase into sustainability indicators, and added to contributions from other phases. Finally, the sustainability metric provides insights into avenues of greening of SHCC through selective adoption of industrial waste streams, without compromising the composite properties necessary to maintain high structural durability. This looping can iterate indefinitely for maximum infrastructure sustainability with the greenest SHCC.

Figure 3 shows the first iteration results of the SIMSS approach applied to a conventional bridge deck and one retrofitted with an SHCC link slab (Lepech and Li, 2009; Keoleian et al, 2005). In this case, the deck with the SHCC link slab shows a reduction in total primary energy (TP Energy) usage 40% lower than that of the conventional deck. A similar level of reduction is found in global warming potential (GWP) measured in equivalent CO_2 emissions. This figure also shows that the major contribution to the energy and carbon footprints occurs in the use phase due to traffic flow alteration (D Traffic) as a result of repair and reconstruction activities. The second most important contribution occurs in the material production phase. This finding emphasizes the importance of structural durability to reducing maintenance requirement, while acknowledging that greening of construction material is necessary to achieve infrastructure sustainability.

3 SHCC MATERIALS AND STRUCTURAL DURABILITY

The durability of many reinforced concrete structures is negatively impacted by the tendency of concrete to crack in tension; tensile stress may be due to live load, restrained shrinkage or thermal loads. Past experiences have demonstrated that cracking and crack widths are difficult to control using steel reinforcements in the field. Given these facts, it would seem natural to look to SHCC for enhancement in structural durability. The tensile ductility characteristics of SHCC should suppress brittle fracture and lead to enhanced structural durability. There is currently only limited field experience (Li and Lepech, 2004; Kunieda & Rokugo, 2006) to support this contention, since SHCC is a relatively new construction material. However, the rapidly increasing volume of laboratory data suggest overwhelmingly that SHCC will contribute to durability of concrete structures, see for example, the State of the Art Report by the RILEM TC HFC (2009). Here we highlight succinctly the material and structural durability of SHCC in the context of prolonging infrastructure service life, reduction in maintenance frequency and enhancement of sustainability.

As a new construction material, SHCC must be carefully scrutinized for its durability in typical infrastructure environments, which may involve a combination of mechanical and environmental loadings. A unique feature of SHCC distinctive from normal or high performance concrete is its tensile strain-hardening behavior. During strain-hardening, the material undergoes controlled multiple microcracking. The strain-hardening stage is expected to be utilized in SHCC infrastructure during service conditions. It is reasonable, therefore, to raise concerns of durability given the expected much higher number of cracks in an SHCC structure in comparison to a normal concrete structure, despite the much tighter crack width in SHCC. Hence it is necessary to experimentally evaluate the durability of SHCC in the elastic state as well as in the strain-hardening state—i.e. when the material has already been loaded to the multiple microcracked state. Further, the potential for changes in transport properties due to the presence of microcracks dictates the need to evaluate structural durability, especially the effect on steel corrosion.

In the following, we highlight durability study results for an Engineered Cementitious Composites (ECC) studied at the University of Michigan and elsewhere. Material durability in the uncracked state under various exposure environments is first summarized. Durability of ECC in the strain-hardening state is then overviewed. Finally, the contribution of ECC to structural durability is highlighted. A typical mix of ECC, labeled M45, is given in table 1. The mix ingredient selection—in terms of ingredient type, proportion, and geometric size—is governed by micromechanical models (Li et al, 2001; Li and Leung, 1992). Specifically, the poly-vinyl-alcohol (PVA) microfibre was coated with a proprietary nanometer scale surface coating to facilitate fibre slippage prior to reaching rupture threshold. Figure 4 shows a typical stress-strain curve of ECC-M45. A tensile strain capacity of 3% can be attained.

Of most importance to the present discussion on durability during the strain-hardening stage is the unique nature of crack development in ECC. Beyond first cracking at about 0.01–0.02%, the number and width of cracks increases with deformation. This continues until the deformation reaches about 1%. Beyond this stage, further deformation is accompanied by increase in crack number but almost constant crack width (figure 4). This constant crack width, termed steady state crack width, is a property of the ECC, analogous to other properties like Young's modulus or compressive strength. In other words, the steady state crack

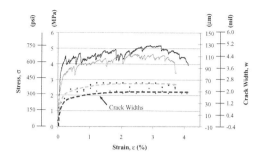

Figure 4. Typical stress-strain-crack width relations of an ECC.

width is independent of specimen or structural size (Lepech and Li, 2004), and is also independent of re-bar size and reinforcement ratio.

3.1 Durability in the uncracked state

Freeze-thaw testing in accordance with ASTM C666, Procedure A, was performed on ECC prism specimens (Li and Lepech, 2004). The dynamic modulus was recorded as a function of freeze-thaw cycles. ECC specimens survived the test duration of 300 cycles with no degradation of dynamic modulus or surface appearance. In addition, ECC coupon specimens after subjected to 300 freeze-thaw cycles were found to retain a tensile strain capacity of about 3%. The observed frost durability of ECC, despite having no deliberate air entrainment, is due to the increase of larger pore volume, and intrinsically high tensile ductility and strength due to the presence of micro PVA fibres (Sahmaran and Li, 2007).

Salt scaling resistance of ECC was evaluated (Şahmaran and Li, 2007) in accordance with ASTM C672. After 50 freeze-thaw cycles in the presence of de-icing salt, the surface condition visual rating and total mass of the scaling residue for ECC prism specimens remain within acceptable limits of ASTM C672. In addition, ECC coupon specimens exposed to freeze-thaw cycles in the presence of de-icing salts for 50 cycles were found to exhibit negligible loss of ductility, and retained a tensile strain capacity of more than 3%. These results confirm that ECC remains durable despite exposure to freeze-thaw cycles in the presence of de-icing salts.

In contrast to freeze-thaw tests designed to simulate temperature changes in winter conditions, hot water immersion tests were conducted to simulate the long-term effects of hot and humid environments (Li et al, 2004). The tensile strain capacity of ECC dropped from 3% at 28 days to 2.75% after 26 weeks of hot water immersion.

Table 1. Typical mix proportion of ECC material.

Cement	Water	Sand	Fly ash	HRWR	Fibre (Vol. %)
1.00	0.58	0.80	1.20	0.013	2.00

HRWR = High range water reducer; all ingredients proportions by weight except for fibre.

Table 2. Durability of uncracked ECC.

Test	Exposure condition	Specimen	Test result	Ref.
ASTM C666, Proc A	Freeze-thaw (300 cycles)	P	Dynamic Modulus retained 100%	Li & Lepech, 2004
Uniaxial tension	Freeze-thaw (300 cycles)	C	Strain capacity retained 3%	Li & Lepech, 2004
ASTM C672	Salt scaling resistance (50 cycles)	P	Visual rating & total mass of scaling residue remain within limits of ASTM C672	Şahmaran & Li, 2007
Uniaxial tension	Salt scaling resistance (50 cycles)	C	Strain capacity retained 3%	Şahmaran & Li, 2007
JIS	Hot water immersion (26 weeks)	C	Strain capacity retained 2.75%	Li et al, 2004
ASTM C1260	High alkaline (NaOH) immersion	B	Expansion within ASTM C1260 limit	Sahmaran & Li, 2008

Specimen Type: P = prisms; C = coupons; B = bars.

While accelerated hot weather testing does result in lower strain capacity of ECC, the strain capacity exhibited after 26 weeks remains over 250 times that of normal concrete.

Another environment that could affect the microstructure and composite properties of ECC is a high alkaline environment. Since ECC has high fly ash content, alkali-silicate reaction (ASR) performance of ECC is expected to be satisfactory. Results of ASTM C1260 test (Sahmaran and Li, 2008) showed no damaging expansion.

Table 2 shows a summary of durability test results for uncracked ECC in a variety of exposure environments.

3.2 Durability in the microcracked strain-hardening state

Many of the same tests conducted on ECC uncracked specimens were also performed on ECC cracked specimens. For example, prism specimens from beams preloaded to as much as 2 mm (equivalent to 1.5% strain on tension side) and then subjected to salt scaling tests in accordance to ASTM C672 showed negligible surface scaling (Sahmaran and Li, 2007). Similarly, preloaded (to 2% strain) coupon specimens retained the same 3% tensile strain capacity after exposure to 50 freeze-thaw cycles in the presence of de-icing salt (Sahmaran and Li, 2007). Exposure of precracked coupon specimens to an alkaline environment (Sahmaran and Li, 2008) up to 3 months at 38°C showed a slight loss of ductility and tensile strength, but retained a strain capacity of more than 2%.

Transport properties of precracked ECC, including permeation, absorption, and chloride ion diffusion, have been measured. Lepech and Li (2008) showed that the tight crack width of less than 100 micron of preloaded (to 3%) ECC maintained its permeability to a level similar to that of normal sound concrete. Based on ponding tests of precracked specimens under high imposed bending deformation, Sahmaran et al (2007) found that the effective chloride diffusion coefficient was linearly proportional to the number of cracks in ECC, whereas the effective diffusion coefficient of reinforced mortar was proportional to the square of the crack width. Therefore, the effect of crack width on chloride transport was more pronounced when compared with that of crack number. This study concludes that controlling crack width is more important than controlling crack number for structural durability associated with chloride ion penetration.

Paradoxically, the tight crack width of SHCC may result in undesirably high water absorption due to capillary suction. Sorptivity test on precracked ECC specimens (Sahmaran and Li, 2009a) indicated that water absorption increased exponentially with the number of surface cracks. Even so, the sorptivity values of pre-loaded ECC specimens up to a 1.5% strain on the exposed tensile face is not particularly high when compared to that of normal concrete. Moreover, in the same study, the addition of water repellent admixture in the ECC mix easily inhibited the sorptivity for the precracked ECC (Sahmaran and Li, 2009a; Martinola et al, 2004).

Table 3 shows a summary of durability test results for pre-cracked ECC in a variety of exposure environments.

3.3 Structural durability

Reinforcing steel bars in concrete structures can be depassivated resulting in corrosion initiation when the chloride concentration reaches threshold levels on the rebar surface (Tuutti 1982). By preserving low chloride ion diffusion rate after cracking as

Table 3. Durability of pre-cracked ECC.

Test	Expo. Condition	Specimen	Test result	Ref
ASTM C672	Salt scaling resist. (50 cycles)	P, Preloaded to 1.5% strain on tension side	Visual rating & total mass of scaling residue within limits of ASTM C672	Sahmaran & Li, 2007
Uniaxial tension	Salt scaling resist. (50 cycles)	C, Pre-loaded to 2% strain	Strain capacity retained 3%	Sahmaran & Li, 2007
ASTM C1260	High alka-line (3 mos at 38°C)	C, Pre-loaded to 2% strain	Strain capacity retained >2%	Sahmaran & Li, 2008
Falling head perm. test	Hydrau. head	C, Pre-loaded to 3% strain	Permeability (8.90 × 10⁻¹² m/sec) similar to sound concrete	Lepech & Li, 2008
AASHTO T259-80.24 Chloride ion diffu-sion	Salt ponding	P, Pre-loaded to 1.5% strain	Effective chloride diffusion coef. lin-early pro-portional to crack nos.	Sahmaran et al, 2007
ASTM C642 & ASTM C1585	Water Absorp-tion & sorption	P, Pre-loaded to 1.5% strain	Water absorption increased exponen-tially with number of surface cracks	Sahmaran & Li, 2009

Specimen Type: P = prisms; C = coupons.

Table 4. Durability of R/ECC in Corrosive Environment.

Test	Exposure condition	Specimen	Test result	Ref.
Macro & Microcell	Wet-dry cycles of acceler-ated chloride	R/ECC beams pre-cracked	Corrosion rate < 0.001 mm/yr in ECC; 0.008 mm/yr in R/C	Miyazato & Hiraishi, 2005
Anti-spalling under accele-rated corrosion	Impressed current (30 V DC) with speci-men in salt bath	Cylinders with embed-ded steel bar (lol-lipop)	No spalling in ECC after 300 hrs; Spalling at 90 hrs for R/mortar	Sahmaran et al, 2008
Corroded steel mass loss	Impressed current (30 V DC) with specimen in salt bath	Cylinders with embed-ded steel bar (lol-lipop)	1% loss in ECC, 12% in concrete specimen; after 75 hrs	Sahmaran et al, 2008
Residual flexural strength	Impressed current (30 V DC) with specimen in salt bath	ECC prisms with embed-ded steel bar	100% in R/ECC, 20% in R/mortar; after 50 hrs	Sahmaran et al, 2008

discussed above, ECC material reduces chloride intrusion to effectively protect reinforcement from corrosion (Sahmaran et al, 2007). Miyazato and Hiraishi (2005) experimentally confirmed that the corrosion rate of steel rebars in preloaded ECC beams was orders of magnitude lower com-pared with those in similarly preloaded concrete beams, when these beams were exposed to wet (saltwater shower 90% RH for 2 days)—dry (60% RH for 5 days) cycles of an accelerated chloride environment.

A second level of protection against steel cor-rosion is the anti-spalling ability of ductile ECC to withstand expansive force generated by steel corrosion, if this ever occurs. In accelerated corro-sion tests in which the embedded steel was forced to corrode by an imposed electro-chemical cell, ECC did not exhibit the severe distress observed in conventional mortar specimens. Expansion of corroding steel reinforcement was absorbed by the inelastic tensile deformation of the surrounding ECC. Corrosion related distress in mortar beams resulted in the reduction of the flexural strength and such reduction was not observed in the ECC beams (Sahmaran et al, 2008).

Finally, it should be pointed out that in many structures, steel reinforcements are used to control concrete crack width. Such reinforcements may be completely eliminated when ECC replaces concrete since the crack width in ECC is self-controlled. The elimination of steel reinforcement renders the cor-rosion related durability issues moot since corro-sion cannot occur without the steel reinforcement present.

Table 4 summarizes the durability test results of reinforced ECC under accelerated corrosive environment.

4 ENGINEERING MATERIALS GREENNESS IN SHCC

One of the shortcomings of SHCC is the envi-ronmental penalty in higher energy and carbon footprints on a unit volume basis, associated with the incorporation of fibres and a typically higher cement content in SHCC compared with normal concrete. The higher cement content in SHCC results from the deliberate elimination of coarse aggregates used as filler material in normal concrete. Figure 5 shows a comparison of the compositions of a typical ECC formulation and a concrete formulation. Figure 6 shows a comparison of the energy consumption per cubic meter of the corresponding formulations (Kandall, 2007). Hence, on a unit volume basis, the primary energy consumption of ECC is a major concern. In the investigation of life-cycle primary energy consumption and equivalent carbon dioxide emission for a bridge deck, Keoleian et al (2005) identified material production as the second largest contributor to these sustainability indicators, behind traffic alterations due to maintenance and reconstruction events. Thus it is important to consider approaches in greening SHCC, even though the enhanced durability of SHCC as discussed in the previous section should drastically reduce repair needs and therefore enhance infrastructure sustainability.

The greening of ECC can target replacement of the virgin fibre and/or the matrix materials with industrial waste stream materials. ECC is optimized for tensile ductility with a minimum amount of fibres. Even so, the typical amount of fibre used is 2% by volume. Attempts at using natural fibre or recycled fibre (e.g. carpet fibre) in SHCC have met with limited success, given the requirement of strong fibre bridges in maintaining composite ten-

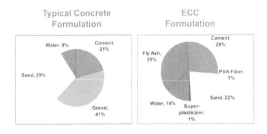

Figure 5. Composition comparison between a typical ECC formulation and a concrete formulation, showing weight percent.

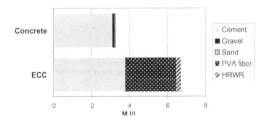

Figure 6. Material energy intensity for ECC and conventional concrete (Kendall, 2007).

sile ductility (Li et al, 2001). However, the greening of the matrix material is promising. We highlight here a few successful cases where Portland cement and silica sand have been replaced by industrial wastes.

Yang et al (2007) investigated green ECCs with Class F fly ash from coal based electric power plants. The "standard" ECC (table 1) already incorporates a relatively high fly ash content of FA/C = 1.2. In the very high fly ash content ECC (HFA-ECC) studied by Yang et al (2007), a FA to cement ratio up to 5.6 was adopted. It was found that at FA/C = 2.8, a 28 day compressive strength of 35 MPa and a 28 day tensile ductility of 3% can be retained. Beyond this FA/C ratio, the compressive strength loss may not be acceptable for many applications, and there may also be durability concerns. Sahmaran and Li (2009b) studied the durability (based on water sorptivity, chloride penetration and accelerated aging tests) of ECC containing FA/C = 2.2, and found that the durability of such HFA-ECC even in the microcracked state to be satisfactory. The use of fly ash actually enhanced composite performance, including reduced shrinkage strain and crack width, and more robust tensile strain capacity (Yang et al, 2007). The improvements in tensile ductility and crack width were attributed to a reduction in chemical bond and an increase in frictional bond due to the presence of the spherical fly ash particles in the interfacial transition zone, thus favoring conditions for tensile strain hardening.

Material sustainability indicators were computed based on life cycle assessment of all material and energy consumption along with water and emission generation associated with raw material extraction and production of ECC and its constituents. For this green ECC, a gain in solid waste reduction (diverting from waste stream) was achieved. With a large amount of cement replaced, this ECC with FA/C = 2.8 now generates a similar amount of CO_2 when compared with normal concrete. Even so, the total energy consumption remains about twice that of normal concrete due to the fibre content.

Another case of successful green ECC development was afforded by Lepech et al (2008), based on replacement of virgin manufactured silica sand by industrial waste sand. The waste foundry sand from the calcinator bag house and green foundry sand from lost foam metal casting were found to have the appropriate particle size distribution as dictated by micromechanics and for rheological control of ECCs. While the bag house sand resulted in an ECC with no loss of performance, the adoption of green sand did lead to a loss of tensile strength and ductility in the composite. It was determined that the fibre/matrix interfacial frictional bond was drastically decreased due to agglomeration of carbon nano-particles that are residuals accompanying the green sand in the loss foam casting process. An oiling agent was originally applied on the PVA fibres to reduce the strong hydrophilicity of the fibre and the excessively strong chemical bond to cement. A reduction of fibre surface oiling agent successfully restored the composite tensile ductility although the tensile strength was still substantially lowered.

Kendall et al (2008) computed material sustainability indicators for the two mixes involving foundry sands. The introduction of foundry sands was found to result in the replacement of an additional 22% of virgin ECC materials, whereas solid waste diverted from landfills increased 93%, when compared with a standard ECC (table 1, see also figure 5).

5 SMART SHCC FOR SUSTAINABLE INFRASTRUCTURE

While durability of SHCC will contribute to longer service life, deterioration of infrastructure will inevitably occur over time, especially in an aggressive environment such as that typical of coastal regions. Two recent advances in SHCC technology should be helpful in this regard—the endowment of self-healing and self-sensing functionalities in SHCC. Some preliminary findings of these functionalities based on ECC material are highlighted below.

Self-healing in an SHCC structure during service can have a profound effect on infrastructure sustainability. Since SHCC are designed to operate in the strain-hardening stage, it is expected that microcracks will be present during normal service conditions. While proven durable as discussed earlier, it is nevertheless advantageous if the material can reheal itself. Self-healing is defined here as the recovery of the undamaged mechanical and transport properties, without external intervention. Recovery of mechanical properties includes the assurance of uncracked tensile and compressive strength and stiffness, and tensile ductility. Recovery of transport properties implies a self-sealing function that prevents intrusion of aggressive agents. Such recoveries suggest a reverse deterioration process over the lifetime of the structure. While mechanical and environmental loading may cause continuous deterioration, self-healing autonomically reverses such deterioration and restores structural health. Ultimately, as structural durability improves and infrastructure maintenance needs are lowered due to self-healing of SHCC materials, the sustainability indicators will also decrease. Within the SIMSS approach (figure 1), self-healing

links material properties to structural properties and infrastructure sustainability performance.

Preliminary studies of self-healing in ECC (Li and Yang, 2007; Yang et al, 2009a) suggest a promising approach towards virtually "crack-free" concrete. The self-healing behavior of ECC has been found to persist in a variety of environments typical of civil infrastructures, including cycles of rain and sunshine (Yang et al, 2009a,b), under continuous water immersion (Yang et al, 2009a,b), under a hydraulic gradient (Lepech and Li, 2008), and in the presence of salt or strong alkali (Sahmaran et al, 2007; Sahmaran and Li, 2008). Self-healing occurs for both young (3 days) (Yang et al, 2009b) and in more mature specimens (6 months) (Yang et al, 2009a). Several characteristics of ECC lend themselves to robust self-healing. Self-healing utilizes unhydrated cement grains in the material for further hydration and for calcium carbonate formation on contact with water and air. This implies that the self-healing process can take place wherever damage occurs since unhydrated cement grains are ubiquitous in the material. The ability of ECC to withstand multiple damage events and to still recover its virgin properties after each self-healing event is less obvious. This issue is currently under investigation at the University of Michigan. Figure 7 shows the full recovery of tensile strength and ductility (3%) and stiffness of an ECC specimen after a 2% damaging tensile pre-load. However, for a specimen with a 3% damaging pre-load, the recovery was incomplete. The material suffered from a reduced first cracking strength. Additional studies of failure mode details and strengthening

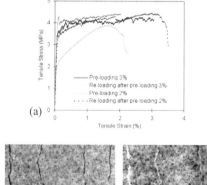

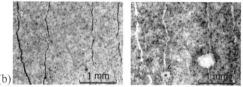

Figure 7. Self-Healing of ECC (a) Recovery of Tensile Properties and (b) Cracks filled with reheal products (Yang et al, 2009a,b).

of the rehealed cracks are needed to advance the concept of "crack-free" concrete.

The current practice of maintenance scheduling for bridges assumes deterioration rates that are based on empirical data of deterioration from similar structures, traffic and environmental exposure conditions. This practice may result in maintenance events that may not be warranted due to better-than-expected performance of the structure, or that may be overdue due to worse-than-expected performance of the structure. An optimized maintenance schedule should be one that is just in time; this would require intimate knowledge of the in-situ surface and internal damage the structure has experienced in real time. An economically feasible technology needed to accomplish this does not currently exist. However, it may be expected that just in time maintenance scheduling can lead to significant savings in materials and reduces unnecessary traffic delays due to minimized reconstruction events. As a result, infrastructure health condition and life cycle sustainability indicators can be significantly and simultaneously improved.

One potential enabling technology to allow real-time in-situ structural health monitoring is a self-sensing SHCC. A self-sensing SHCC will serve both as a structural load-bearing damage tolerant material by virtual of its tensile ductility, while at the same time, be able to self-measure its strain state and crack width. By coupling with wireless communication technologies and cyber infrastructure, it is possible to report to remote monitoring stations the condition of any part of the structure in real time. Such a technology is currently being researched at the University of Michigan (Lynch et al, 2009).

Self-sensing SHCC for structural health monitoring fits seamlessly with the SIMSS design approach (figure 1). Preliminary investigation (Hou and Lynch, 2005) demonstrated that ECC exhibited piezoresistive properties. Specifically, changes in material deformation, especially in the inelastic stage, are accompanied by changes in electrical resistivity (figure 8). At the materials level, manipulation of the piezoresistive sensitivity can be attained by control of the composite ingredients, such as doping the composite with carbon nanotubes or carbon blacks. At the structural scale, strategic location of electrodes within the structure allows current injection and voltage measurements from which data a spatial damage map can be extracted via electrical impedance tomographic approaches (Hou and Lynch, 2009). Unlike other sensing approaches, self-sensing ECC provides direct damage sensing. System scale deterioration information is communicated via a cyber-network to structural health monitoring stations where the

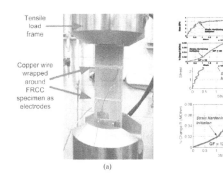

Figure 8. (a) ECC plate element loaded in unaxial tension as resistivity is measured by two-point probing, and (b) Resistivity versus strain plots (Hou and Lynch, 2005).

data can be assimilated with risk and life cycle analyses tools. In this scheme, data collection is autonomous and low cost, and the data is objective. The result is optimized maintenance decision-making with regard to when, what and where repairs should be performed. While attractive, there remain many technical challenges that need to be overcome to turn this concept into a deployable technology. A multi-disciplinary team involving specialists in wireless sensors, materials design, power harvesting, radio technology, cyber infrastructure, structural analyses, human-structure interaction, and industrial ecologists have been assembled (Lynch et al, 2009) to tackle this complex problem.

6 CLOSURE

In this paper, SHCC is depicted as a new class of concrete materials that can drive the environmental sustainability of the next generation of reinforced concrete structures. The most direct contribution to infrastructure sustainability comes from the enhanced durability through the unique tensile ductility and tight crack width control of SHCC. These properties assist in suppressing many common deterioration mechanisms of current reinforced concrete structures. As a result, prolonged service life and reduction of maintenance needs substantially curtail energy input and carbon dioxide output over the life cycle of an infrastructure. It should be noted that even though the energy and carbon footprints are emphasized in this paper, other resource input (e.g. water) and emission output (e.g. NOx and SOx) will also be positively impacted by replacing concrete with SHCC. Further, the magnitude of reduction in environmental burden is infrastructure and geographically

189

dependent. Transportation infrastructure such as bridges and roads that suffers rapid deterioration especially in cold weather or coastal regions will likely have the best gain in environmental sustainability with the use of SHCC.

The greening of SHCC has made important advances in recent years. It appears that green SHCC with carbon footprint similar to that of normal concrete is on the horizon, although higher energy content is almost inevitable due to the use of fibres. However, as suggested in the case study result shown in figure 3, the total primary energy and global warming potential over the service life of the bridge deck due to materials are actually lower than that of a deck built with normal concrete. This is so because the total volume of materials used over the life cycle is lessened due to the reduced maintenance events when SHCC is adopted. In that example, the calculation was done with a standard ECC (table 1). With a greener version of ECC adopting a high volume of fly ash partially replacing cement and recycled industrial sand replacing virgin silica sand, the improvements in sustainability indicators should be even more favorable.

While green and durable SHCC have been demonstrated, its self-healing and self-sensing functionalities hold an even more significant amount of untapped potential for major gains in infrastructure sustainability. Self-sensing of SHCC provides a means to continuously track not only material and structural damage, but also the amount of self-healing that takes place over time. Further research will lead to intelligent infrastructure with abilities to monitor its own health in real time, and also self diagnose its recovery should damage occur; or provide meaningful information to responsible agencies for repair assistance. While substantial technical challenges remain, the worldwide efforts in multi-functional SHCC development and field-testing offer infrastructure sustainability a promising future.

ACKNOWLEDGEMENTS

The author acknowledges support of the National Institute of Science and Technology (NIST) Technology Innovation Program (TIP) for support of this work via Cooperative Agreement 70NANB9H9008 to the University of Michigan. The program manager is Dr. Jean-Louis Staudenmann. Acknowledgement is also due to Dr. G. Keoleian for his many contributions to life cycle modeling of infrastructure and to Dr. J. Lynch for his many contributions to self-sensing ECC materials.

REFERENCES

ASCE, Report Card for America's Infrastructure. http://www.infrastructurereportcard.org/ Accessed, Aug., 2009.

Hou, T. and Lynch, J.P. 2005, Monitoring Strain in Engineered Cementitious Composites using Wireless Sensors, *Proceedings of the International Conference on Fracture (ICF XI)*, Turin, Italy, March 20–25.

Hou, T. and Lynch, J.P. 2009, Electrical Impedance Tomographic Methods for Sensing Strain Fields and Crack Damage in Cementitious Structures, in J. of Intelligent Material Systems and Structures, Sage Publications, Vol. 20, No. 11, pp. 1363–1379.

Kendall, A. 2007, Concrete Infrastructure Sustainability: Life Cycle Metrics, Materials Design, and Optimized Distribution of Cement Production, PhD Thesis, University of Michigan.

Kendall, A., Keoleian, G.A. and Lepech, M. 2008, Material Design for Sustainability through Life Cycle Modeling of Engineered Cementitious Composites, Materials and Structures, Vol. 41, No. 6, pp. 1117–1131.

Keoleian, G.A., Kendall, A., Dettling, J.E., Smith, V.M., Chandler, R., Lepech, M.D. and Li, V.C. 2005, Life Cycle Modeling of Concrete Bridge Design: Comparison of Engineered Cementitious Composite Link Slabs and Conventional Steel Expansion Joints, ASCE J. Infrastructure Systems, Vol. 11, No. 1, pp. 51–60.

Kunieda, M. and Rokugo, K. Recent Progress on HPFRCC in Japan—Required Performance and Applications, 2006, J. of Advanced Concrete Technology, Vol. 4, No. 1, pp. 19–23.

Lepech, M. 2006, A Paradigm for Integrated Structures and Materials Design for Sustainable Transportation Infrastructure, PhD Thesis, University of Michigan.

Lepech, M.D. 2009, Sustainable Infrastructure Systems using Engineered Cementitious Composites, Proc. of the American Society of Civil Engineers Structures Congress. Austin, Texas, USA.

Lepech, M. and Li, V.C. 2004, Size Effect in ECC Structural Members in Flexure, in Proc., FRAMCOS-5, Colorado, Eds. V.C. Li et al, pp. 1059–1066.

Lepech, M.D. and Li, V.C. 2008, Water Permeability in Engineered Cementitious Composites, Accepted, J. of Cement and Concrete Composites, Dec., 2008.

Lepech, M.D. and Li, V.C. 2009, Application of ECC for Bridge Deck Link Slabs, RILEM J. of Materials and Structures, DOI 10.1617/s11527-009-9544-5.

Lepech, M.D., Li, V.C., Robertson, R.E. and Keoleian, G.A. 2008, Design of Ductile Engineered Cementitious Composites for Improved Sustainability, ACI Materials J., Vol. 105, No. 6, pp. 567–575.

Li, V.C., Horikoshi, T., Ogawa, A., Torigoe, S. and Saito, T. 2004, Micromechanics-Based Durability Study of Polyvinyl Alcohol Engineered Cementitious Composite (PVA-ECC), ACI Materials Journal, Vol. 101, No. 3, pp. 242–248.

Li, V.C. and Lepech, M. 2004, Crack Resistant Concrete Material for Transportation Construction, in Proc., Transportation Research Board 83rd Annual Meeting, Washington, D.C., Compendium of Papers CD ROM, Paper 04-4680.

Li, V.C. and Leung, C.K.Y. 1992, Steady State and Multiple Cracking of Short Random Fiber Composites, ASCE J. of Engineering Mechanics, Vol. 118, No. 11, pp. 2246–2264.

Li, V.C., Wang, S. and Wu, C. 2001, Tensile Strain-Hardening Behavior of PVA-ECC, ACI Materials J., Vol. 98, No. 6, pp. 483–492.

Li, V.C. and Yang, E.H. 2007, Self-Healing in Concrete Materials, in *Self Healing Materials: An Alternative Approach to 20 Centuries of Materials Science*, Ed. S. van der Zwaag, Springer, pp. 161–193.

Lynch, J.P., Kamat, K., Li, V.C., Flynn, M.P., Sylvester, D., Najafi, K., Gordon, T., Lepech, M., Emami-Naeini, A., Krimotat, A., Ettouney, M., Alampalli, S. and Ozdemir, T. 2009, Overview of a Cyber-enabled Wireless Monitoring System for the Protection and Management of Critical Infrastructure Systems, SPIE Smart Structures and Materials, San Diego, CA.

Martinola G., Baeuml M.F. and Wittmann, F.H. 2004, Modified ECC by means of internal impregnation. J. Adv Concrete Tech. Vol. 2, No. 2, pp. 207–212.

Miyazato, S. and Hiraishi, Y. 2005, Transport Properties and Steel Corrosion in Ductile Fiber Reinforced Cement Composites, Proc. of the Eleventh International Conference on Fracture, Turin, Italy.

Sahmaran, M., Lachemi, M. and Li, V.C. 2009, Assessing the Durability of Engineered Cementitious Composites (ECC) Under Freezing and Thawing Cycles, In print, J. of ASTM Int'l, Vol. 6, No. 7.

Sahmaran, M., Li, M. and Li, V.C. 2007, Transport Properties of Engineered Cementitious Composites Under Chloride Exposure, ACI Materials J., Vol. 104, No. 6, pp. 604–611.

Sahmaran, M. and Li, V.C. 2007, De-icing Salt Scaling Resistance of Mechanically Loaded Engineered Cementitious Composites, J. Cement and Concrete Research, Vol. 37, pp. 1035–1046.

Sahmaran, M. and Li, V.C. 2008, Durability of Mechanically Loaded Engineered Cementitious Composites under Highly Alkaline Environment, J. Cement and Concrete Composites, Vol. 30, No. 2, pp. 72–81.

Sahmaran, M. and Li, V.C. 2009a, Influence of Microcracking on Water Absorption and Sorptivity of ECC, RILEM J. of Materials and Structures, Vol. 42, No. 5, pp. 593–603.

Sahmaran, M. and Li, V.C. 2009b, Durability Properties of Micro-Cracked ECC Containing High Volumes Fly Ash, J. Cement and Concrete Research, DOI information: 10.1016/j.cemconres.2009.07.009.

Sahmaran, M., Li, V.C. and Andrade, C. 2008, Corrosion Resistance Performance of Steel-Reinforced Engineered Cementitious Composites Beams, ACI Materials J., Vol. 105, No. 3, pp. 243–250.

Tuutti, K. 1982, Corrosion of Steel in Concrete, CBI Swedish Cement and Concrete Research Institute, Stockholm, Sweden, 1982, 159 pp.

RILEM TC HFC SC 2 STAR, 2009, Durability of Strain-Hardening Fiber Reinforced Cement-Based Composites (SHCC)—State-of-The-Art, Wittmann, F., and G. Van Zijl (eds.), in preparation.

United States Environmental Protection Agency (USEPA), 2000, Sources of dioxin-like compounds in the United States. Draft exposure and human health reassessment of 2,3,7,8-tetrachlorodibenzo-pdioxin (TCDD) and related compounds. Rep. No. EPA/600/P-00/001Bb, Environmental Protection Agency.

van Oss, H.G. and Padovani. A.C. 2002, Cement Technology, Journal of Industrial Ecology, Vol. 6, No. 1, pp. 89–105.

Vaysburd, A.M., Brown, C.D., Bissonnette, B. and Emmons, P.H. 2004, "Realcrete" versus "Labcrete", Concrete International, Vol. 26, No. 2, pp. 90–94.

World Business Council on Sustainable Development (WBCSD), 2002, Toward a Sustainable Cement Industry, Draft Rep., Battelle Memorial Institute/WBCSD, Geneva.

World Commission on Environment and Development (WCED), 1987, *Our Common Future*, Oxford University Press, New York.

Yang, E.H., Yang, Y. and Li, V.C. 2007, Use of High Volumes of Fly Ash to Improve ECC Mechanical Properties and Material Greenness, ACI Materials J., Vol. 104, No. 6, pp. 620–628.

Yang, Y., Lepech, M.D., Yang, E.H. and Li, V.C. 2009a, Autogenous Healing of Engineering Cementitious Composites under Wet-Dry Cycles, J. Cement and Concrete Research, Vol. 39, pp. 382–390.

Yang, Y., Yang E.H. and Li, V.C. 2009b, Autogenous Healing of Engineered Cementitious Composites at Early Age, Submitted, ACI Materials J.

Advances in Cement-Based Materials – van Zijl & Boshoff (eds)
© 2010 Taylor & Francis Group, London, ISBN 978-0-415-87637-7

Modelling the influence of cracking on chloride ingress into Strain-Hardening Cement-based Composites (SHCC)

F. Altmann & V. Mechtcherine

Institute of Construction Materials, TU Dresden, Germany

ABSTRACT: Strain-hardening cement-based composites (SHCC) are high-performance fibre-reinforced composites characterised by their high ductility under tensile load. This is achieved through the formation of multiple cracks with limited crack width. It is expected that structural SHCC members will typically contain steel reinforcement (R/SHCC), thus steel corrosion due to chloride ingress will often be critical to the durability of such members. Due to the limited crack width cracked SHCC provides greater resistance against the ingress of chloride than cracked ordinary concrete. This is especially true under high tensile strains. To fully utilise this advantageous property of SHCC in the design of durable R/SHCC members and structures, a performance-based durability design concept is required that is applicable to both crack-free and cracked SHCC. In this paper experimental investigations on the influence of cracking on chloride ion ingress in SHCC are presented. It was found that for expected strain levels under working loads of up to 1% cracking was not evenly distributed. This resulted in highly variable chloride diffusion coefficients for cracked SHCC. At the design stage the actual crack distribution of an SHCC member will not be known, thus the influence of cracking must be quantified based on the expected strain under working loads. The observed spatial distribution of cracks was taken into account using fuzzy probability theory. This approach allowed a realistic quantification of the crack distribution and its influence on chloride ingress based on the limited data available from the experimental investigations. That definition of the chloride diffusion coefficient was then integrated into the fuzzy-probabilistic model for chloride ingress in SHCC developed previously by the authors to forecast chloride ingress in SHCC.

1 INTRODUCTION

Strain-hardening cement-based composites (SHCC) are a group of materials exhibiting superior ductility and promising durability properties (Li 2003; Mechtcherine & Schulze 2005). These materials are characterised by a high tensile strain capacity due to the gradual formation of a large number of fine, well-distributed cracks. Figure 1 shows a typical stress-strain curve for SHCC with 2.25 Vol.-% of PVA-fibres and corresponding average crack widths as a function of the induced strain.

This group of materials is suitable for non-structural and structural applications and in case of structural applications will likely be used in combination with ordinary steel reinforcement (Mechtcherine & Altmann 2009). According to Li (2003) one target application for SHCC are structures requiring durability under severe environmental loading. In such cases the protection of steel reinforcement from corrosion, for instance due to chloride ingress, is a key durability requirement. Additionally, chloride has a detrimental effect on the fibre-matrix bond properties of the composite

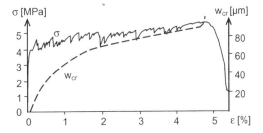

Figure 1. Typical stress-strain and strain-crack width curves for SHCC.

and consequently the mechanical properties of SHCC (Kabele et al. 2006).

The design approach for elements made of SHCC implies that multiple cracking will occur under operational conditions. As shown by Mechtcherine et al. (2007) and Yang et al. (2009), cracking has a significant influence on the transport parameters of the material. Thus it is necessary to explicitly consider the effect of cracking on chloride ingress into the material and subsequently on its durability.

It is reasonable to expect that steel-reinforced SHCC members (R/SHCC) will generally have a superior durability compared to conventional RC members subjected to the same tensile str ain (Li & Stang 2004) due to their limited crack widths (cf. Figure 1). However, the material's durability can only be fully utilised based on an accurate prediction of the time until chloride-induced corrosion begins. Such a model for chloride ingress in cracked and crack-free SHCC has to take into account the uncertainties of the model and parameters. Altmann et al. (2008) developed a fuzzy-probabilistic model for chloride ingress in SHCC based on the DuraCrete model.

In this paper results of a preliminary study on the effect of cracking on chloride ingress in SHCC are presented. Based on these results the aforementioned fuzzy-probabilistic model for chloride ingress is used to predict the chloride profiles in the specimens after 50 years. Furthermore, the suitability of the test methods, the evaluation procedure and the model are discussed and the need for further research will be highlighted.

2 MATERIALS AND METHODS

2.1 *Material composition*

An SHCC developed by Mechtcherine & Schulze (2005) with 2.25% by volume of PVA (Polyvinyl-Alcohol) fibres with a length of 12 mm was used for the experiments. Its composition is given in Table 1. Jun & Mechtcherine (2007) investigated the stress-strain behaviour of this SHCC. As can be seen from Table 2, the material behaves very similar under monotonic and cyclic load and the material response for small and large specimens is comparable. Furthermore, the authors found the crack development under monotonic and cyclic loading to be very similar for strain levels between 0.5 and 2.0%.

2.2 *Specimen preparation and crack generation*

Two prisms P1 and P2 with a cross-section of 100 mm × 100 mm and a length of 600 mm and two large dumbbell shaped tensile specimens A and B (cf. Figure 2) and were produced. All specimens were cast horizontally in steel forms and demoulded after two days. While in the mould and until tensile loading the specimens were wrapped in plastic foil and stored in a climate chamber at 20°C and a relative humidity of RH = 65%.

Prior to chloride immersion the tensile specimens A and B were loaded in uniaxial tension at an age of 24 days using the cyclic load regime developed by Jun & Mechtcherine (2007). A deformation-controlled cyclic load was applied via

Table 1. SHCC composition [kg/m³].

Portland cement	321
Fly ash	749
Fine sand	535
Water	335
Super plasticizer	16.6
Viscosity agent	3.2
PVA fibre	29.3

Table 2. Mechanical properties of dumbbell-shaped SHCC specimens under tensile load (Jun & Mechtcherine 2007).

Loading regime	Stress at first cracking σ_1 [MPa]	Tensile strength f_t [MPa]	Strain capacity ε_{tu} [MPa]
Monotonic	3.2	3.8	5.0
Cyclic	3.2	3.4	3.8

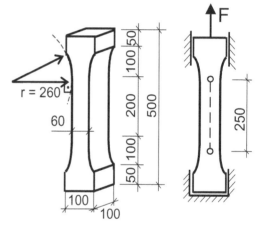

Figure 2. Dumbbell-shaped tensile specimens, 250 mm gauge length.

non-rotatable steel plates at a strain rate of $4 \cdot 10^{-5}$ s^{-1} with a constant strain increment $\Delta\delta = 0.1\%$ from cycle to cycle. When the preset value for $\Delta\delta$ was reached for a cycle, the specimen was unloaded until the lower reversal point, defined as zero load ($F_{min} = $ const. $= 0$ N), was reached. This was repeated until a maximum strain of 0.76% and 0.60% was reached for specimens A and B respectively, which corresponded to a residual strain of $\varepsilon_{res,A} = 0.40\%$ and $\varepsilon_{res,B} = 0.36\%$, cf. Figure 3. As the crack widths at the residual strain level are close to those observed under load for the same strain (Mechtcherine et al. 2007), the chloride immersion tests on unloaded specimens allow the determination of the chloride diffusion coefficient for cracked SHCC under load.

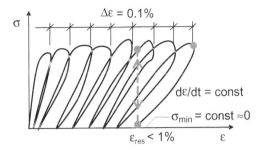

Figure 3.   Tensile load regime for crack generation.

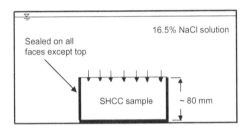

Figure 4.   Test setup for the chloride immersion test according to NT BUILD 443 (1995).

It should be noted that at the investigated relatively low residual strain levels the cracks were not distributed evenly along the length of the specimens (cf. Figure 5). This is in line with the findings by Mechtcherine et al. (2007).

2.3   Chloride immersion tests

The chloride diffusion coefficient was determined according to NT BUILD 443 (1995), an accelerated immersion test. In this test the specimens are coated with epoxy resin on all but one side and then immersed in an aqueous NaCl solution (cf. Figure 4).

Immediately after tensile pre-loading of the specimens A and B all specimens A, B, P1 and P2 were stored in water until fully saturated. Subsequently the outermost 10 mm of SHCC on the face to be exposed to the NaCl solution were sawn off all specimens in accordance with NT BUILD 443 and all other faces were coated with epoxy resin: Two layers of SikaCor 277 were applied on a base coat and a levelling layer of Sikafloor 156. Once the resin had fully hardened, the specimens were again stored in water until fully saturated and then immersed in a 16.5% aqueous solution of NaCl for 48 to 72 days.

For both crack-free specimens P1 and P2 one chloride profile was determined, while for each of the pre-cracked specimens A and B two profiles were determined, one in the area with the lowest

crack concentration (A-2, B-3.2), the other in the area with the highest concentration of cracks (A-4, B-4). To this end the specimens were dry-sawn into samples with a length of 60 mm (cf. Table 4 and Figure 5) before grinding off material in layers of 1.0 to 3.8 mm parallel to the exposed surface. To avoid edge effects, the epoxy resin and 10 mm of

Table 3.   Procedure of chloride immersion tests.

Step	Age [d]	Time [d]
Concreting	0	–
De-moulding	2	–
Tensile pre-loading	24	–
Water storage	24	6
Epoxy-coating	30	
Coating		7
Hardening		7
Water storage	44	5
Chloride immersion	49	48–72
Chloride profiling	97–121	–

Table 4.   Samples for chloride immersion test.

Designation	Immersion period [d]	Tensile strain ε_{res} [%]	Number of cracks
A-2	71	0.40	1
A-4	71	0.40	9
B-3.2	72	0.36	3
B-4	71	0.36	8
P1	48	–	0
P2	70	–	0

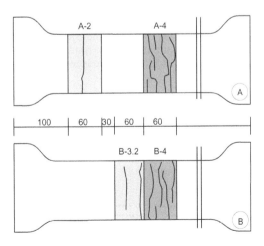

Figure 5   Schematic view of the dumbbell-shaped tensile specimens A and B showing the location at which chloride profiles were determined. The shown cracks are only indicative of the actual crack status.

SHCC at the flanks perpendicular to the exposed surface were removed before the grinding.

For each profile the acid soluble chloride concentration in percent of SHCC mass was determined in a minimum of eight layers by ion-specific photometric analysis. The sample material obtained from each layer was dried at 105°C until constant mass was achieved. Subsequently the chloride was extracted from the sample using an 18% nitric acid (HNO₃) solution. After filtering, the clear solution was analysed using a Hach Lange DR 2800 eco Spectrophotometer.

3 EXPERIMENTAL RESULTS

The pre-cracked specimens A and B exhibited significant self-healing during the water storage and later during the immersion in the saline solution (cf. Figure 6). However, the experimentally determined chloride diffusion coefficient is calculated based on the assumption of being constant over the immersion period. Thus the determined diffusion coefficients for cracked SHCC include the effects of self-healing over the immersion period.

Using the least squares method the chloride diffusion coefficient $D_{ex,\xi}$ at the age t'_ξ and the surface chloride concentration C_s were determined from these results by fitting Crank's (1979) solution (cf. Equation 1) of Fick's 2nd law to the measured chloride concentrations.

$$C(x,t) = C_s \, erfc\left(\frac{x}{2\sqrt{Dt}}\right) \quad (1)$$

where $C(x, t)$ = the total chloride concentration in the SHCC, C_s = the surface chloride concentration, x = the distance from the concrete surface, D = the diffusion coefficient t = the time and $erfc$ is the complement to the error function. The result of the surface layer was omitted in the regression analysis.

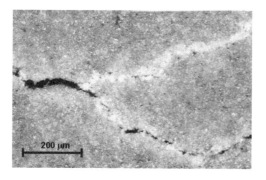

Figure 6. Partially healed cracks after immersion in saline solution.

Table 5. Results of the chloride immersion tests.

Specimen	t'_ξ [d]	$D_{ex,\xi}$ [10^{-12}m²/s]	C_s [mass % of SHCC]
A-2	120	4.58	2.09
A-4	120	13.0	2.10
B-3.2	121	5.09	2.19
B-4	120	8.90	1.98
P1	97	5.71	1.88
P2	119	4.54	2.04

* t'_ξ = age at which $D_{ex,\xi}$ was determined; ** Experimentally determined diffusion; *** Experimentally determined surface chloride concentration

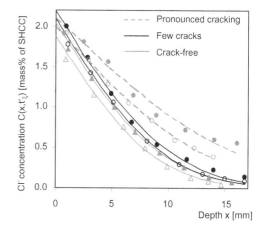

Figure 7. Experimentally determined chloride profiles.

As can be seen from Table 5 and in Figure 7, the surface chloride concentration C_s did not vary significantly between all investigated specimens, irrespective of residual strain or number of cracks.

Similarly, the diffusion coefficients $D_{ex,\xi}$ for the pre-cracked specimens A-2 and B-3.2, exhibiting one and three cracks respectively, did not vary significantly from the coefficients determined for the crack-free specimens P1 and P2. In contrast to that, $D_{ex,\xi}$ increased significantly with the presence of more cracks, as evidenced by the results obtained for the specimens A-4 and B-4.

4 FUZZY-PROBABILISTIC MODELLING OF CHLORIDE INGRESS

4.1 The fuzzy-probabilistic model

To fully utilise the advantageous durability properties of SHCC a performance-based durability design approach is required that takes the cracked state into account. Altmann et al. (2008) took a first step in that direction with a fuzzy-probabilistic model for chloride ingress in SHCC based on the DuraCrete model (2000) for crack-free ordinary concrete.

This fuzzy-probabilistic model allows the quantification of parameters based on limited information and expert knowledge. If insufficient data for a stochastic parameter description is available, this lack of knowledge may be interpreted as fuzzy uncertainty. Fuzziness can be quantified with a member-ship function μ (see Figure 8). Random and fuzzy uncertainties exist simultaneously, resulting in fuzzy-randomly distributed parameters. Fuzzy random parameters may be quantified by defining the variables of their probability functions as fuzzy variables. Figure 9 shows a fuzzy probability distribution function for a normally distributed parameter with a fuzzy mean value.

Using tildes to denote fuzzy and fuzzy-stochastic variables, the DuraCrete model for chloride ingress can be expressed according to Equation 2:

$$\tilde{C}(x,t) = \tilde{C}_s \, erfc\left(\frac{x}{2\sqrt{\tilde{D}_a t}}\right) \tag{2}$$

where $\tilde{C}(x,t)$ = the total chloride concentration in the SHCC, $\tilde{C}_s$ = the (constant) surface chloride concentration, x = the distance from the concrete surface, $\tilde{D}_a$ = the apparent diffusion coefficient, t = the duration of exposure and $erfc$ is the complement to the error function.

Instead of the function for $\tilde{D}_a$ prescribed in the DuraCrete model, a time-integrated average value in accordance with Tang & Gulikers (2007) is used:

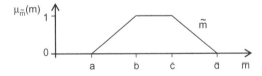

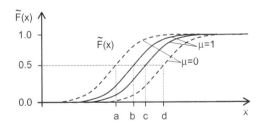

Figure 8. Membership-function of fuzzy mean value.

Figure 9. Fuzzy probability distribution function.

$$\tilde{D}_a = \tilde{D}_0 \frac{1}{1-\tilde{n}}\left[\left(1+\frac{t'_{ex}}{t}\right)^{1-\tilde{n}} - \left(\frac{t'_{ex}}{t}\right)^{1-\tilde{n}}\right]\left(\frac{t'_0}{t}\right)^{\tilde{n}} \tag{3}$$

where $\tilde{D}_0$ = the diffusion coefficient at the time, t'_0, t = duration of exposure, t'_0 = reference time, t'_{ex} = concrete age at first exposure, $\tilde{n}$ = age factor. For maximum flexibility in the modelling of the time-dependent influence of cracks on the diffusion coefficient, $\tilde{D}_0$ is described somewhat differently from the description by Altmann et al. (2008) as

$$\tilde{D}_0 = \tilde{k}_e \tilde{k}_t \tilde{k}_c \, (\tilde{D}_{ex,0,cf} + \tilde{k}_{cr} \tilde{D}_{ex,0,cr}) \tag{4}$$

with $\tilde{k}_e$ = environment parameter, $\tilde{k}_t$ = test method parameter, $\tilde{k}_c$ = curing parameter according to the DuraCrete model, $\tilde{D}_{ex,0,cf}$ = the experimentally determined diffusion coefficient for crack-free SHCC at the time t'_0, and $\tilde{D}_{ex,0,cr}$ = the contribution of cracks to the diffusion coefficient of cracked SHCC at the time t'_0. Furthermore, a crack intensity factor $\tilde{k}_{cr}$ is introduced to account for the spatially variable influence of cracks for a given strain level.

4.2 Modelling of chloride ingress in investigated specimens

To forecast future chloride profiles for similar specimens exposed to the same conditions it is necessary to quantify the parameters given in Equations 2 to 4 based on the presented experimental results and expert knowledge.

The diffusion coefficients in $D_{ex,\xi}$ Table 5 were determined at different times t'_ξ. From these known pairs of diffusion coefficient and time it is possible to calculate the diffusion coefficient $D_{ex,0}$ at the time $t'_0 = 28$ d for each experimentally determined chloride profile:

$$D_{ex,0} = \left(\frac{t'_0}{t'_\xi}\right)^{\tilde{n}} D_{ex,\xi} \tag{5}$$

with $\tilde{n}$ = age factor according to Equation 3.

Based on these results the fuzzy mean values $\tilde{m}$ of the parameters $\tilde{D}_{ex,0,cf}$, $\tilde{D}_{ex,0,cr}$ and $\tilde{D}_0$ according to Equation 4 can be quantified. As the diffusion coefficients $D_{ex,\xi}$ for the cracked specimens exhibiting few cracks do not vary much from those of the crack-free specimens, $\tilde{D}_{ex,0,cf}$ can be quantified using the results of the specimens $P1$, $P2$, A-2 and B-3.2. Using the expression $\tilde{p} = < a; b; c; d >$ for the membership function of a fuzzy parameter $\tilde{p}$ (cf. Figure 8) this yields:

$$\mu_{\tilde{m}}(\tilde{D}_{ex,0,cf}) = \langle D_{ex,0,P2}; D_{ex,0,A-2}; D_{ex,0,P1}; D_{ex,0,B3.2} \rangle \tag{6}$$

Table 6. Fuzzy-stochastic parameters for modelling chloride ingress in SHCC*.

Parameter	Dimension	Type of uncertainty	Distribution	Mean a**	b	c	d	Standard Deviation*** a**	b	c	d
$\tilde{D}_{ex,0,cf}$	[10^{-12} m²/s]	Fuzzy Random	Normal	10.8	11.0	12.0	12.3	2.17	2.19	2.41	2.45
$\tilde{D}_{ex,0,cr}$	[10^{-12} m²/s]	Fuzzy Random	Normal	9.05	14.1	15.2	20.2	1.81	2.83	3.04	4.04
n	[–]	Random	Beta, a = 0, b = 1	0.6	–	–	–	0.15	–	–	–
$\tilde{k}_{cr}$	[–]	Fuzzy	–	0.16	0.19	0.22	0.27	–	–	–	–
$\tilde{C}_s$	[mass % SHCC]	Fuzzy Random	Log normal	1.88	1.98	2.10	2.19	0.38	0.40	0.42	0.44
k_e	[–]	Crisp	–	1	–	–	–	–	–	–	–
k_t	[–]	Crisp	–	1	–	–	–	–	–	–	–
k_c	[–]	Crisp	–	1	–	–	–	–	–	–	–

* For t'_0 = 28 d & n = 0.6; ** For random or crisp (neither random nor fuzzy) parameters this column specifies the crisp mean value/standard deviation; *** As defined in the DARTS model (2004) for n and NT BUILD 443 (1995) for all other parameters.

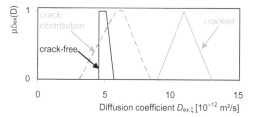

Figure 10. Fuzzy quantification of mean values for the diffusion coefficient $D_{ex,\xi}$ (cf. Table 5).

with $\mu_{\tilde{m}}(\tilde{D}_{ex,0,cf})$ = the membership function for fuzzy mean value of $\tilde{D}_{ex,0,cf}$, and $D_{ex,0,\zeta}$ the diffusion coefficient $D_{ex,0}$ for specimen ζ.

The contribution of cracks to chloride diffusion $\tilde{D}_{ex,0,cr}$ can be determined according to Equation 7:

$$\mu_{\tilde{m}}(\tilde{D}_{ex,0,cr}) = \tilde{m}(\tilde{D}_{ex,0}) \oplus (-\tilde{m}(\tilde{D}_{ex,0,cf})) \quad (7)$$

$\tilde{\mu}_{\tilde{m}}(\tilde{D}_{ex,0,cr})$, $\tilde{\mu}_{\tilde{m}}(\tilde{D}_{ex,0,cf})$ and $\tilde{\mu}_{\tilde{m}}(\tilde{D}_{ex,0})$ = the membership functions of the fuzzy mean values of the parameters $\tilde{D}_{ex,0,cr}, \tilde{D}_{ex,0,cf}$ according to Equation 4 and $\tilde{D}_{ex,0}$, the diffusion coefficient for the cracked specimens A-4 and B-4. For $\tilde{\mu}_{\tilde{m}}(\tilde{D}_{ex,0})$ it is

$$\mu_{\tilde{m}}(\tilde{D}_{ex,0}) = \langle D_{ex,0,B-4}; b; c; D_{ex,0,A-4} \rangle \quad (8)$$
$$b = c = 0.5 (D_{ex,0,B-4} + D_{ex,0,A-4})$$

The age factor $\tilde{n}$ was modelled as a random parameter n in accordance with the DARTS model (2004). It should be noted that the results of the

laboratory experiments presented in this paper do not allow the definition of an age factor for SHCC. A fuzzy parameter quantification was not performed. The considerable uncertainty regarding its value would have reduced the informative value of the results with regards to the influence of cracking on the diffusion coefficient.

As noted previously, the crack intensity factor $\tilde{k}_{cr}$ was introduced to account for the fact that the cracks in the pre-cracked specimens were not equally distributed. This factor quantifies the likelihood that the chloride profile at an arbitrary location is influenced by the presence of cracks. To quantify $\tilde{k}_{cr}$ the specimens A and B were sub-divided into crack-free areas and areas deemed to be influenced by cracking. It was assumed that the influence of cracks extends up to 1 mm into the matrix from to crack surface. Then three lines were drawn along the length of both specimens. For each line the ratio of the length cutting through cracked areas over the gauge length (cf. Figure 2) was determined. Using the expression $\tilde{k}_{cr} < a; b; c; d >$ for the membership function, the values for a and b were defined as the two lowest measured ratios, those for c and d as the two highest ratios.

The surface chloride concentration $\tilde{C}_s$ was quantified using the experimental values given in Table 5, while all further parameters were quantified as crisp with the value 1 (cf. Table 6).

4.3 Results

Figures 11 to 14 show chloride profiles for SHCC exposed to a 16.5% NaCl-solution for 50 years. Figure 11 gives the fuzzy mean values for cracked and crack-free material, while Figure 12 shows the 95% quantiles for the same conditions. The influence of

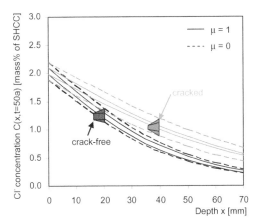

Figure 11. Fuzzy mean values of the chloride profiles for cracked and crack-free SHCC exposed to 16.5% NaCl solution for 50 years.

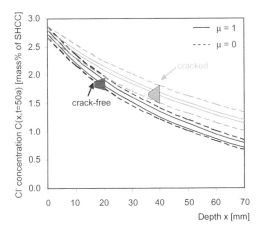

Figure 12. Fuzzy 95% quantiles of the chloride profiles for cracked and crack-free SHCC exposed to 16.5% NaCl solution for 50 years.

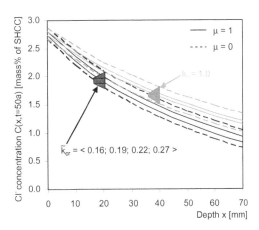

Figure 13. Fuzzy 95% quantiles of the Cl⁻ profiles for SHCC with $\varepsilon \leq 0.40\%$ at the location of cracks ($k_{cr} = 1$) & at arbitrary locations (unknown crack-status, $\tilde{k}_{cr} = <0.16;0.19;0.22;0.27>$) exposed to 16.5% NaCl solution for 50 years.

To account for this fact, self healing may be modelled with a different age factor $\tilde{n}$ for the cracked material. For $\tilde{k}_e = \tilde{k}_t = \tilde{k}_c = 1$ it is

$$\tilde{D}_a = \tilde{D}_{ex,0}\,\tilde{T} + \tilde{k}_{cr}\,\tilde{D}_{ex,0,cr}\,\tilde{T}_{cr} \qquad (7)$$

with

$$\tilde{T} = \frac{1}{1-\tilde{n}}\left[\left(1+\frac{t'_{ex}}{t}\right)^{1-\tilde{n}} - \left(\frac{t'_{ex}}{t}\right)^{1-\tilde{n}}\right]\left(\frac{t'_0}{t}\right)^{\tilde{n}} \qquad (8)$$

$$\tilde{T}_{cr} = \frac{1}{1-\tilde{n}_{cr}}\left[\left(1+\frac{t'_{ex}}{t}\right)^{1-\tilde{n}_{cr}} - \left(\frac{t'_{ex}}{t}\right)^{1-\tilde{n}_{cr}}\right]\left(\frac{t'_0}{t}\right)^{\tilde{n}_{cr}} \qquad (9)$$

For mathematical reasons the age factor may not assume values greater than one. Thus there is a limitation to the rate of self-healing that can be modelled using $\tilde{n}_{cr}$, and it remains to be seen if Equations 7 to 9 allow a realistic representation of self-healing.

To demonstrate the potential and the limitation of this approach, two different non-fuzzy age factors n_{cr} have been defined. The resulting 95% quantiles of the chloride profiles for (i) $\mu(n_{cr}) = 0.6$, $\sigma(n_{cr}) = 0.15$ and (ii) $n_{cr} = 0.999$ (crisp) after 50 years of immersion in 16.5% NaCl solution can be seen in Figure 14. For the age factor $n_{cr} = 0.999$ a new correction of $\tilde{D}_{0,cr}$ as described before was required, yielding the fuzzy mean value $\mu(\tilde{D}_{0,cr}) = <3.81; 4.68; 4.68; 5.55>*10^{-11}$ m²/s and the fuzzy standard deviation $\sigma(\tilde{D}_{0,cr}) = <7.62; 9.35; 9.35; 11.1>*10^{-12}$ m²/s.

the crack intensity factor $\tilde{k}_{cr}$ on the 95% quantiles of the chloride profile is shown in Figure 13.

Small cracks in cementitious materials may close fully or partially due to self-healing depending on the crack width, matrix composition, age, loading history and other factors. The experimental results presented in this paper indicate significant self-healing but do not allow a quantification of this process. However, while bearing in mind the different composition of the matrix, the results reported by Yang et al. (2009) for self-healing of SHCC suggest that it is a quicker process than the reduction of D in crack-free material.

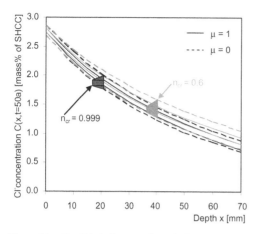

Figure 14. Possible influence of modelling self-healing using the age factor $n_{cr} = 0.999$ on the fuzzy 95% quantiles of the chloride profile for cracked SHCC exposed to 16.5% NaCl solution for 50 years.

4.4 *Discussion*

When interpreting the above results it must be born in mind that the exposure to a 16.5% NaCl solution is much more severe than any exposure that can be expected under field conditions. Thus the chloride concentrations according to Figures 11 to 14 are much higher than they would be under field conditions.

With the fuzzy chloride profiles in Figures 11 to 14 it is possible to determine (i) the depth at which a certain chloride concentration is reached with a given likelihood and (ii) the chloride concentration at a given depth. In Table 6 the results for the upper boundaries of the membership functions ($\alpha = 0$, cf. value d in Figure 8) for the mean values and 95% quantiles of the chloride concentration according to Figures 11 and 14 are tabulated.

According to Table 7 the mean value of the chloride concentration at $x = 50$ mm after 50 years of exposure according to NT BUILD 443 is not higher than 0.55 mass% of SHCC for crack-free and 0.99 mass% of SHCC for cracked SHCC. Furthermore, the likelihood that the chloride concentration exceeds 1.5 mass% of SHCC is not higher than 5% at depths between 37 mm for crack-free SHCC and 59 mm for cracked SHCC. If the crack-status of a member with a strain $\varepsilon \leq 0.40\%$ is unknown, the 95% quantile of the penetration depth is reduced to 45 mm or as little as 38 mm under the assumption of an age factor $n_{cr} = 0.999$.

A comparison of the last value with the penetration depth for crack-free SHCC highlights the urgent need for a quantification of the age factor of SHCC in crack-free material and especially of the self-healing of cracks in SHCC.

Table 7. Upper fuzzy boundaries ($\alpha = 0$, cf. value d in Figure 8) of chloride profiles after 50 years exposure to conditions according to NT BUILD 443 (1995).

Crack status	$\tilde{x}$ (C = 1.5 mass % of SHCC)[mm]	$\tilde{C}$ (x = 50 mm) [mass% of SHCC]
Mean		
Crack-free	17	0.55
Cracked	27	0.99
95% quantile		
Crack-free	37	1.19
Cracked	59	1.62
Unknown	45	1.38
Unknown, $n_{cr} = 0.999$	38	1.23

5 CONCLUSIONS AND OUTLOOK

Chloride ingress and subsequent corrosion of the steel reinforcement is a critical corrosion process for R/SHCC members. To fully utilise material ductility and durability, a performance-based durability design concept that is applicable to both crack-free and cracked SHCC is required. In this paper the results of experimental investigations on the influence of cracking on chloride ion ingress were presented and subsequently used to model chloride profiles exposed to these conditions for extended periods of time.

Significant self-healing of cracked specimens immersed in water and saline solution was observed. Nonetheless the experimentally determined chloride diffusion coefficient was greatly influenced by the presence of cracks. For these reasons and as it is well-known that cracks of limited widths as present in SHCC may fully heal over time (Yang et al. 2009, Edvardsen 1996), further investigations into self-healing of SHCC are required to accurately model chloride ingress in the cracked material.

Furthermore, it was found that at the investigated strain levels cracks were not evenly distributed. As was shown in this paper, such a localisation has a major influence on the quantification of an expected chloride diffusion coefficient at an arbitrary location in a member under tensile strain. Thus the spatial distribution as well as further characteristic crack properties such as length and width as a function of the induced load or strain need to be investigated further.

According to Gehlen (2000) the age factor n of crack-free material has a greater impact on the chloride ingress over the lifetime of a structure than the diffusion coefficient D_0. For SHCC no information on the age factor is available, and thus significant research effort in this area is required.

While the potential of the presented fuzzy-probabilistic chloride ingress model is obvious, a significant effort is still required to reliably quantify the model parameters. Ongoing experimental investigations will provide more information that will help in the refinement of the model and the quantification of its parameters.

ACKNOWLEDGEMENT

The authors would like to acknowledge the ongoing support of Dr.-Ing. Uwe Reuter of the Department of Civil Engineering, TU Dresden, in the development of the fuzzy-probabilistic model.

REFERENCES

Altmann, F., Mechtcherine, V. & Reuter, U. 2008. A novel durability design approach for new cementitious materials: Modelling chloride ingress in strain-hardening cement-based composites. In M.G. Alexander et al. (eds.), *2nd Int'l Conference on Concrete Repair, Rehabilitation and Retrofitting II (ICCRRR 2008)*, Cape Town, South Africa. CRC Press/Balkema.

Crank, J.C. 1979. *The mathematics of diffusion.* Oxford: Clarendon Press.

DARTS 2004. *Data.* The European Union—GROWTH 2000, Project GRD1–25633.

DuraCrete 2000. *Probabilistic performance based durability design of concrete structures.* The European Union—Brite Euram III, Project BE95-1347 R15.

Edvardsen, C.K. 1996. *Wasserundurchlässigkeit und Selbstheilung von Trennrissen in Beton.* Berlin: Beuth.

Gehlen, C. 2000. *Probabilistische Lebensdauerbemessung von Stahlbetonbauwerken.* Berlin: Beuth.

Jun, P. & Mechtcherine, V. 2007. Stress-strain behaviour of strain-hardening cement-based composites (SHCC) under repeated tensile loading. In A. Carpinteri et al. (eds.), *6th Int'l Conference on Fracture Mechaincs of Concrete and Concrete Structures (FraMCoS-6),* Catania, Italy. London: Taylor & Francis.

Kabele, P., Novák, L., Němeček, J. & Kopecký, L. 2006. Effects of chemical exposure on bond between synthetic fiber and cementitious matrix. *1st Int'l RILEM Conference on Textile Reinforced Concrete (ITCR'2006),* Aachen, Germany.

Li, V.C. 2003. On Engineered Cementitious Composites (ECC): A review of the material and its applications. *Journal of Advanced Concrete Technology* 1(3): 215–230.

Li, V.C. & Stang, H. 2004. Elevating FRC material ductility to infrastructure durability. In M.d. Prisco et al. (eds.). *BEFIB'2004,* Varenna, Italy. RILEM Publications

Mechtcherine, V. & Altmann, F. 2009. Durability of structural elements and structures. In F.H. Wittmann et al. (eds.), *Durability of Strain-Hardening Fibre-Reinforced Cement-Based Composites (SHCC)—State-of-the-Art (in preparation).*

Mechtcherine, V., Lieboldt, M. & Altmann, F. 2007. Preliminary tests on air permeability and water absorption of cracked and uncracked strain hardening cement-based composites. In K. Audenaert et al. (eds.), *Int'l RILEM Workshop on Transport Mechanisms in Cracked Concrete,* Ghent. Acco.

Mechtcherine, V. & Schulze, J. 2005. Ultra-ductile concrete—Material design concept and testing. *CPI Concrete Plant International* 5: 88–98.

NT Build 443. 1995. Concrete, hardened: Accelerated chloride penetration.

Tang, L. & Gulikers, J. 2007. On the mathematics of time-dependent apparent chloride diffusion coefficient in concrete. *Cement and Concrete Research* 37(4): 589–595.

Yang, Y., Lepech, M.D., Yang, E.-H. & Li, V.C. 2009. Autogenous healing of engineered cementitious composites under wet-dry cycles. *Cement and Concrete Research* 39(5): 382–390.

Advances in Cement-Based Materials – van Zijl & Boshoff (eds)
© *2010 Taylor & Francis Group, London, ISBN 978-0-415-87637-7*

Physical and mechanical properties of Strain-Hardening Cement-based Composites (SHCC) after exposure to elevated temperatures

M.S. Magalhães, R.D. Toledo Filho & E.M.R. Fairbairn
PEC/COPPE, Universidade Federal do Rio de Janeiro, Brasil

ABSTRACT: An experimental program was carried out to study the physical and mechanical properties of strain-hardening cement-based composites (SHCC) after exposure to high temperatures. The SHCC were prepared using Portland cement (CP), fly ash (FA), silica sand (S), admixtures, and PVA fibers. The thermal stability of the fibers was determined by thermal analysis up to the temperature of 800°C whereas the composites were subjected to the temperatures of 120°C and 150°C before being tested at room temperature (22°C). The specimens were heated at a constant heating rate of 1°C/min with residence time of 1 h. The residual stress-strain behavior under compression and direct tension loads of the SHCC was determined and the change in the water porosity of the specimens with heating was measured by water absorption tests. Thermal analysis of the PVA fibers indicated that above 256°C the process of fiber degradation becomes very intense until the temperature of 470°C when the fiber loses about 70% of its initial mass. The results of direct tension tests indicated that the specimens exposed to the temperature of 120°C presented the same behavior of the non heated control specimens. Some degradation in the strain capacity of the composite was observed, however, for the specimens submitted to 150°C. The results of the compression tests indicated that occur a reduction in the elastic modulus and an increase in the compressive strength of the composites after being exposed to the temperatures of 120°C and 150°C.

1 INTRODUCTION

Several types of high performance fiber-reinforced cement-based composites (HPFRCC) have been developed since 1970 (Naaman 2008). Among the various classes of HPFRCC, a particular one is of interest here: the strain hardening cement-based composites (SHCC). SHCC is a type of high performance fiber reinforced concrete that has superior tensile ductility, with a high tensile strain capacity (up to 5.0%). This ductility is achieved by the formation of multiple cracks, with generally less than 80 µm in width. This exceptional ductility has resulted in more durable structures (Li 1997).

SHCC is designed with micromechanical principles to achieve high damage tolerance under severe loading and high durability under normal service conditions. Micromechanical parameters associated with fiber, matrix, and interface are combined to satisfy a pair of criteria, the first crack stress criterion and steady state cracking criterion to achieve the strain hardening behavior (Li 1998, Li 2003, Li et al., 2001).

SHCC have been studied in various aspects, such as: mechanical behavior (Li 1998; Li et al., 2001, Li et al., 1994, Fukuyama et al., 1999; Kesner et al., 2003, Suthiwarapirak et al., 2002), creep (Billington & Rouse 2003), shrinkage (Weimann & Li, 2003), durability (Lepech & Li, 2006). However nothing is known about its thermal durability.

Concrete structures, when exposed to thermal load gradients or high temperatures can present modifications in their properties committing their efficiency and, consequently, their durability. The mechanism of concrete deterioration, due to thermal loads, consists of the appearance of differential internal stresses that will promote and spread cracks, forming rupture surfaces or a preferential net of connectivity among the pores increasing the concrete permeability, making it more susceptible to attack of aggressive agents.

In this way, the determination of the thermal properties is very important to predict the behavior of the structure when submitted to thermal loads. The aim of the present work is therefore to study the effect of high temperatures on the mechanical and physical behavior of strain hardening cement-based composites (SHCC).

2 MATERIALS AND METHODS

2.1 Materials

The materials used in the mixtures were: Portland cement (PC) CPII F-32 defined by the Brazilian standard (NBR 11578 1991) composed with filler with 32 MPa of compressive strength at 28 days, fly ash (FA); silica sand with a maximum diameter of 0.21 mm and a density of 2.60 g/cm^3; water; polyvinyl alcohol (PVA) fibers; and a melamine superplasticizer (CC583). The physical and chemical characteristics of the cementitious materials are presented in Table 1 and the properties of PVA fiber (PVA REC 15—Kuraray Japan) are listed in Table 2. Details of the mixture, used in this investigation, are given in Table 3.

2.2 Casting, curing and heating of specimens

The mixture was prepared in a mechanical mixer with capacity of 20 liters. The cementitious materials and sand were dry mixed for 3 minutes (for homogenization) with the subsequent addition of superplasticizer. The powder material was mixed for

Table 1. Chemical and physical compositions of cementitious materials.

	PC	FA
Chemical composition (%)		
Na$_2$O	0.331	0.26
MgO	1.344	0.50
Al$_2$O$_3$	3.706	28.24
SiO$_2$	15.326	57.78
P$_2$O$_5$	0.101	0.06
SO$_3$	3.327	–
Cl	0.086	–
K$_2$O	0.189	2.54
CaO	71.476	1.26
MnO	0.045	0.03
Fe$_2$O$_3$	3.777	4.76
ZnO	0.034	–
SrO	0.257	–
TiO$_2$	–	0.95
BaO	–	<0.16
Density (g/cm^3)	3.08	2.35

Table 2. Properties of fiber PVA.

Properties	
Length (mm)	12.0
Diameter (μm)	40.0
Density (g/cm^3)	1.3
Modulus (GPa)	40.0
Elongation at break (%)	7.0
Tensile strength (MPa)	1600

Table 3. Mixture properties of SHCC.

Properties	
FA/C	1.20
W/CM*	0.36
Cement (PC) (Kg/m^3)	505.00
Fly Ash (FA) (Kg/m^3)	606.00
Sand (S) (Kg/m^3)	404.00
Water (W) (Kg/m^3)	404.00
Superplasticizer (SP) (Kg/m^3)	15.15
Fiber (PVA) (Kg/m^3)	26.00
Air content (%)	2.80

* CM—cementitious materials (cement + fly ash).

more 30 seconds and when the water was added, it was mixed for more 5 minutes. The specimens were cast in steel molds and demolded 24 hours after casting. During this period the specimens were kept covered with damp cloths and a polythene sheet. After this time the specimens were cured until the age of 28 days in a cure chamber with 100% relative humidity (RH) and 22 ± 1°C of temperature. Three samples were tested immediately after the curing, and the results will be referred to as those obtained under normal curing (control tests carried out at room temperature (22 ± 1°C)).

The samples submitted to high temperatures were maintained in ambient temperature (22 ± 1°C) for 24 h before they were heated. The specimens were heated in an electric furnace to peak temperatures of 120°C and 150°C at a slow increasing rate of 1°C/min. When the target peak temperature was reached, the furnace temperature was maintained for 60 minutes and then the furnace was turned off and specimens were allowed to cool down to room temperature.

2.3 Testing procedure and methods

2.3.1 Thermal analyses of PVA fiber

Thermogravimetric analysis (TG) were performed from 35 to 800°C in N$_2$ (53 mL/min) and in O$_2$ (8 mL/min) at a heating rate of 10°C/min) in a simultaneous TG/DTA Rigaku, model TAS 100.

2.3.2 Physical properties

The physical properties were studied by total porosity, density and total absorption tests according to NBR 9778 (2005). The samples were cylindrical with 50 × 25 mm (diameter × height).

2.3.3 Compressive strength

The axial compression tests were performed on cylindrical specimens of 50 mm × 100 mm (diameter × height). The cylinders were tested in a Shimadzu universal testing machine (UH-FI-1000KN) controlled by computer under strain

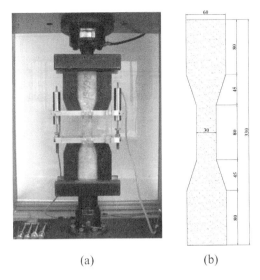

(a)   (b)

Figure 1.   (a) Tension test set-up, and (b) the dimensions of the tensile specimen (all dimensions in mm).

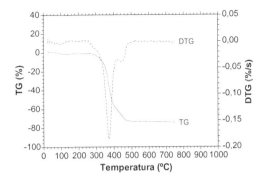

Figure 2.   Thermogravimetric analysis (TG) and differential thermogravimetric analysis (DTG) of PVA fibers.

Table 4.   Residual physical properties of SHCC.

T (°C)	TP (%)	WA (%)	D (g/cm³)
22	23.97(1.74)	14.54(1.65)	1.67(0.52)
120	23.10(1.82)	13.66(2.32)	1.69(0.62)
150	22.08(1.72)	13.08(3.62)	1.68(1.11)

control at a loading rate of 0.2% FS/min. The complete load—displacement date was recorded. The reported data are the average values from three specimens. The Young's modulus was calculated according to the ABNT NBR 8522 (2003).

2.3.4   *Direct tension tests*

In order to determine the direct tension behavior of the studied composites, a Shimadzu AGX 100 kN was used. The tests were carried out at a crosshead rate of 0.1 mm/min.

Three specimens with the dimensions of 330 mm × 30 mm × 30 mm (length × width × thickness) were tested under direct tension. The tensile specimen test set-up is shown in Figure 1. Two LVDTs were used to measure displacements between two points on the specimen at a gauge length of 80 mm. Loads and corresponding displacements were continuously recorded during the tests.

3   EXPERIMENTAL RESULTS AND ANALYSIS

3.1   *Thermal analyses of PVA fiber*

The thermogravimetry (TG) and differential thermal analysis (DTA) test results are shown in Figure 2. At the temperature of 256°C the PVA fiber began to decompose slowly until the temperature of 300°C. Until this temperature the fiber lost approximately 3.30% of its mass. After that, the process of fiber degradation is faster, until the temperature of 470°C, when the fiber lost approximately 73% of its initial mass.

3.2   *Residual physical properties*

The residual physical properties of the composites namely total porosity (TP), density (D) and water absorption (WA) are presented in Table 4. The obtained results indicate that when the temperature is increased the total porosity and water absorption is reduced and the density is slightly increased. For example, when the samples were heated to 120°C and 150°C the total porosity was reduced by, respectively, 3.6 and 7.9% whereas the water absorption was reduced by 6% and 10%, respectively. The density, on the other hand, was not influenced (an increase of about 0.5–1% was observed).

3.3   *Stress-strain behavior under compression*

Typical residual stress-strain curves obtained for specimens tested at room temperature and after exposure to 120°C and 150°C are presented in Figure 3. The compressive strength and elastic modulus are presented in Table 5.

The results indicate that compressive strength of the composites was increased by about 9% when the specimens were heated to the temperatures of 120°C and 150°C. This strength gain may probably be related with the reaction between unhydrated fly ash particles and calcium hydroxide at elevated temperatures.

Regarding to the results of elastic modulus of the composite, it was observed that it decreased by about 24% when the temperature was increased to the temperatures of 120°C and 150°C.

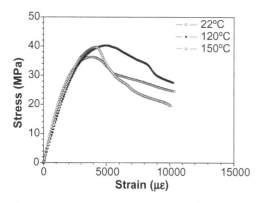

Figure 3. Typical compression stress-strain curves of SHCC after exposure to high temperatures.

Table 5. Residual compressive strength and elastic modulus.

T (°C)	Compression strength (MPa)	Elastic modulus (GPa)
22	36.27(0.41)	19.00(3.52)
120	39.58(1.76)	14.57(1.98)
150	39.44(3.86)	14.48(23.8)

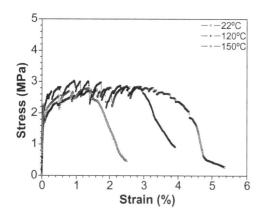

Figure 4. Direct tension stress-train behavior of SHCC after exposure to high temperatures.

3.4 Stress-strain behavior under direct tension

Typical residual stress-strain curves obtained for specimens tested at room temperature and after exposure to 120°C and 150°C are presented in Figure 4 and the results of first cracking strength, ultimate tensile strength and tensile strain capacity are shown in Table 6.

The results indicated that the ultimate tensile strength and the residual strain capacity of the composite were affected by the heating only for

Table 6. Experimental results of the direct tension tests.

T (°C)	First cracking strength, MPa	Ultimate tensile strength, MPa	Tensile strain capacity, %
22	2.12(11.19)	2.90(6.47)	2.98(14.02)
120	2.10(6.73)	3.01(1.18)	2.91(3.40)
150	2.04(6.98)	2.61(5.64)	1.40(3.98)

(a)

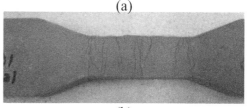

(b)

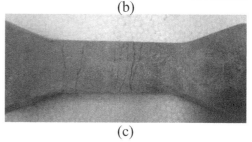

(c)

Figure 5. Typical crack pattern observed after direct tension tests. (a) Typical specimen without heating, (b) typical specimen heated to 120°C and (c) typical specimen heated to 150°C.

the temperature of 150°C. In this case, the strain capacity was reduced by about 53.0% and the ultimate tensile strength by about 10%. When exposed to the temperature of 120°C the tensile strength of the composite was slightly increased (by about 4%) when compared to the result presented by the unheated reference specimens. It is important to mention, however, that the composites still presented a strain hardening behavior with a pronounced multiple cracking pattern (see Fig. 5) after being heated to 150°C. The first cracking strength of the specimens heated to 120°C and 150°C was not modified significantly.

4 CONCLUSIONS

On the basis of the experimental results the following conclusions can be drawn:

- The temperature has affected the physical properties of SHCC. The total porosity was reduced by 3% and 8% when the specimens were heated to temperatures of 120°C and 150°C, respectively, whereas the water absorption was reduced by 6% and 10%, respectively.
- An increase of about 9% in the residual compressive strength the SHCC samples exposed to the temperatures of 120°C and 150°C was observed in the present study. The elastic modulus, on the other hand, was decreased by about 24% when the temperature increased up to 150°C.
- The results of direct tension tests indicated that up to the temperature of 150°C the behavior of SHCC is still ductile but it has experienced decrease in both ultimate tensile strength (by about 10%) and strain capacity (by about 53%). The first cracking strength of SHCC was not significantly modified with the exposure to the temperatures studied.

REFERENCES

Billington, S.L., Rouse, M. 2003. "Time—dependent response of highly ductile fiber—reinforced cement-based composites". Proceedings of the BMC-7, Warsaw, Poland, eds. A.M. Bramn, V.C. Li and I.M. Marshall, pp. 47–56.

Fukuyama, H., Matsuzaki, Y., Nakano, K. et al, 1999, "Structural performance of beam elements with PVA-ECC." in: Proc. of High Performance Fiber Reinforced Cement Composites 3 (HPFRCC 3), Ed. Reinhardt and A. Naaman, Chapman & Hull, 531–542.

Kesner, K.E., Billington, S.L., Douglas, K.S., 2003, "Cyclic response of highly ductile fiber-reinforced cement-based composites". ACI Material Journal. V. 100 n° 5, pp. 381–390.

Lepech, M.D., Li, V.C., 2006, "Long Term Durability Performance of Engineered Cementitious Composites", Journal of Restoration of Buildings and Monuments, Vol. 12, No. 2, pp. 119–132.

Li, V.C., Mishra, D.K., Naaman, A.E. et al, 1994, "On the Shear Behavior of Engineered Cementitious Composites," Journal of Advanced Cement Based Materials, Vol. 1, No. 3, pp. 142–149.

Li, V.C., 1997, "Damage tolerance of engineered cementitious composites." in: Advances in Fracture Research, Proc. 9th ICF Conference on Fracture, Sydney, Australia, Ed. B.L. Karihaloo, Y.W. Mai, M.I. Ripley and R.O. Ritchie, Pub. Pergamon, UK, 619–630.

Li, V.C., 1998, "Engineered Cementitious composites (ECC)—Tailored Composites through Micromechanical Modeling", in Fiber Reinforced Concrete: Present and Future. Eds: N. Banthia, A. Bentur, and A. Mufti, Canadian Society of Civil Engineers, Montreal, 64–97.

Li, V.C., Wang S., Wu, C., 2001, "Tensile Strain-hardening Behavior of PVA-ECC," ACI Materials Journal, Vol. 98, n° 6, Nov.–Dec., pp. 483–492.

Li, V.C., 2003, "On Engineered Cementitious Composites (ECC). A Review of the Material and its Applications," Journal Advanced Concrete Technology, Vol. 1, n° 3, pp. 215–230.

Naaman, A.E., 2008, "Development and Evolution of Tensile Strain-Hardening FRC Composites," In: Seventh RILEM International Symposium on Fibre Reinforced Concrete: Design and Applications—BEFIB2008, 2008, Indian. Seventh RILEM Symposium on Fibre Reinforced Concrete BEFIB2008. v. 1. pp. 1–28.

NBR 9778, 2005, Argamassa e concreto endurecidos—Determinação da absorção de água por imersão—Índice de vazios e massa específica. Associação Brasileira de Normas Técnicas (ABNT).

NBR 8522, 2003. Determinação dos módulos estáticos de elasticidade e de deformação e da curva tensão-deformação.

Suthiwarapirak, P., Matsumoto, T., Kanda, T., 2002, "Flexural Fatigue Failure Characteristics of an Engineered Cementitious Composite and Polymer Cement Mortars", proceedings of JSCE, n° 718, pp. 121–134.

Weimann, M.B., LI, V.C., 2003, "Drying Shrinkage and Crack Width of ECC," BMC-7, Poland, October, pp. 37–46.

Advances in Cement-Based Materials – van Zijl & Boshoff (eds)
© *2010 Taylor & Francis Group, London, ISBN 978-0-415-87637-7*

Determining the durability of SCC using the South African early-age Durability Index tests

B.G. Salvoldi, M.E. Gillmer & H. Beushausen
Department of Civil Engineering, University of Cape Town, South Africa

ABSTRACT: Advantages associated with self compacting concrete (SCC) commonly relate to improved productivity on site, a better working environment, and enhanced homogeneity of the mix. The original objective of developing SCC was to increase the durability of concrete structures by making the concrete less susceptible to poor workmanship during placement and compaction. Given that SCC is a relatively new material, practical experience and research results on durability characteristics of SCC are still limited. Differences in pore structure and paste content between SCC and conventional vibrated concrete suggest that different transport characteristics may exist for aggressive agents such as carbon dioxide or chlorides. Research on durability reported in the literature however indicates that the durability characteristics of SCC are similar to those of vibrated concrete produced with the same binder type and water/binder ratio. Contradictory opinions still exist about whether or not common early age durability prediction test, which were developed and calibrated for conventional concrete, can equally be applied to SCC.

This research aimed at identifying if the South African Durability Index approach is applicable for the prediction of the potential durability of SCC. Durability Index values of both SCC and conventional concrete samples were compared to accelerated carbonation and chloride diffusion characteristics. The test results indicate that similar durability characteristics exist for SCC and vibrated concrete and that the relationship between DI values and carbonation or chloride ingress are comparable.

1 INTRODUCTION

As a result of its rheological properties self-compacting concrete is intrinsically distinguishable from conventional vibrated concrete (normal concrete, NC). It is defined as concrete that can completely fill formwork, encapsulate reinforcement and de-aerate under the influence of gravity alone. The mix design for SCC is different from the mix design of NC as it focuses on rheological properties such as flowability and passing ability.

SCC has commonly a higher paste content and a lesser amount of coarse aggregates, compared to NC, which can have an effect on the durability characteristics. The ingress of aggressive agents, such as carbon dioxide or chlorides depends to a large extend on the pore structure and penetrability of the paste. On the one hand a higher paste content may result in increased carbonation or chloride ingress, considering that the paste is the main path way for the ingress of aggressive agents. On the other hand, SCC has been reported to have a refined pore structure and more uniform and denser ITZ, which results in lower permeability [1]. For example, Boel et al. tested the microstructure of SCC and determined that the critical pore sizes of SCC were much smaller than those of NC

and that the ITZ exhibited greater adhesion and in some instances the concrete matrix partly penetrated the porous aggregates [2, 3].

In general, the factors influencing transport properties of SCC appear to be the same as those for conventional concrete, i.e. water/binder ratio, degree of hydration and mineral additions. Research on carbonation resistance of SCC still yields contradictory results but it appears that carbonation of SCC is not significantly different from that of conventional concrete. In terms of chloride resistance, the high powder content and admixture dosage in SCC may result in different pore volumes, different ionic compositions in the pore solution and perhaps different chloride binding characteristics, compared with conventional concrete [1]. In principle, current experience and knowledge suggests that the chloride resistance of SCC is to a large degree dependent on the type of cement and additional powder material used, similar to conventional concrete. Common test methods for chloride resistance include electrochemical test methods, which allow a fast assessment of chloride transport characteristics, and methods that measure actual chloride ingress of concrete samples immersed in a salt solution. De Shutter et al. indicate that the relationship between accelerated test results and actual chloride ingress may be different for SCC, compared to NC [1]. They stress that when electrochemical test methods, which were developed and calibrated for conventional concrete, are applied to characterise SCC for chloride penetration, experimental results need to be carefully analysed and interpreted.

The South African early age Durability Index tests were designed to predict the resistance of concrete against carbonation and chloride ingress and have successfully been applied for specification and quality control of concrete structures. The DI approach addresses both the transport of aggressive agents within the concrete matrix and the correlation between lab and practice. In an effort to check if the South African Durability Index Approach could be used to predict Self-Compacting Concretes potential resistance to deterioration this study was conducted. SCC mixes were prepared with different binders and different water/binder ratios and compared to normal concrete mixes of similar composition.

2 EXPERIMENTAL PROGRAMME

2.1 Objectives and test procedures

Self compacting concrete mixes and conventional concrete mixes were designed and tested for their durability characteristics. The durability index (DI) parameters investigated in this research include the Oxygen Permeability Index (OPI) and the Chloride Conductivity Index. A detailed discussion of the South African DI approach, test methods and application principles can be found in the literature [4, 5]. In addition, the concrete mixes were tested for accelerated chloride ingress using the Bulk Diffusion Test [6] and accelerated carbonation. The experimental programme had the following main objectives:

- Develop self compacting concrete mixes using locally available materials
  Compare the potential durability of SCC and NC
- Compare the relationship between OPI values and accelerated carbonation for SCC and NC
- Compare the relationship between chloride conductivity values and accelerated chloride ingress for SCC and NC.

The South African DI approach links OPI and chloride conductivity parameters to service life prediction models and performance specifications. The service life models used in this approach are based on the relevant DI parameter, depending on whether the design accounts for carbonation-induced or chloride-induced corrosion. Designers and constructors can use the approach to optimize the balance between required concrete quality and cover thickness for a given environment and binder system. One of the main advantages of the DI test methods is that the concrete can be tested at young ages and adjustments to the mix design and construction procedures be made during the planning phase or execution of the project.

The OPI test method consists of measuring the pressure decay of oxygen passed through a 25 to 30 mm thick slice of a (typically) 68 mm diameter core of concrete placed in a falling head permeameter. The oxygen permeability index is defined as the negative log of the coefficient of permeability. Common OPI values for South African concretes range from 8,5 to 10,5, a higher value indicating a higher impermeability and thus a concrete of potentially higher quality.

The accelerated carbonation method consists of a chamber in which the RH, temperature and CO_2 concentration can be kept constant. For this research these parameters were kept at 65% RH, 32°C and 5% CO_2 concentration. The carbonation depths are measured by applying phenolphthalein on freshly cut surfaces perpendicular to the carbonation front and measuring with vernier calipers.

The chloride conductivity test apparatus consists of a two cell conduction rig in which preconditioned concrete core samples are exposed on either side to a 5M NaCl chloride solution. The movement of chloride ions occurs due to the application of a 10 V potential difference and the chloride conductivity is determined by measuring the current flowing through the concrete specimen.

In the bulk diffusion test [6], samples are saturated with limewater, sealed on all sides except the top face, and submerged in a 2.8 M NaCl solution, for a minimum of 35 days. After this time chloride profiling is done, by grinding 0.5 mm layers of the sample and measuring the chlorides at different depths. The chloride concentration profile is then used to determine the diffusion coefficient and surface concentration values.

2.2 Mix design

The mix design of SCC focuses mainly on the fresh properties of the concrete to ensure flowability and passing ability. Once these rheological parameters are achieved the hardened properties can be controlled by adjusting the powder proportioning of cement, cement extenders and inert fines called fillers.

The fresh properties can principally be obtained by limiting the coarse aggregate content, which results in a higher paste volume, and reducing water/powder ratios. Limiting mix design values were presented in [1].

The water/binder ratios chosen for this research were 0.4 and 0.5. The mix designs were based on methods suggested by Audenaert, et al. [7].

One of the most important parameters when designing SCC is the paste content of the mix as a high paste content ensures good flowability and viscosity. SCC requires about 40% paste content and 60% aggregate content depending on aggregate size, shape and grading. Based on the selected water/binder ratios, a paste content of 43% for the 0.4 mixes and 39% for the 0.5 mixes were chosen. The fine and coarse aggregates were calculated as the remaining volume with the fine aggregates making up 55% of that volume. During the preparation of trial mixes, Superplasticiser was added until the desired rheological parameters were achieved. An optimization of the mix designs, which would have been the next step towards production of economical, high performance SCC was not undertaken due to time constraints.

The viscosity and flowability of the SCC mixes were tested with the slump flow test. This test is carried out with a slump cone and a large plate where the centre and the circumference of a 500 mm diameter circle are indicated. The cone is inverted, placed in the centre and held down steadily while it is being filled. Once the cone is lifted the spread of the SCC is measured as well as the time it takes for the SCC to reach the 500 mm diameter line (T_{500} time).

The mix designs for the NC mixes were based on the respective SCC mixes to allow a good comparison. The paste content was lowered and the fine and coarse aggregate proportioning adjusted. Mix designs and selected hardened properties for the SCC and NC mixes are presented in Tables 1 and 2, respectively. All specimens were water-cured at 23°C for 28 days and subsequently prepared for the tests.

Table 1. Mix designs for all SCC mixes.

	Units	SCC 0.4 CEM1	SCC 0.4 CS	SCC 0.4 FA	SCC 0.4 SF	SCC 0.5 CEM1	SCC 0.5 CS	SCC 0.5 FA	SCC 0.5 SF
CEM 1 42.5 N	(kg/m³)	600	360	420	580	480	288	336	464
Corex slag	(kg/m³)	–	240	–	–	–	192	–	–
Fly ash	(kg/m³)	–	–	180	–	–	–	144	–
Silica fume	(kg/m³)	–	–	–	20	–	–	–	16
Water	(kg/m³)	240	240	240	240	240	240	240	240
Water:binder ratio	–	0.4	0.4	0.4	0.4	0.5	0.5	0.5	0.5
Phillipi dune sand	(kg/m³)	781	781	781	781	455	455	455	455
Crusher sand	(kg/m³)	–	–	–	–	454	454	454	454
Coarse aggregate	(kg/m³)	639	639	639	639	606	606	606	606
Fine aggregate %	–	55	55	55	55	60	60	60	60
Superplasticiser:	(l/m³)	8.7	4.7	4.0	7.3	4.7	1.3	0.7	4.0
Achieved flow:	(mm)	540	550	560	550	540	520	560	550
T_{500} time	(s)	≤2	≤2	≤2	≤2	≤2	≤2	≤2	≤2
Slump flow class/ Viscosity class	–	SF1/ VS1	SF1/ VS1	SF1/ VS1	SF1/ VS1	SF1/ VS1	SF1/ VS1	SF1/ VS1	SF1/ VS1
28d comp. str.	(MPa)	57	63	60	57	41	57	38	41

Table 2. Mix designs for all NC mixes.

	Units	NC 0.4 CEM1	NC 0.4 CS	NC 0.4 FA	NC 0.4 SF	NC 0.5 CEM1	NC 0.5 CS	NC 0.5 FA	NC 0.5 SF
CEM 1 42.5 N	(kg/m³)	500	300	350	483	400	240	280	386
Corex slag	(kg/m³)	–	200	–	–	–	160	–	–
Fly ash	(kg/m³)	–	–	150	–	–	–	120	–
Silica fume	(kg/m³)	–	–	–	17	–	–	–	14
Water	(kg/m³)	200	200	200	200	200	200	200	200
Water:binder ratio	–	0.4	0.4	0.4	0.4	0.5	0.5	0.5	0.5
Phillipi dune sand	(kg/m³)	640	640	640	640	840	840	840	840
Coarse aggregate	(kg/m³)	960	960	960	960	840	840	840	840
Fine aggregate %	–	40	40	40	40	50	50	50	50
Superplasticiser:	(l/m³)	0.0	0.0	0.0	0.0	0.3	0.0	0.0	0.3
Slump:	(mm)	120	100	75	80	50	65	75	60
28d comp. str.	(MPa)	65	71	54	60	41	58	38	41

2.3 Test results: Oxygen permeability and accelerated carbonation

The test results for Oxygen Permeability Index values are presented in Figures 1 and 2 for mixes with water/binder ratios of 0.4 and 0.5 respectively.

When interpreting the results of the OPI tests it is important to keep in mind that the OPI scale is a logarithmic one and therefore a sample with an OPI value of 10 would be twice as permeable as a sample with an OPI value of 10.3 as the corresponding permeability values (k) would be $1 * 10^{-10}$ and $5 * 10^{-11}$ respectively. The OPI values of all mixes are well above 10.0, indicating high quality concrete with low permeability.

In general, OPI values obtained on SCC samples were similar to the respective NC samples. Exceptions to this observation were specimens made with plain CEM I and Corex Slag (CS) blend and a water/binder ratio of 0.4. For these specimens, a noticeably lower OPI value was obtained on the SCC mixes, which could not be explained. Comparing the OPI values of the SCC mixes to their corresponding NC mixes it can be seen that the CEM I and CS binders had higher OPI values

for the NC mixes than for the SCC mixes and vice versa for the FA and SF binders. A reason for this could be found in the bleeding characteristics of the SCC mixes. SCC specimens made with CEM I and CS showed significantly higher bleeding than mixes made with fly ash or silica fume. This may have contributed to the comparatively higher permeability observed on those specimens. In practice, this would be of lesser significance as SCC mixes would commonly be optimized to reduce bleeding.

The carbonation depths measured after exposure of 56 days were generally very small (Figures 3 and 4), indicating high quality concrete with good carbonation resistance, which corresponds to the high OPI values measured for the same mixes. Specimens made with fly ash had significantly higher carbonation depths than specimens made with the other binder types. Specimens made with Corex slag and silica fume generally outperformed the plain CEM I concrete. All this was expected. Similar to OPI results, carbonation depth results indicate a similar performance of SCC and NC

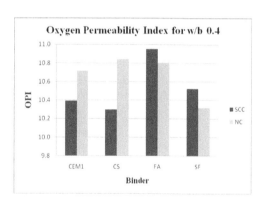

Figure 1. Results for OPI 0.4 w/b mixes.

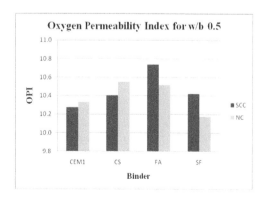

Figure 2. Results for OPI, 0.5 w/b mixes.

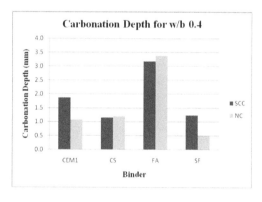

Figure 3. Results for accelerated carbonation, 0.4 w/b mixes.

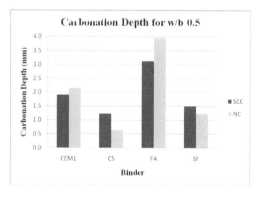

Figure 4. Results for accelerated carbonation, 0.6 w/b mixes.

samples of same binder type and water/binder ratio.

2.4 *Test results: Chloride conductivity and bulk diffusion*

The Chloride Conductivity Index results obtained for the different mixes are presented in Figures 5 and 6. The values obtained correspond well to values reported in the literature for similar mixes (same water/binder ratio and binder type) [8]. The only exception are the mixes made with silica fume (SF). Commonly, the use of silica fume results in significantly lower chloride conductivity, as compared to plain CEM I mixes. However, in this research the silica fume content was limited to 3% in order to achieve the required workability, which is much lower than the usual 7–10%. The replacement of a mere 3% of CEM I with silica fume therefore did not have a positive effect on the test results.

Similarly to the chloride conductivity results, chloride diffusion coefficients obtained with the Bulk Diffusion Test were virtually the same for

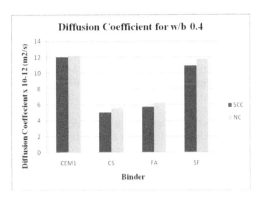

Figure 7. Results of the bulk diffusion test, 0.4 w/b.

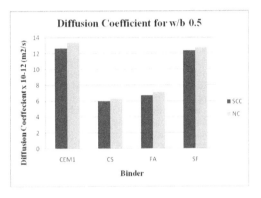

Figure 8. Results of the bulk diffusion test, 0.5 w/b.

SCC and their respective NC mixes. The relationships between chloride conductivity values and chloride diffusion coefficients were as expected and correlate well with previous work done at the University of Cape Town.

The above indicates that, as far as the resistance against chloride ingress is concerned, the SCC mixes used in this research correspond well to their equivalent conventional concrete mixes. The use of the chloride conductivity test to estimate the chloride diffusion coefficient appears applicable for NC and SCC alike. This allows modelling the durability of SCC in the marine environment using the South African Durability Index approach that was developed and calibrated for conventional vibrated concrete.

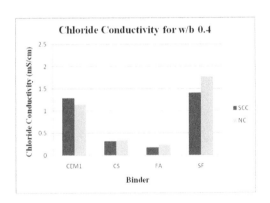

Figure 5. Results chloride conductivity, 0.4 w/b.

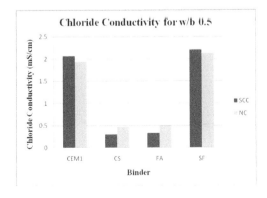

Figure 6. Results chloride conductivity, 0.5 w/b.

3 CONCLUSIONS

Due to different mix compositions (mainly relating to differences in paste content, fines content and amount of coarse aggregates), and different pore structure characteristics of normal concrete

(NC) and self compacting concrete (SCC) (refined pore structure of SCC due to improved ITZ and increased fines content), different durability characteristics may be expected. However, the preliminary investigation discussed in this paper indicates that the durability characteristics of SCC closely compare to the durability characteristics of equivalent NC.

The South African Durability Index test methods, originally developed for and calibrated against NC, were found to be also applicable for SCC.

REFERENCES

[1] De Schutter, G. and Audenaert, K. (editors), 'Durability of self-compacting concrete', *RILEM TC 205-DSC: State-of-the-art report*, Paris: RILEM, April 2007.

[2] Boel V. and De Schutter G. 'Determination of the porosity and the critical pore size of hardened self-compacting cement paste' [Conference] // 5th International RILEM Symposium on Self-Compacting Concrete—Ghent, Belgium: RILEM Publications, 2007a. pp. 571–576.

[3] Boel V. and De Schutter G. 'Optical and electron microscopy on the microstructure of traditional and self-compacting concrete' [Conference] // 5th International RILEM Symposium on Self-Compacting Concrete—Ghent, Belgium: RILEM Publications, 2007b. pp. 577–582.

[4] Alexander, M.G., Ballim, Y. and Stanish, K., 'A framework for use of durability indexes in performance-based design and specifications for reinforced concrete structures', *Materials & Structures*, Vol. 41, No. 5, June 2008, pp. 921–936.

[5] Alexander, M.G. & Beushausen, H., 'Performance-based durability testing, design and specification in South Africa: latest developments', Proceedings: *International Conference on Excellence in Concrete Construction through Innovation*, London, 9–10 September 2008, CRC Press, pp. 429–434

[6] ASTM C1556-04, Standard Test Method for Determining the Apparent Chloride Diffusion Coefficient of Cementitious Mixtures by Bulk Diffusion, Philadelphia: American Society for Testing and Materials, 2004.

[7] Audenaert K. and De Schutter G. 'Modelling the carbonation process of self-compacting concrete' [Conference] // 5th International RILEM Symposium on Self-Compacting Concrete—Ghent. Belgium: RILEM Publications, 2007. pp. 689–694.

[8] Alexander, M.G., Ballim, Y. & Beushausen, H., 'Concrete durability', Fulton's Concrete Technology, 9th edition, Cement and Concrete Institute, Midrand, South Africa, 2009, pp. 155–188.

Advances in Cement-Based Materials – van Zijl & Boshoff (eds)
© *2010 Taylor & Francis Group, London, ISBN 978-0-415-87637-7*

Characterisation of crack distribution of Strain-Hardening Cement Composites (SHCC) under imposed strain

C.J. Adendorff, W.P. Boshoff & G.P.A.G. van Zijl
Stellenbosch University, South Africa

ABSTRACT: The formation of multiple cracks under tensile load is believed to be a durability enhancing mechanism in strain-hardening cement composites (SHCC). The mechanism associated with multiple crack formation is control of individual cracks to widths in the micro-range, whereby new cracks form at close spacing, rather than old cracks widening. This is achieved by effective fibre bridging of the matrix cracks, enabling resistance to higher tensile loads without significant crack widening. To link the cracking characteristics of SHCC to durability, it is required to study crack formation and widening under various loading regimes, and relate them to physical and chemical processes of degradation. In this endeavour it is useful to find elegant ways to observe and characterise crack patterns, considering the high cracking densities, or small spacing of cracks at roughly 1–10 mm, combined with the complexities of crack tortuousity, branching and coalescence. In this paper crack observation with an ARAMIS non-contact, optical, 3D digital deformation observation device is reported. The system combines microscopy and digital image photography with speckle pattern recognition software, whereby local deformations can be computed with high resolution. For the results reported here, an observation area of 70 mm × 30 mm on a tensile specimen was used, whereby up to 60–70 cracks could be observed simultaneously. Crack patterns determined in this manner are reported for loaded and unloaded states. The methodology is used here to study the influence of cyclic loading and loading rate on crack width in SHCC.

1 INTRODUCTION

Strain-hardening cement composites (SHCC) derive their pseudo strain-hardening response to tensile load from steady-state cracking. This means that crack profiles are controlled by effective fibre bridging to allow crack lengthening and the formation of new cracks in the cement-based matrix, as opposed to significant widening of existing cracks. The so-called multiple cracking has been reported by several authors (Li et al. 2001, Weimann and Li 2003, Wang and Li 2006, Boshoff and van Zijl 2007, Jun and Mechtcherine 2009) and shown to afford SHCC deformation compatibility with steel reinforcement in steel reinforced SHCC (Fischer and Li 2004), as well as resistance to moisture and chlorides ingress (Sahmaran et al. 2007).

Several research groups have embarked on studies to establish whether the crack control is retained under general loading conditions, including sustained tensile load (Boshoff et al. 2009, Jun and Mechtcherine 2009), cyclic tensile load (Jun and Mechtcherine 2009) and at various tensile loading rates (Boshoff 2007). The crack pattern characterisation is tedious, as it involves the initiation and evolution of several cracks. In this paper, a methodology of crack pattern characterisation is proposed

and used for investigating crack width evolution with imposed tensile strain under monotonic loads at various loading rates, covering 4 orders of magnitude, as well as under cyclic tensile loading at two different deformation-controlled cycle increments.

2 CRACK CHARACTERISATION METHOD

The crack width measurement used in this research was performed by speckle pattern recognition technology and relative displacement calculation from digital images taken of the specimen gauge area throughout mechanical testing in a Universal Materials Testing Machine. The ARAMIS 3D photographic system was used, from which high resolution images could be traced at specified stages of deformation. In the research reported here, direct tensile tests were performed on dumbbell specimens shown in Figure 1.

An observation area of 70 mm × 30 mm was used. During the tests digital photo's were taken at a frequency of 10 images per second. From these images local deformations were calculated in a post-processing phase. A typical set of average tensile stress-strain responses of SHCC developed

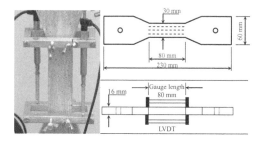

Figure 1. Tensile test setup, showing specimen geometry, LVDT displacement measurement and three lines for crack width evaluation along the central gauge area.

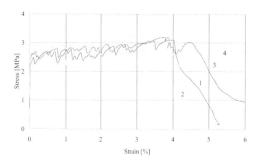

Figure 2. Typical direct tensile responses, with average strain determined from LVDT measurements over the 80 mm gauge length.

at Stellenbosch University is shown in Figure 2. The strains in Figure 2 were calculated by division of the averaged LVDT displacement measurements with the 80 mm gauge length. These LVDT measurements were synchronized with the digital photography, to allow identification of specific stages of imposed strain.

For the purpose of crack identification and crack width calculation, a set of three lines was selected along the specimen gauge area, as shown schematically in Figure 1 (top, right). A grid of roughly 70 points along each line was selected by the ARAMIS system from the speckle pattern applied to the specimen surface by spraying—see Figure 3a.

Referring to the reference state, i.e. the speckle pattern image photographed in the undeformed state, the ARAMIS software analyses each stage to determine the local deformations. In Figures 3b,c the visual crack pattern and contours of local deformation produced by the ARAMIS system are shown for a particular loading stage. It is clear that the crack pattern is traced accurately. From these local deformations crack widths are calculated by assuming that the deformation between cracks is negligible, i.e. the local deformation between speckles is assumed to be the width of a

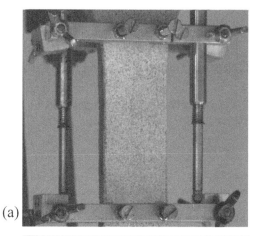

(a)

(b)

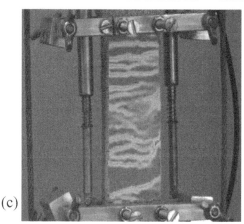

(c)

Figure 3. Tensile specimen showing (a) Speckle pattern, (b) crack pattern under imposed tensile strain, and (c) ARAMIS visualisation of deformations.

crack between such speckles. The possibility exists to adjust the crack widths by subtraction of average deformations observed in uncracked regions. By neglecting such 'uncracked' deformation a

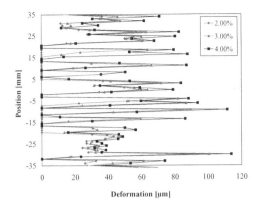

Figure 4. Typical incremental deformation along gauge length, for tensile strain levels 2%, 3% and 4%.

Table 1. Typical SHCC mix proportions.

Ingredient	kg/m³
Cement (CEM I 42.5)	550
Fly Ash	650
Water	395
Sand ($\varphi_s < 0.21$ mm)	550
Fibre: PVA RECS15	26
$L_f = 12$ mm, $d_f = 0.04$ mm	

Table 2. Direct tensile test series.

Test series	No. of specimens	Displacement rate (mm/s)	Average strain rate (s⁻¹)
Rate 1	4	0.0008	10^{-5}
Rate 2	4	0.008	10^{-4}
Rate 3a	6	0.08	10^{-3}
Rate 3b	4		
Rate 4	4	0.8	10^{-2}
Rate 5	4	8	10^{-1}
Cyclic 1: Large strain cycle	3	0.08	10^{-3}
Cyclic 2: Small strain cycle	3	0.08	10^{-3}

conservative crack width is reported. Furthermore, to observe and 'measure' individual cracks, a speckle spacing smaller than the crack spacing is required.

In Figure 4 a typical graph of local deformation, or crack width is shown along the 70 mm centre line of the specimen gauge area at various stages of imposed average strain.

3 EXPERIMENTAL PROGRAM

Several test series were executed in order to study crack patterns, including spacing and width evolution in SHCC under uniaxial tension. A typical SHCC mix was used for the whole test series, with ingredients and mass proportions listed in Table 1. Note that a superplasticiser and a viscous modifying agent chemical additives were used to obtain appropriate levels of workability and consistency.

Tensile dumbbell specimens as shown in Figure 1 were produced by mixing in a Hobart type mixer of 8 liter capacity. The mix was cast into steel moulds, protected by coverage until stripping after one day. The specimens were tested at the age of 14 days, until which date they were cured in water at a controlled temperature of 23°C.

The specimens were tested in direct tension in a Zwick Z250 Universal Materials Testing Machine under cross-head displacement control of 4.8 mm/minute. Special clamps were manufactured to grip the specimen by friction and bearing. Deformation of the gauge length was measured with a set of two 10 mm HBM LVDT's, while an HBM U2a 500 kg load cell measured the force. The LVDT's and load cell were connected to a Spider8 high frequency data logging system.

The specimens were observed throughout the test by the ARAMIS 3D photographic system described in the previous section.

Table 2 summarises the test series reported here, including monotonic tensile tests at various loading rates, as well as two series of cyclic tests. The latter series allowed crack with measurement in the loaded as well as unloaded state at various imposed strain levels.

These test series were performed to characterise crack width evolution under various loading conditions. An important further condition is sustained tensile load, or creep, whereby crack width evolution under sustained tensile load can be measured. Such tests are currently performed and will be reported in later publications. The purpose of such crack characterisation is to establish strain limits at which acceptable levels of crack width and associated durability may be ensured.

Note that for series Rate 1–Rate 5 (excluding Rate 3b) crack measurements were taken only along the central line of the specimens. For Rate 3b and the Cyclic 1 and 2 series, cracks widths were measured along all three lines shown in Figure 1 (top, right).

4 CRACK PATTERN AT VARIOUS LOADING RATES

The results of series Rate 1 to Rate 5 are presented in Figures 5–7, in the form of stress-strain responses and average crack widths and spacing as function of strain level.

From Figure 5 is appears that no significant loading rate dependence acts for the ultimate tensile resistance for this SHCC material. However, in agreement with the findings of Boshoff and van Zijl (2007b), an increased first cracking stress is observed with increased loading rate. A slight reduction in ductility with loading rate increase over 5 orders of magnitude is also observed. Nevertheless, the ultimate tensile strain remains more than 3%.

The crack width evolution with imposed strain is illustrated in Figure 6, by plotting the number of cracks for each specimen loaded at Rate 3. The observed trend of increased crack density, and stabilisation at a tensile strain level of roughly 3% is typical of all loading rates. Also shown for each specimen is the average crack width as function of

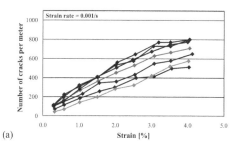

(a)

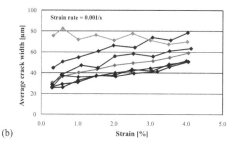

(b)

Figure 6. (a) Number of cracks and (b) average crack width as function of imposed strain level at loading Rate 3, for each of the 6 specimens.

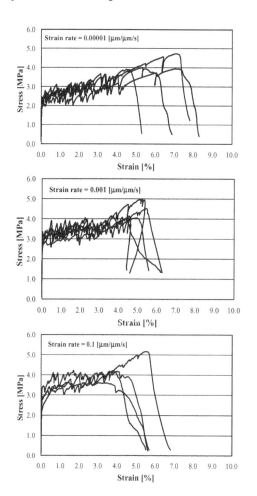

Figure 5. Stress-strain responses at loading Rate 1, Rate 3a and Rate 5 (Table 2).

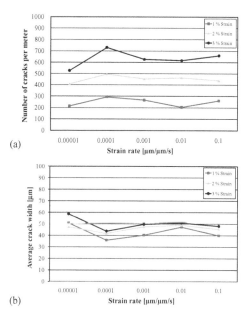

(a)

(b)

Figure 7. Average (a) number of cracks and (b) crack width at various imposed strain levels.

imposed tensile strain. It appears that increased tensile deformation is dominated by new crack formation, while crack widening with increased strain contributes to a lesser degree.

218

Figure 7 illustrates the influence of loading rate on the average crack density and width. These values are plotted for tensile strain levels of 1%, 2% and 3%, for each loading rate. Figure 7a confirms the increase in the number of cracks as the imposed strain level is increased. However, an increase in average crack width of only 10–15 μm is observed in Figure 7b for an increase in imposed strain from 1% to 3%.

At these imposed strain rates, the average crack width appears to be insensitive to the loading rate.

5 CRACKING UNDER CYCLIC LOAD

The responses of test series Cyclic 1 and Cyclic 2 are shown in Figure 8 in terms of averaged stress-strain values calculated from the load cell force and LVDT displacement measurements over the 80 mm gauge length. The test was conducted under displacement control, governed by the Universal Materials Testing Machine crosshead displacement. The cycles were set to 0.5% (Cyclic 1) and 0.1% (Cyclic 2) average strain respectively. Due to the fact that the crosshead displacement and the deformation of the gauge area do not match leads to the variation in strain cycles shown in Figure 8. In the tests roughly 8–10 (Cyclic 1) and 25–50 (Cyclic 2) cycles were applied.

From Figure 8 it appears that the ductility of these SHCC specimens reduced with increased number of load cycles. More tests are envisaged to study this phenomenon carefully, and to build a more substantial data base. From the current responses the crack width evolutions of these specimens are studied next, based on the ARAMIS crack width measurements.

The average crack widths for the test series Rate 3b, (see Table 2) are shown for the four individual specimens as function of strain level in Figure 9. The trends seen in Figure 6b (for series Rate 3a) are confirmed, although crack measurement was only done to a strain level of 3% as shown in Figure 9.

5.1 *Crack width in monotonic vs cyclic load*

Figure 10 compares the crack widths for the cases of monotonic loading (Rate 3b) and cyclic loading (Cyclic 1) to see the influence of cyclic loading on crack width in SHCC. The averaged crack widths for all specimens of each series are shown, for ease of comparison. From Figure 10 the difference in crack widths under these loading conditions appear to be negligible.

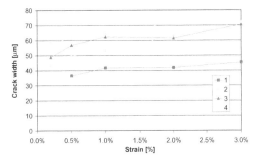

Figure 9. Average crack widths for the four specimens of test series Rate 3b, i.e. for monotonic tensile loading.

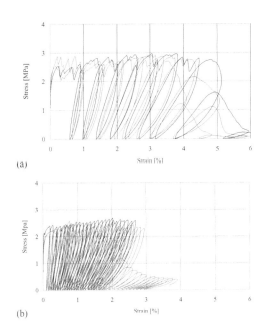

(a)

(b)

Figure 8. Responses to cyclic tensile tests at (a) large load cycles and (b) small load cycles.

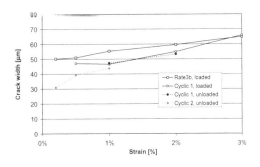

Figure 10. Average crack widths for the test series Rate 3b (monotonic) and Cyclic 1 and 2 (cyclic tensile loading).

5.2 *Crack width in loaded vs unloaded state*

Also shown on Figure 10 are the average crack widths in the unloaded state, for the test series Cyclic 1 and Cyclic 2. No significant difference in average crack widths can be seen for the loaded vs unloaded state at coinciding imposed or residual tensile strain levels.

5.3 *Average vs maximum crack width*

Note that the average strains are shown in Figures 9 and 10. While these values remain within a tight range, below 0.1 mm, the maximum crack widths at localised positions are significantly larger. The maximum crack widths found along the three lines shown in Figure 1 in the gauge area in each series of tests are shown in Figure 11. Clearly, cracks exceeding 0.1 mm in width appear already at a relatively low strain.

It remains to be determined whether the average crack width or the maximum crack width is the most reliable indicator of SHCC durability.

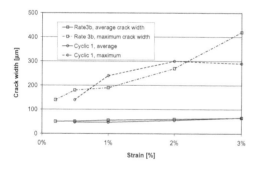

Figure 11. Maximum crack width in series Rate 3b and Cyclic 1, shown in comparison with the averaged values.

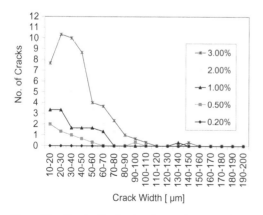

Figure 12. Crack distribution in specimen 1 of test series Rate 3b.

A crack distribution graph, such as presented in Figure 12, may be useful in combination with durability testing related to moisture, gas and chloride ingress in SHCC. Note that the average number of cracks is shown per crack width category. A unit value may indicate a crack which passes through all three lines (shown in Figure 1), or crosses a single line at three positions.

6 CONCLUSION

A non-contact method for accurate measurement of crack widths has been presented, and argued to be suitable for elegant crack measurement in the presence of multiple cracks particular to SHCC. By combination of microscopic enlargement and high resolution and frequency digital photography, it is possible to extract crack width distributions in a semi-automated post-processing phase.

With aid of this technology, several series of SHCC tests were analysed to study the crack width evolution as a function of imposed tensile strain, tensile loading rate and loading path. The following conclusions can be drawn about SHCC specimens tested in this project in direct tension:

- In the strain-hardening regime of SHCC the average crack width remains below 80 μm in SHCC in direct tension.
- This average crack width is reached at a relatively low tensile strain and new cracks form instead of significant widening of existing cracks.
- Negligible difference exists between average crack widths under tensile load compared with average crack widths in the unloaded state at the same strain level.
- Average crack widths in specimens loaded in monotonic tension, and those that arise from cyclic tensile loading to the same imposed strain level, differ insignificantly.
- While average crack widths are kept within 80 μm over a large tensile strain regime, the maximum crack width is significantly higher already at relatively low tensile strains. This occurs in localised areas only, which can be detected best by from crack width distribution graphs, whereby the low frequency of such wide cracks are visible at a glance.

ACKNOWLEDGEMENT

The support by the Ministry of Trade and Industry through the Technology and Human Resources for Industry Programme (THRIP), as well as the industrial partners of the THRIP project ACM-S, is gratefully acknowledged.

REFERENCES

Boshoff, W.P. 2007. *Time-dependant behaviour of ECC*, PhD dissertation, Stellenbosch University, South Africa.

Boshoff, W.P. and Van Zijl, G.P.A.G. 2007. Tensile creep of SHCC. *Proc. International RILEM CONFERENCE on High performance fibre reinforced cement composites*, 10–13 July 2007, Mainz, Germany, pp. 87–95.

Boshoff, W.P. and van Zijl, G.P.A.G. 2007b. Time-dependent response of ECC: Characterisation of creep and rate dependence. *Cement and Concrete Research*, 37, pp. 725–734.

Boshoff, W.P., Adendorff, C. and van Zijl, G.P.A.G. 2009. Creep of pre-cracked strain-hardening cement composites (SHCC), *Proceedings of 8th International Conference on Creep, Shrinkage and Durability of Concrete and Concrete Structures (CONCREEP 8)*, Oct 2008, Ise-Shima, Japan, pp. 723–728.

Fischer, G. and Li, V.C. 2004. Effect of fiber reinforcement on the response of structural members, *Fracture mechanics of concrete and concrete structures*, Vail, USA, April 2004, pp. 831–838.

Jun, P. and Mechtcherine, V. 2009. Deformation behaviour of cracked SHCC under sustained and repeated tensile loading, *Proceedings of 8th International Conference on Creep, Shrinkage and Durability of Concrete and Concrete Structures (CONCREEP 8)*, Oct 2008, Ise-Shima, Japan, pp. 487–493.

Li, V.C., Wang, S. and Wu, C. 2001. Tensile strain-hardening behaviour of Polyvinyl Alcohol Engineered Cementitious Composites (PVA-ECC). *ACI Materials Journal*, Nov.–Dec. 2001, pp. 483–492.

Sahmaran, M., M. Li, and V.C. Li 2007. Transport Properties of Engineered Cementitious Composites Under Chloride Exposure, *ACI Materials J.* 104(6), 604–611.

Wang, S. and Li, V.C. 2006. Polyvinyl Alcohol Fiber reinforced engineered cementitious composites: material design and performances. *Proceedings of International workshop on HPFRCC in structural applications*, Honolulu, Hawaii, USA, May 23–26, 2005.

Weimann, M.B. and Li, V.C. 2003. Hygral Behavior of Engineered Cementitious Composites (ECC), *International Journal for Restoration of Buildings and Monuments* Vol. 9, No. 5, 513–534.

Advances in Cement-Based Materials – van Zijl & Boshoff (eds)
© *2010 Taylor & Francis Group, London, ISBN 978-0-415-87637-7*

Shrinkage of highly flowable cement paste reinforced with glass fiber

A. Sangtarashha, A. Shafieefar, N.A. Libre & M. Shekarchi
Construction Material Institute, Department of Civil Engineering, University of Tehran, Tehran, Iran

I. Mehdipour
Construction and Concrete Research Center, Islamic Azad University, Qazvin, Iran

ABSTRACT: Self-consolidating concrete (SCC) has a high flowability and a moderate viscosity, and no blocking may occur during flow. SCC is prone to considerable shrinkage, due to the high amount of paste and reduced aggregate content. The addition of fibers may take advantage of its high performance in the fresh state to achieve a more uniform dispersion of fibers, which can help to mitigate the shrinkage of the self-compacting composite. This research focuses on the development of shrinkage in highly flowable cement paste (HFCP), which plays an important role in shrinkage of SCC, reinforced with glass fibers. The effect of fiber length and fiber content on total shrinkage of the cementitious material is investigated in this research using shrinkage curves over 190 days. Nine cement paste mixtures were prepared containing 0 to 5 percent of glass fibers. Besides, the rheological and mechanical properties of fiber reinforced cement paste are investigated by pertinent examinations.

1 INTRODUCTION

Nowadays the use of highly-flowable or self-compacting cement based mixtures is common in building industries. This is due to high workability and elimination of vibration in these composites. The self-compacting trait of these composites helps in placing by their own weight. Due to higher volume of paste in these mixtures than ordinary mixtures the cracking susceptibility of them are often raised (Hammer 2003). The increase in the volume of paste improves the workability but it may decrease the mechanical and time-dependent deformations properties. For example, drying shrinkage and non-restrained shrinkage increases with an increase in volume of paste (Bissonnette et al. 1999, Rozie're et al. 2007). Thus, shrinkage of the cement based mixtures is mostly related to the cement paste content in them.

By the process of curing and drying of the mixture, tensile stresses occur due to hydration and loss of moisture. The tensile stresses result in surface cracking of the mixture. If these cracks that develop as a result of shrinkage remain unnoticed, they become channels for passage of external deteriorating agents and reduce long-term durability (Johansen & Dahl 1993).

There are four main types of shrinkage mentioned for cement based materials: plastic shrinkage, carbonation shrinkage, autogenous shrinkage, and drying shrinkage. The shrinkage of fresh concrete due to early age moisture loss from the concrete before, or shortly after, the concrete sets is referred to as plastic shrinkage while carbonation shrinkage is caused by the chemical reaction of various cement hydration products with carbon dioxide in the air. Carbonation shrinkage is limited to the surface of the low-permeable concrete. Drying shrinkage is used to describe the volumetric change of the mixture due to drying. This change in volume of the mixture is not equal to the volume of water lost. Autogenous shrinkage, which occurs when a concrete can self-desiccate (losing water from the capillary pores) during hydration of cement, and which becomes more significant as the strength of concrete is increased, is analogous to drying shrinkage (Carlton & Mistry 1991, Neville 1997). Shrinkage of concrete in its hardened state can be divided in two different processes of drying and autogenous shrinkage (Neville 1995). The shrinkage of concrete during its hardening is mainly related to autogenous shrinkage, but long-term shrinkage of the concrete is a combination of drying and autogenous shrinkage.

Reinforcing concrete with fibers is an impressive method for controlling plastic shrinkage. Randomly distributed fibers of steel, polypropylene, glass, etc. provide bridging forces across cracks and thus prevent them from growing (Grzybowski & Shah 1990, Qi et al. 2003). Some researchers have shown that incorporating non-metallic fibers such as polypropylene, glass, polyethylene, etc will reduce drying shrinkage crack widths of concrete at later ages (Nanni et al. 1993). Polypropylene fibers can be used to decrease long-term shrinkage of highly flowable mortar (Shekarchi et al. 2008). Investigations made by Romauldi et al. (1963, 1964) clear that the fibers act as crack arrestors, and that shear stresses developed at the fiber-hardened cement paste (HCP) inter-faces tend to close the cracks in the vicinity of the fiber.

The addition of glass fibers to the cement matrix significantly improves the flexural and tensile behavior of cement matrix. However, the performance of these composites with aging depends on the matrix mix ingredients (Marikunte et al. 1997). Reinforcement with long glass fibers is more effective, but the use of these long fibers introduces new problems: fiber breakage, large viscosity increases and yield stresses (Thomasset et al. 2005). Rapid loss of surface bleed water on evaporation results in the rapid drawdown in pore water level, causing an increase in pore water pressure, which tends to bring the neighboring solid particles closer. All this leads to shrinking of cement paste (Banthia & Gupta 2006, Wittmann 1976). The mechanism of controlling shrinkage by fibers is known to be by gradually discharging absorbed water of mixture in fibers and compensating the drop in pore water level due to hydration and loss of moisture.

By increasing the amount of non metallic fibers even though the crack characteristics are significantly improved, the concrete mixtures lose their workability (Sivakumar & Santhanam 2007) which results in high volumes of entrapped air in concrete, and causes reduction in mechanical properties of the composites.

This paper, investigates the shrinkage of a highly-flowable cement paste (HFCP) reinforced with glass fiber. Nine cement paste mixtures were prepared containing 0, 1, 2, 3, 4 and 5 percent of 6 and 12 mm long glass fibers with respectively 150 and 200 aspect ratios. Shrinkage curves over 190 days, mini-slump flow diameter and mini V-funnel measurements and the 28 days compressive strength obtained from the tests are used in this study. Besides the effect of glass fiber on short-term shrinkage strains of the specimens over 7 days is compared to that on long-term shrinkage strains over 190 days, where the short-term shrinkage strains which develop during hardening of the composites mostly represent the properties of autogenous shrinkage and long-term shrinkage is the result of a combination of autogenous and drying shrinkage.

2 EXPERIMENTAL PROGRAM

2.1 Characterization of materials

ASTM Type II, normal Portland cement was used in all mixtures. To obtain sufficient consistency in mixtures a novel polycarboxylic acid based superplasticizer was used in all mixtures. Optimum dosage of these admixtures is about 0.5% of cement weight as stated by the admixtures suppliers.

Reference cement composites were modified with low volumetric fractions of two different dispersible Alkali-resistant (AR) glass fibers with 6 and 12 mm length and aspect ratios of 150 and 200 respectively supplied by the manufacturer Saint Gobain-Vetrotex España. Both AR-glass fiber types which are shown in Figure 1, have the ability of dispersing homogenously in fresh composite when introduced in the mixer. Other nominal properties of AR-glass fibers are a Young modulus of 70 GPa and a density of 2580 kg/m^3. In all mixtures water-cement ratio is 35%. The cement content is fixed at 700 kg/m^3, and the SP content is 0.5% of cement mass.

2.2 Test methods

All specimens were fabricated according to ASTM C 192/C 192M-02, Standard Practice for Making and Curing Concrete Test Specimens in the Laboratory.

2.2.1 Mini-slump test
The mini-slump test is based on the measurement of the spread of mortar placed into a cone-shaped mould. The truncated cone (diameters: 100 and 70 mm, height: 60 mm) is placed on a smooth and non-absorbing plate, filled with paste and lifted. The resulting final diameter of the fresh paste sample is the mean value of two measurements made in two perpendicular directions as shown in Figure 2.

2.2.2 Mini V-funnel test
This test consists of measuring the time required for a given volume of mortar (1 liter) to flow through the nozzle. This test is often used to

Figure 1. Glass fibers with 12 mm and 6 mm length.

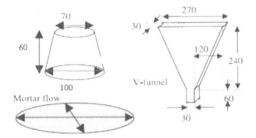

Figure 2. Dimensions of mini-slump cone and mini V-funnel.

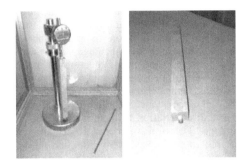

Figure 3. Measurement of shrinkage strain.

measure the viscosity of the mortar which may be related to properties such as cohesiveness, pumpability and finishability. Note that due to instability or inadequate flowability of mortars, the V-funnel values of some mixtures could not be measured. The instrument used in this test can be seen in Figure 2.

2.2.3 *Shrinkage measurement*

After fabricating the shrinkage specimens, they were covered with wet burlap. The plastic sheet was not used for covering the specimens because there was no concern about keeping the burlap wet in the first 24 hours. After 24 hours of wet curing, the specimens were removed from the steel molds.

The test method involves measuring the length change of 25 mm × 25 mm × 285 mm specimens using a length comparator shown in Figure 3 in accordance with C 490-00a, Standard Practice for Use of Apparatus for the Determination of Length Change of Hardened Cement Paste, Mortar, and Concrete.

When initial readings were taken, the specimens were placed freely on a table and covered with a plastic sheet for keeping the temperature and humidity in a relative range. The length change measurements were conducted in the 1st, 2nd, 3rd,

Table 1. Mix proportions.

Label	Size of glass fiber (mm)	Fiber content (%)	Fiber content (kg/cm³)
NF	–	0	0
SG1	6	1	26
SG2	6	2	52
SG3	6	3	78
SG4	6	4	104
SG5	6	5	130
LG1	12	1	26
LG2	12	2	52
LG3	12	3	78

5th and the 7th day of the first week. Subsequent length change measurements were conducted every week up to 28 days, and then every month up to 190 days.

2.3 *Mixture properties*

The performance of HFCPs incorporating glass fibers is investigated by the following experiments. First, the optimum admixture dosage and maximum possible fiber content are determined. Then, the mechanical performances of HFCPs reinforced by the maximum possible amount of fibers (with 2 aspect ratios) are compared. Nine cement paste mixtures are prepared containing 0, 1, 2, 3, 4, 5 volumetric percent of 6 mm and 1, 2, 3 volumetric percent of 12 mm long glass fibers. The 12 mm long glass fiber mixtures contained at most 3 percent of them, because beyond this fiber content the rheological properties of mixtures were decreased in a way that workability of the mixtures was questionable.

Mix proportions are summarized in Table 1. The mini-slump test, mini V-funnel and T20 measurement were carried out to evaluate yield stress and viscosity.

3 RESULTS AND DISCUSSION

3.1 *Rheology of fresh HFCPs*

Fibers can have rheological effects and the optimized fiber combinations can better increase mechanical performance while maintaining adequate flow properties for fiber-reinforced highly flowable cement paste. In this research, mini-slump and mini V-funnel tests were conducted as the rheological tests and results of these experiments are presented in Table 2.

Table 2. Properties of fresh paste.

Label	Mini-slump (cm)	T20 (s)	Mini V-funnel (s)
NF	46	0.3	1.1
SG1	40	0.34	2.8
SG2	39	0.5	3.65
SG3	38	0.6	4
SG4	35	0.6	4
SG5	29	0.9	Blocked
LG1	34	0.7	5.3
LG2	31	0.7	Blocked
LG3	19	–	Blocked

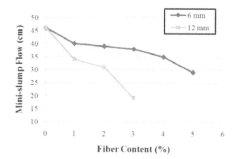

Figure 4. Result of mini-slump test.

These results show that with constant water-cement ratio, the use of fibers result in lower mini-slumps, as can be seen in Figure 4. It means that by increasing the fiber content, increase in yield stress or reduction in flowability of fresh cement paste may occur. Additionally, increase in fiber content leads to higher mini V-funnel results or higher viscosity of the mixture.

It is observed that addition of fiber content of lower aspect ratio up to 4% does not make considerable changes in mini-slump flow and mini V-funnel test results, as it is evident in SG1 to SG4 test results. On the other hand, increasing fiber content of higher aspect ratio up to 3% makes considerable changes in test results. Thus increasing the aspect ratio of fibers, with the same fiber content affects the mini-slump flow and mini V-funnel test results. In other words fiber geometry affects the rheological properties of the cement paste such as viscosity and yield stress.

Based on our experiments, the minimum acceptable spread of slump test is about 35 cm. Thus, reaching to the desirable rheology properties, using 6 mm glass fibers is permitted up to 4% but 12 mm glass fibers permitted only up to 1%. So, the acceptable volumetric fiber content ratio for maintaining workability of the cement paste is

about 1% for both sizes of fiber and 2, 3 and 4 percent for shorter fibers with lower aspect ratio.

3.2 Non-restrained shrinkage of hardened HFCPs

High flowability of HFCPs causes fibers to uniformly disperse in the mixture. The use of fibers in such composites can lead to lower shrinkage strains. In this study, shrinkage of the specimens was measured over 190 days. The shrinkage curves are presented in Figures 5a, b.

Figure 5a pertains to shrinkage strains of HFCPs reinforced with glass fibers of 6 mm length with lower aspect ratio, while Figure 5b represents that of the 12 mm length and higher aspect ratio. It's clear that with constant water-cement ratio, the addition of glass fibers of both sizes decreases the shrinkage strains of the cement paste. But increasing the amount of fiber content does not necessarily lead to lower shrinkage strains. Considering the information given in Table 2, it can be concluded that loss of flowability has a direct relation with control of long-term shrinkage. Therefore, in order to obtain good workability and the best control of shrinkage, 1% of 12 mm long glass fiber with the aspect ratio of 200 and 4% for fibers with 6 mm length and

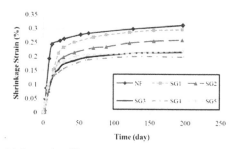

(a) 6 mm glass fiber

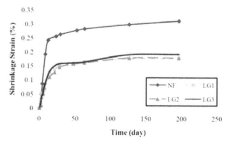

(b) 12 mm glass fiber

Figure 5. Shrinkage curves of glass fiber reinforced HFCPs.

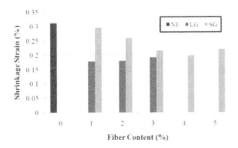

(a) Shrinkage strains

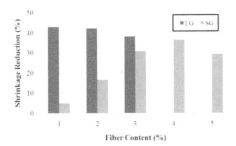

(b) Shrinkage reduction

Figure 6. Effect of glass fibers on shrinkage after 190 days.

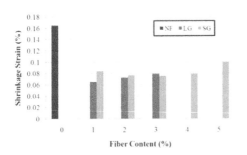

(a) Shrinkage strains

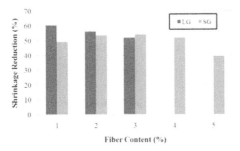

(b) Shrinkage reduction

Figure 7. Effect of glass fibers on shrinkage after 7 days.

aspect ratio of 150 would be the most appropriate amount of glass fiber in HFCPs.

Figures 6a, b show the shrinkage strain information of the specimens after 190 days, which mostly is referred to as autogenous and drying shrinkage. In Figure 6a, shrinkage strains of HFCPs reinforced with different contents of glass fibers of both sizes are shown. Figure 6b illustrates the percentage of success of the mixtures concerned with control of shrinkage strains. The obvious conclusion of this figure is that fibers with higher aspect ratio have significantly greater ability in decreasing long-term shrinkage strains.

In Figure 7, the shrinkage strain information of the specimens after 7 days, which can be considered as the result of autogenous shrinkage, are presented in the same way. According to these bar charts, glass fibers of both sizes have great ability to mitigate the short-term shrinkage of HFCPs. So, it can be concluded that the effect of glass fiber in control of autogenous shrinkage is more significant than that in drying shrinkage. Furthermore, over increasing the fiber content of each size, decreases their ability of controlling shrinkage strains.

3.3 Mechanical properties of hardened HFCPs

Cement paste samples were cured in the water for 28 days at the water temperature of $23 \pm 1°C$.

Fibers with 6 mm length seem to have no effect on compressive strength of cement paste. In fact excessive fibers decrease the sample strength. In this research the compressive 28 days strength of samples made with 1% to about 4% of 6 mm fiber show little difference from the control sample. Increase in amount and meaningfully aspect ratio of fiber causes the compressive strength to drop considerably, as it can be seen in Figure 8. This could be due to balling of fibers or fibers coagulation in the mixtures. By choosing appropriate length and amount of fibers, convenient flowability and control of shrinkage can be obtained without sacrificing much strength.

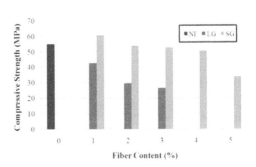

Figure 8. Compressive strength of glass fibers reinforced HFCP.

4 CONCLUSIONS

The investigation compiles rheological and mechanical test results with experimental shrinkage strain curves obtained from specimens of highly-flowable cement paste (HFCP) fabricated by two sizes of glass fibers. The main conclusions of the study are as follows:

1. Although this research shows that using fibers reduces the composite workability and compressive strength, choosing fibers with appropriate quality, length and quantity improves the time-dependent shrinkage properties of composite. Workability and compressive strength can be reduced to a minimum as well.
2. In constant water-cement ratio, increasing the aspect ratio of glass fibers causes rheological properties such as viscosity and yield stress to increase considerably.
3. The results show that loss of flowability of the cement paste samples due to high fiber content, leads to lower shrinkage control and drop in compressive strength. So, it can be concluded that there is a direct relationship between flowability of the samples in fresh state and time-dependent properties of hardened cement paste.
4. There seems to be a certain limit of fiber content, beyond which the reduction in shrinkage continues with a lower rate.
5. The measurements in this research illustrate that fibers with higher aspect ratio have greater ability in decreasing long-term shrinkage strains than those with lower aspect ratio.
6. The comparison between long-term and short-term shrinkage strains of highly-flowable cement pastes shows that the effect of glass fiber in control of autogenous shrinkage is more significant than that in drying shrinkage.
7. According to the test results, the optimum fiber content to achieve appropriate workability and mechanical properties, with suitable autogenous and drying shrinkage control is about 1% of 12 mm long fibers with the aspect ratio of 200 and 4% for fibers with 6 mm length and aspect ratio of 150.

REFERENCES

Banthia, N. & Gupta, R. 2006. Influence of polypropylene fiber geometry on plastic shrinkage cracking in concrete. *Cement and Concrete Research* 36: 1263–1267.

Bissonnette, B. et al. 1999. Influence of key parameters on drying shrinkage of cementitious materials. *Cem Concr Res* 25(5): 1075–1085.

Carlton, D. & Mistry, P.J.M. 1991. Thermo-elastic-creep analysis of maturing concrete. *Comput. Struct.* 40(2): 293–302.

Grzybowski, M.C. & Shah, S.P. 1990. Shrinkage cracking of fiber reinforced concrete. *ACI Materials Journal* 87(2): 138–148.

Hammer, T.A. 2003. Cracking susceptibility due to volume changes of self-compacting concrete. In O. Wallevik & I. Nielsson (eds), *Proceedings of third RILEM international symposium on self-compacting concrete*: 553–557. Reykjavik, Iceland: RILEM Publications.

Johansen, R. & Dahl, P.A. 1993. Control of plastic shrinkage in concrete at early ages. *Proceedings of the eighteenth conference on our world in concrete and structures*: 149–154. Singapore.

Marikunte, S. et al. 1997. Durability of glass fiber reinforced cement composites-effect of silica fume and metakaolin. *Adv. Cem. Bas. Mat.* 9(3/4): 100–108.

Nanni, A. et al. 1993. Plastic shrinkage cracking of restrained fiber-reinforced concrete. *Transport Res Rec* (1382): 69–72.

Neville, A.M. 1995. *Properties of concrete*. New York: John Wiley & Sons.

Neville, A.M. 1997. *Properties of concrete Fourth Edition Reprinted*: 884. Harlow, UK: Longman Group.

Qi, C. et al. 2003. Characterization of plastic shrinkage cracking in fiber reinforced concrete using image analysis and a modified Weibull function. *Materials and Structures* 36(260): 386–395.

Romualdi, J.P. & Batson, G.B. 1963. Mechanics crack arrest in concrete. *J. Eng. Mech. Div. ASCE* 89: 147–167.

Romualdi, J.P. & Mandel, J.A. 1964. Tensile strength of concrete afforded by uniformly distributed and closely spaced short lengths of wire reinforcement. *J. Amer. Concrete Inst.* 61(6): 657–672.

Rozie`re, E. et al. 2007. Influence of paste volume on shrinkage cracking and fracture properties of self-compacting concrete. *Cement & Concrete Composites* 29: 626–636.

Shekarchi, M. et al. 2008. Shrinkage of highly flowable mortar reinforced with polypropylene fiber. *Proceedings of the third ACF international conference*: 154–160. Ho Chi Minh, Vietnam.

Sivakumar, A. & Santhanam, M. 2007. A quantitative study on the plastic shrinkage cracking in high strength hybrid fiber reinforced concrete. *Cement & Concrete Composites* 29: 575–581.

Thomasset, J. et al. 2005. Rheological properties of long glass fiber filled polypropylene. *J. Non-Newton Fluid Mech.* 125: 25–34.

Wittmann, F.H. 1976. On the action of capillary pressure in fresh concrete. *Cem. Concr. Res.* 6(1): 49–56.

Advances in Cement-Based Materials – van Zijl & Boshoff (eds)
© *2010 Taylor & Francis Group, London, ISBN 978-0-415-87637-7*

Early age cracking and capillary pressure controlled concrete curing

V. Slowik & M. Schmidt
Leipzig University of Applied Sciences, Leipzig, Germany

ABSTRACT: Due to the evaporation of water at fresh concrete surfaces, a capillary pressure is built up in the pore system of the material leading to shrinkage deformations, i.e., to the so-called capillary or plastic shrinkage, and possibly to cracking. By influencing the capillary pressure build-up, the risk of cracking in the very early age, i.e., within the first few hours after casting, may be reduced significantly. A method of controlled concrete curing is proposed. It is based on in situ measurements of the capillary pressure by using wireless sensors. If the measured pressure reaches a previously defined threshold value, the concrete surface is rewetted. Experimental and numerical results concerning the physical behavior of drying suspensions are presented and observations made during on-site capillary pressure measurements are discussed.

1 INTRODUCTION

The presented work is aimed to the reduction of the early age cracking risk in concrete construction. Cracks in concrete structures may occur already within the first about four hours after casting, i.e., when the material is still in its plastic stage and before it has reached a significant tensile strength. Similar phenomena as those leading to concrete cracking in this age may also be observed in drying suspensions with inert solid particles, for instance in silt. Physical processes rather than chemical reactions are the predominant reason for volume changes and cracking in plastic concrete (Wittmann 1976, Cohen et al. 1990, Schmidt et al. 2007, Slowik et al. 2008a). Therefore, drying suspensions consisting of fly ash and water may serve as model materials for studying these processes (Slowik et al. 2008a, 2009). Such suspensions are characterized by a cement-like particle size distribution and by spherical particle shapes.

Planar concrete structures like floors and roads are in particular prone to shrinkage and cracking in the plastic stage. This may be attributed to their comparably large surface subjected to the evaporation of water. After casting, the fine solid particles, i.e., those of the cement and of additives, are covered by a plane water film at the top face of the concrete member where usually evaporation takes place. The self-weight of the solid particles may lead to a settlement of the same and, accordingly, to the transport of additional water towards the upper surface, i.e., to the bleeding of the concrete. Due to evaporation, the thickness of the water film reduces and, eventually, the near-surface particles

are no longer covered by a plane water film. Due to adhesive forces, a curved water surface with so-called menisci in the interparticle spaces, see Figure 1, is then formed. Accordingly, a negative capillary pressure is built up in the water. Its magnitude may be calculated by using the Gauss-Laplace equation.

$$p = -\gamma\left(\frac{1}{R_1} + \frac{1}{R_2}\right) \qquad (1)$$

The pressure p depends on the surface tension γ of the liquid phase and on the main radii R of its curved surface. It has to be considered that in cement paste water loss is not only caused by evaporation but also by the cement hydration beginning later.

The negative capillary pressure results in inward forces on the particles at the surface. The microstructure is compacted resulting initially in a measurable settlement or vertical shrinkage of the material. After the material has been separated

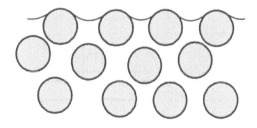

Figure 1. Curved water surface between near-surface solid particles in a drying suspension.

from the side faces of the mould or after cracks were formed, the contracting capillary forces lead also to a shrinkage strain in the horizontal direction. Up to this stage, the volume change of the material is approximately equal to the volume of the evaporated water (Grube 2003, Slowik et al. 2008a). The expansion of entrapped air might cause a deviation, however. While water is evaporating, the curvature of the menisci is increasing, the absolute capillary pressure value continues to rise, and the material volume is getting smaller. Since this volume change is caused by capillary forces, plastic shrinkage of concrete is also referred to as capillary shrinkage.

At a certain material specific pressure, air penetrates suddenly into the pore system (Wittmann 1976, Slowik et al. 2008a). This is the case when the curvature of the water surface is too large for bridging all the spaces between the particles at the surface where evaporation takes place. According to the terminology used in soil physics, this pressure value is referred to as air entry value (Slowik et al. 2008a).

Air entry appears to be a local event. Because of the irregular particle arrangement it does not occur simultaneously in all pores. The latter are drained successively starting with the largest ones. The air entry value marks the first instance of air penetrating the pore system. If the capillary pressure is measured at a location where air entry takes place, the pressure "breaks through" (Wittmann 1976), i.e., it drops down to zero. Figure 2 shows measured curves of the capillary pressure versus time in a cement paste sample as well as in a sample made of fly ash and water. Each of the samples was instrumented with two capillary pressure sensors at different locations, but at the same depth of 4 cm. Details of the experimental set-up are described by Slowik et al. (2008a). It may be seen that the curves obtained for different sensor positions follow the same path. The maximum absolute pressure values, however, are different. This may be attributed to the air entry occurring not simultaneously into all pores. For this reason, the maximum absolute pressure value measured at

a certain location can not be considered to be a material property. It depends on the sensor location. In addition, the pressure might "break down" locally due to air bubbles reaching the sensor tip.

When the concrete is in its plastic stage, all pores are interconnected and the capillary pressure is almost constant in the vicinity of the surface. According to the authors' experimental observations, this is the case up to a depth of at least 10 cm in common cementitious materials. Hydrostatic pressure differences are much smaller than the absolute value of the capillary pressure being built up.

When the air entry value is reached, the cracking risk increases significantly. The pores where air has penetrated into are weak spots at the material surface and origins of crack initiation. This has been shown experimentally by electron microscopic observations (Slowik et al. 2008a, Schmidt et al. 2007) and by force measurements (Slowik et al. 2008b). In suspensions consisting of fly ash and water it was observed that shortly after air entry cracks were formed along a line connecting the weak spots mentioned above. Crack initiation requires air entry, whereas air entry does not necessarily result in cracking. Strain localization and crack formation might be prevented by a limited mobility of the solid particles, although air entry takes place.

The causality between air entry and plastic shrinkage cracking led to an idea for a concrete curing concept. If the absolute capillary pressure being built up in the plastic concrete is permanently kept below the air entry value of the material, cracking can not occur. A corresponding curing method is described in section 4.

Early age cracks are not only an aesthetic problem. They also may degrade the durability of the structures. Even if they are not visible or if they have been temporarily closed during surface finishing, they do have an influence on damage processes taking place during the service life of the structure. Numerical studies into the drying shrinkage cracking of hardened concrete have shown that the obtained crack patterns are strongly influenced by pre-existing early age cracks (Slowik et al. 2008b). The latter may lead to distinct damage localization and larger crack widths. Thus, the concrete permeability will be increased and the durability of the structure might be unfavorably affected.

2 MEASUREMENT OF THE CAPILLARY PRESSURE IN PLASTIC CONCRETE

In laboratory experiments, the capillary pressure build-up in drying suspensions has been studied (Slowik et al. 2008a). The test materials included cement paste as well as suspensions made of fly ash and water. In addition to the capillary pressure,

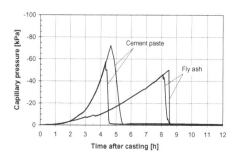

Figure 2. Capillary pressure versus time, measured at a depth of 4 cm, specimen height 6 cm.

deformations, the specimen temperature as well as the electrical conductivity in different depths were measured. For the capillary pressure measurement, miniature pressure transducers were installed outside the forms the specimens were cast into. The connection to the water-filled pore system of the material was provided by metallic tubes having an inside diameter of 3 mm. The location of the tip of the respective tube is regarded as sensor location. In the following, some experimental observations are described.

The slope of the capillary pressure versus time curve depends on the evaporation rate and on the material characteristics. The higher the evaporation rate and the smaller the particle sizes, the steeper the increase of the absolute capillary pressure value will be. The last mentioned effect results from the smaller surface pores in the case of smaller particle sizes. Moreover, for smaller particle sizes higher absolute pressure values are reached. Air entry into the smaller pores requires a higher curvature of the water surface and, accordingly, a higher capillary pressure. Therefore, high-performance concrete compositions tend to be more vulnerable to early age shrinkage and cracking when compared to conventional structural concrete. This is due to the small particle sizes, the high binder contents and the low water-cement ratios. These characteristics lead to comparably high absolute capillary pressure values and shrinkage strains. Consequently, the early age cracking risk is higher. In addition, a more intense self-desiccation increases the water loss rate.

The slope of the capillary pressure versus water loss curve depends on the specimen height of the drying suspension. This effect has been extensively investigated by Radocea (1992). The higher the specimen, the higher the potential for material consolidation. In other words, out of a higher specimen more water can be transported to the surface where evaporation is taking place.

As mentioned before, the maximum absolute capillary pressure value, the so-called "break-through" pressure, measured in a drying suspension depends on the sensor location and may not be regarded as a material property. However, the pressure reached at the first instance of air entry into the pore system should be specific for a certain material. This pressure is called air entry value and mainly depends on the pore structure at the specimen surface and on the mobility of the particles. It has been found that under certain experimental conditions, reaching the air entry value is accompanied by a temporary maximum of the settlement, i.e., of the vertical shrinkage, by an increasing deviation between specimen volume change and evaporating water volume as well as by a sudden drop of electrical conductivity (Slowik et al. 2008a). On the basis of these observations, the air entry value may be identified

in laboratory experiments as a material parameter for a drying suspension.

Capillary pressure measurements were also undertaken under site conditions. Figure 3 shows capillary pressure versus time curves measured on a road construction site. Two sensors were applied to an actual concrete member which was treated by using a curing agent. A third sensor was applied to an uncured reference specimen which had the same height and was cast simultaneously. It may be seen that the capillary pressure rise was much faster in the case of the uncured concrete specimen. Obviously, the curing agent reduced the evaporation rate. Consequently, the capillary pressure increase was retarded, even though not prevented. This retardation reduces the cracking risk. Due to the beginning cement hydration, the particle mobility starts to decrease resulting in an increasing mechanical resistance to the capillary pressure.

Figure 3 also shows that at two different sensor locations in the cured concrete member almost the same capillary pressure values were measured. This confirms the results obtained under laboratory conditions.

In view of a capillary pressure based concrete curing, the corresponding measuring technique had to be improved. Since cable connections usually cause technical problems during the construction process, a wireless capillary pressure sensor was developed, see Figure 4. It consists of a conic plastic tip which is plunged into the concrete, a small cubic box containing the transducer as well as the radio module, and an antenna. The range of wireless transmission amounts to approximately 50 m. Prior to the sensor application, the plastic tip has to be filled with water in order to connect the sensor to the pore system of the material. After the measurement, usually after the final setting of the concrete, the sensor is simply extracted from the concrete surface and the remaining hole is closed. The plastic sensor tip may be replaced, if necessary.

Figure 5 shows the complete measuring system. It consists of the wireless sensors, the radio

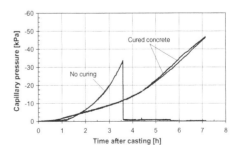

Figure 3. Capillary pressure versus time, measured at a road construction site.

Figure 4. Wireless capillary pressure sensor to be used for on-site measurements.

Figure 5. Measuring system for on-site capillary pressure measurements.

receiver box and a contactless battery charger. The receiver box may serve up to eight wireless sensors. In Figure 5, it is connected to a netbook via USB.

3 NUMERICAL SIMULATION OF CAPILLARY PRESSURE BUILD-UP AND CRACK INITIATION

The described capillary pressure build-up and the resulting crack initiation have been numerically simulated by using a 2D particle-based model (Slowik et al. 2009). It represents a drying suspension consisting of inert solid particles surrounded by water. Circular particles of different size are generated and placed in a rectangular specimen. The top face of the specimen is assumed to be open; i.e., evaporation may take place. In order to simulate the loss of water, the absolute capillary pressure value is incrementally increased and the according course of the water front is calculated under the assumption of constant curvature of the water surface.

In addition to forces resulting from the capillary pressure, the solid particles are subjected to gravitational and interparticle forces. The latter

mainly include electrostatic and van-der-Waals forces. By superposition of these forces, a resulting interparticle force is obtained, see Figure 6. In the short-distance range, however, a simplified force-distance function has been adopted for computational reasons (Slowik et al. 2009).

Figure 7 contains the results of a numerical simulation of the capillary pressure build-up in a drying suspension with circular solid particles (dark color) having a size ranging from 4 μm to 32 μm. The absolute capillary pressure value in the water (gray color) is incrementally increased. The top figure shows the particle arrangement under zero pressure and the bottom figure for the maximum absolute pressure value reached. The latter amounted to 88 kPa. During the capillary pressure build-up, i.e., under decreasing water content of the system, menisci are formed between the solid particles at the surface leading to downward forces and a settlement of the material. It may also be seen that the air entry into the material does not occur uniformly. Under certain conditions, strain localization and crack initiation may be observed in such simulations. In the simulation shown in Figure 7, a "gap" is formed on the right side of the specimen and widened by the rising capillary pressure. This phenomenon of strain localization and separation is regarded as crack initiation.

By using the described 2D model several influences on the capillary pressure build-up and on the early age cracking risk were investigated. The cracking risk is increased if the particle sizes are decreased. Furthermore, the capillary pressure versus water loss curve becomes steeper with

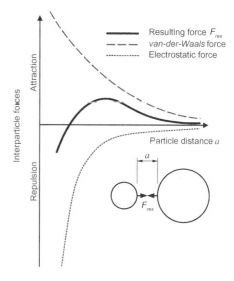

Figure 6. Interparticle forces versus particle distance (not to scale).

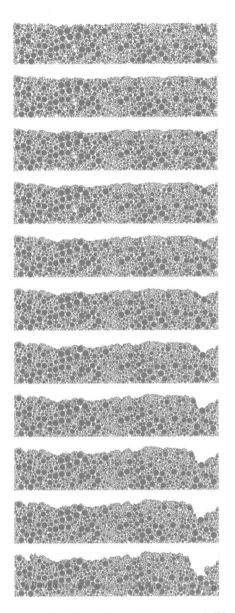

Figure 7. Simulation of the capillary pressure build-up in a drying suspension (absolute capillary pressure value increasing from top down, particles sizes ranging from 4 μm to 32 μm).

decreasing particle sizes. The same effects could be observed when the portion of fine particles was increased. It was also found that the material appears to be more vulnerable to cracking in this case. These numerical results are in good accordance with experimental observations.

The capillary pressure values obtained in the simulations are in the same order of magnitude as those determined experimentally. The ongoing work on the simulation of capillary shrinkage cracking is aimed to a better understanding of the material based influences on the air entry value and on the early age cracking risk. Possibly, the numerical results can help to select an appropriate threshold pressure value for a closed-loop controlled concrete curing.

4 CONCRETE CURING BASED ON CAPILLARY PRESSURE MEASUREMENT

If the air entry value of a certain cementitious material is known, it is possible to define a critical absolute capillary pressure value which should not be exceeded during concrete processing. This value should be smaller than the absolute air entry value. In this way, the early age cracking risk may be significantly reduced since cracking requires air entry.

A method of closed-loop controlled concrete curing is proposed. It is based on the in situ measurement of the capillary pressure. If the measured absolute value reaches a previously defined threshold, the concrete surface is rewetted. This results in a temporary reduction of the capillary pressure. The rewetting is terminated when a lower limit is reached. It is recommended to always maintain a negative capillary pressure in order to prevent the formation of a water film on the concrete surface which might have an unfavorable effect on the near-surface material properties. Furthermore, a moderate capillary pressure leads to an advantageous compaction of the microstructure.

Figure 8 shows the capillary pressure versus time curve measured in a concrete slab subjected to a controlled concrete curing. The corresponding curve measured in an uncured reference slab is also shown. It may be seen that the capillary pressure could be kept within a certain range between two limit values. The evaporation rate was monitored by using curing meters (Jensen 2006). An initial value of about 0.5 kg/(m²h) has been measured, see Figure 9.

For the rewetting of the concrete surface, a commercially available fogging device was used, see Figure 10. Experience has shown that only a few seconds of fogging are required for reducing the absolute capillary pressure value down to its lower limit. The major advantage of the closed-loop control is that only the amount of water actually needed for preventing early age cracks is added to the concrete surface. Hence, the surface quality is not degraded by the rewetting.

In field experiments, the effect of the proposed curing method on the crack pattern formed in the plastic stage could be demonstrated (Slowik et al.

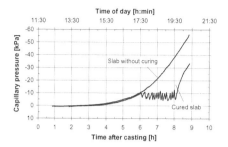

Figure 8. Capillary pressure versus time in a concrete slab under controlled curing.

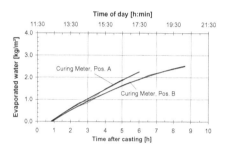

Figure 9. Evaporated water mass versus time, measured by using curing meters (Jensen 2006).

Figure 10 Fogging device.

2008b). The number of cracks as well as the total crack length was significantly reduced.

5 CONCLUDING REMARKS

Concrete in its plastic stage may be regarded as a drying suspension consisting of solid particles and water. The loss of water results in the build-up of a negative capillary pressure in the liquid phase of the material. Capillary pressure-induced local air entry into the material may then lead to crack initiation.

By keeping the absolute capillary pressure value below the value at which air entry takes place,

crack initiation may be prevented. This concept requires the in situ measurement of the capillary pressure and the closed-loop controlled rewetting of the plastic concrete.

The material dependent critical capillary pressure value and the required duration of the controlled curing will be subject of further research.

The application of the proposed curing method might be too expensive for many ordinary projects in concrete construction. Even in these cases, the wireless capillary pressure sensors might be utilized in order to obtain valuable information on the actual cracking risk and on the efficiency of curing measures. The sensors are comparably inexpensive and do not disturb the construction process.

REFERENCES

Cohen, D.C., Olek, J. & Dolch, W.L. 1990. Mechanism of plastic shrinkage cracking in Portland cement and Portland cement-silica fume paste and mortar. *Cement and Concrete Research* 20(1): 103–119.

Grube, H. 2003. Definition der verschiedenen Schwindarten, Ursachen, Größe der Verformungen und baupraktische Bedeutung (Definition of the different types of shrinkage. Causes, magnitude of deformations and practical relevance). *beton* 12: 598–603.

Jensen, O.M. 2006. Monitoring water loss from fresh concrete. In Proceedings of the International RILEM-JCI Seminar Concrete Durability and Service Life Planning—ConcreteLife '06; Ein-Bokek, Israel, March 14–16, 2006, 197–202. Bagneux, France: RILEM Publications SARL.

Radocea, A. 1992. A study on the mechanism of plastic shrinkage of cement-based materials. PhD thesis. Göteborg: Chalmers University of Technology Göteborg.

Schmidt, D., Slowik, V., Schmidt, M. & Fritzsch, R. 2007. Auf Kapillardruckmessung basierende Nachbehandlung von Betonflächen im plastischen Materialzustand (Early age concrete curing based on capillary pressure measurement). *Beton- und Stahlbetonbau* 102(11): 789–796.

Slowik, V., Schmidt, M. & Fritzsch, R. 2008a. Capillary pressure in fresh cement-based materials and identification of the air entry value. *Cement & Concrete Composites* 30(7): 557–565.

Slowik, V., Schmidt, M., Neumann, A. & Dorow, J. 2008b. Early age cracking and its influence on the durability of concrete structures. In Tanabe, T., Sakata, K., Mihashi, H., Sato, R., Maekawa, K., Nakamura, H. (eds.). Proceedings of the 8th International Conference on Creep, Shrinkage and Durability Mechanics of Concrete and Concrete Structures (CONCREEP 8); Ise-Shima, Japan, Sept. 30–Oct. 2, 2008, Vol. 1: 471–477. London: Taylor & Francis Group.

Slowik, V., Hübner, T., Schmidt, M. & Villmann, B. 2009. Simulation of capillary shrinkage cracking in cement-like materials. *Cement & Concrete Composites* 31(7): 461–469.

Wittmann, F.H. 1976. On the action of capillary pressure in fresh concrete. *Cement and Concrete Research* 6: 49–56.

Computational modelling of advanced cement-based materials

Advances in Cement-Based Materials – van Zijl & Boshoff (eds)
© 2010 Taylor & Francis Group, London, ISBN 978-0-415-87637-7

Finite element fracture analysis of reinforced SHCC members

Petr Kabele

Faculty of Civil Engineering, Czech Technical University in Prague, Prague, Czech Republic

ABSTRACT: After a brief review of various existing approaches to FEM modeling of fracture in R/SHCC structural members, a methodology based on a treatment of individual cracks is proposed. This method is shown to provide more accurate reproduction of experimental results, especially when SHCC material is exposed to non-proportional loading. Analytical results also indicate that when a multiply cracked SHCC is exposed to shearing, it may lose some of its strength and deformational capacity, probably due to shearing-induced damage of fiber bridging.

1 INTRODUCTION

Fiber reinforced strain hardening cementitious composites (SHCC) have a characteristic ability to sustain increasing tensile load with increasing overall elongation even after cracking. The inelastic deformation is mostly attributed to formation and opening of a large number of distributed fine cracks, which are bridged by fibers—so-called multiple cracking. When the tensile capacity of the material is exhausted (typically at the overall strain on the order of 10^{-1} to $10^{0}\%$), fracture localizes into one of the cracks and the overall response becomes softening. A typical representative class of SHCC are Engineered Cementitious Composites (ECC), in which the superior fracture performance is achieved by conscious composite design even with a small amount (up to 2% by volume) of short randomly orientated polymeric fibers (Li 2003). SHCC have been used, or their use is being considered, in various structural applications, which require that the cementitious material can accommodate relatively large deformations without losing macroscopic integrity. These applications include ductile energy-dissipating elements for earthquake-resistant structures, e.g. short columns, coupling beams, shear walls (Fukuyama & Suwada 2003, Maruta *et al.* 2005, Fukuyama *et al.* 2006), anchorage of steel studs (Qian & Li 2009), bridge deck link slabs (Lepech & Li 2009) and many others. The composites are often used in combination with conventional steel reinforcing bars (reinforced SHCC or R/SHCC).

The growing use of SHCC and R/SHCC for novel structural solutions implies that there is a need for their reliable numerical modeling. Numerical analysis can serve not only as an aid for designing structural elements, but it can also complement experimental verification of their performance.

To this end, the finite element method (FEM) has been mostly used in conjunction with constitutive models that represent the specific mechanical behaviors of SHCC materials, namely the tensile pseudo strain-hardening and subsequent localized fracture and softening.

In most cases, constitutive models simplify the tensile response of SHCC in multiple-cracking state by monotonous hardening (linear or piecewise-linear), for example as shown in Figure 1. The hardening is usually calibrated from a uniaxial tension test, although Kabele (2007) also outlined a multiscale framework, which allows determining the tensile stress-strain relationship on the basis of micro-scale properties of the composite and its constituents. Uniaxial behavior is then generalized for a multiaxial stress state by various ways. For example, Kabele & Horii (1997) employed the incremental theory of plasticity with Rankine yield function. Other widely used approaches have been derived by adapting models originally developed

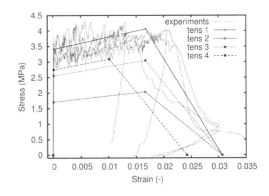

Figure 1. Various approximations of uniaxial response of PVA-ECC. (Kabele & Kanakubo 2007).

for concrete, e.g. the smeared crack model (Kabele *et al.* 1999, Takeuchi 1999), the modified compression field theory (Xoxa 2003), the total strain-based rotating crack model (Han *et al.* 2003) or an orthotropic model (Suwada & Fukuyama 2006). Kabele (2002, 2007) derived the constitutive law using the homogenization methods of micromechanics of solids with defects. Some of the above-mentioned models also deal, to various extent, with the effect of transversal stress (compression or tension) on the strain-hardening response. In models, where the crack direction is fixed once multiple cracking process starts (e.g. in the fixed smeared crack model), it is also necessary to represent possible sliding of cracks, which may occur if the direction of principal stress changes.

A common feature of these models is that the behavior of SHCC in multiple cracking state is represented by a relationship between stress and strain. Individual cracks are not explicitly modeled; instead, the constitutive law describes the overall behavior of a representative volume element (RVE), which contains a large number of cracks. This approach has proven itself as computationally effective and capable of reasonably well reproducing global load-displacement response and load carrying capacity of various R/SHCC structural members. However, results reported by several authors show that, for example, in members failing in shear-tension mode, this approach leads to unrealistically high stiffness just before the peak load is attained (Kanda 1998, Suwada & Fukuyama 2006, Kabele & Kanakubo 2007). In the present paper we demonstrate, that in such situations, more accurate results can be achieved if the formation of individual cracks, as opposed to the homogenized approach, is captured in the model.

First, a homogenization-based model is reviewed. Then a modification of the model to account for individual cracks is outlined and the results of the individual-crack-based model are compared with the results of the former one.

2 HOMOGENIZATION-BASED MODELING

Kabele & Kanakubo (2007) used the finite element program Atena (Červenka *et al.* 2008) to perform numerical reproductions of experiments on conventionally reinforced ECC beams, which were designed and loaded so as to fail by diagonal cracks in a shear-tension mode (so-called Ohno shear beams—Fig. 2).

ECC material was represented with constitutive model proposed by Kabele (2002). This constitutive law is obtained as a relation between overall stress and overall strain of an RVE, which contains

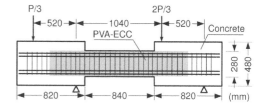

Figure 2. Configuration of Ohno shear beam.

up to two mutually orthogonal sets of multiple parallel cracks. It is assumed that a crack set, perpendicular to the principal stress direction, forms once the maximum principal stress attains the first crack strength σ_{fc}. Consequently, the overall response in the crack-normal direction is governed by a linear hardening relationship until the ultimate tensile strength σ_{tu} is reached at overall strain ε_{tu}. If the principal stress direction changes, cracks' direction remains fixed, but they may slide. It is assumed that crack sliding is resisted by bridging fibers acting as elastic beams. The corresponding tangential traction is obtained by homogenizing the effect of individual fibers over a crack area. Localized softening cracks, which form after the strain capacity ε_{tu} is attained, are treated by the crack band approach. The fracture model is combined with a plasticity model for compressive behavior—see Červenka & Papanikolaou (2008).

Conventional reinforcement was modeled by elastic-perfectly plastic truss elements. Its interaction with surrounding cementitious material was modeled by perfect bond or by bond-slip relationship.

The finite element mesh typically consisted of about 3200 elements and the mesh was highly refined (with element size around 2 cm) in the sheared central part of the beams. This refinement was chosen so as to accurately capture the anticipated variation of stress field, especially with regard to the closely spaced reinforcing bars.

The authors conducted a parametric study focusing mainly on the effects of:

– the assumed uniaxial hardening behavior,
– the assumed resistance of fiber-bridged cracks against sliding (represented by the fiber shear modulus G_{fib}), and
– the bond between ECC and steel reinforcing bars.

The effect of considering perfect bond or bond-slip between ECC and steel reinforcement was found to be negligible. However, Kabele & Kanakubo (2007) found that in order to match the beam load carrying capacity observed in experiments, the tensile hardening relation had to be

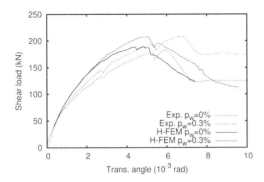

Figure 3. Experimental (Exp.) and analytical (H-FEM) results of tests on beams with stirrup reinforcement ratios $p_w = 0$ (PVA20-00) and $p_w = 0.3\%$ (PVA20-30). The analytical results were obtained with the homogenization-based model (Kabele & Kanakubo 2007).

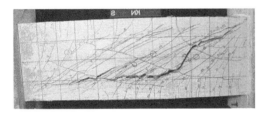

Figure 4. Cracking pattern of beam PVA20-00 (Kabele & Kanakubo 2007).

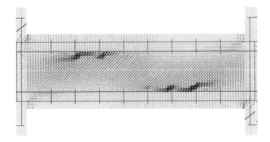

Figure 5. Cracking of beam PVA20-00 obtained with homogenization-based model (Kabele & Kanakubo 2007).

determined as a lower bound fit to the results of uniaxial tests and the strain capacity had to be reduced by about 40% (line "tens4" in Fig. 1). In addition, the value of the fiber shear modulus G_{fib} had to be considerably reduced. The authors concluded that not only opening but also sliding of cracks had a dominant effect on the overall load and deformational capacity of the beams. They hypothesized that in the experiment, bridging

fibers were damaged by the relative slip of crack surfaces and consequently only a fraction of the ECC's uniaxial tensile strength and strain capacity was actually utilized in the R/ECC member. Also, this mechanism may explain why it was necessary to use a very low apparent shear modulus of fibers to match the experimental results.

Figure 3 shows that despite capturing the peak loads, the analyses reproduced much stiffer pre-peak response. It is also seen, when comparing Figures 4 and 5, that the major direction of cracks in the central part of the beam was not correctly represented by the model, with the model predicting cracks inclined at about 45° and the experiment revealing cracks at much lower angle.

3 INDIVIDUAL-CRACK-BASED MODEL

When we closely inspect the finite element model used for the foregoing analyses and the photographs of the fractured experimental specimen, it is obvious that the element size (~2 cm) is actually less than or on a par with the spacing of most of the distributed multiple cracks. This implies that the finite elements de facto do not correspond to an RVE of a composite in multiple cracking state, which was used as a basis for formulating the constitutive law in terms of overall stress and overall strain. On the contrary, it appears as more appropriate to represent the behavior of each individual crack by a traction-separation law, that is, by a relation between the bridging traction and relative displacement of the crack surfaces.

3.1 Traction-separation law

Since, as in the previous model, the fixed crack concept is used, the traction separation law in 2-D relates two components of traction $\{\sigma_{nn}, \tau_{tn}\}$ (normal and tangential with respect to a crack surface) to the corresponding components of relative displacement $\{\delta_n, \delta_t\}$ (crack opening displacement—COD and crack sliding displacement—CSD). We also identically adopt the assumption that opening and sliding of a bridged crack can be modeled independently.

3.1.1 Crack opening mode
The dominant phenomena governing the response of an opening crack are the cohesive effect of matrix and the bridging effect of fibers. Since the cementitious matrix itself is quite brittle, it is represented by a linear tension-softening relationship, which is determined by two parameters: the cracking strength σ_{Mc} and the COD at which the matrix cohesion is completely lost δ_{M0}:

$$\sigma_{nn,M}(\delta_n) = \sigma_{Mc}\left(1 - \frac{\delta_n}{\delta_{M0}}\right) \quad \text{for} \quad \delta_n \le \delta_{M0} \qquad (1)$$

Note that the matrix fracture energy is $G_M = \frac{1}{2}\sigma_{Mc}\delta_{M0}$.

The hardening part of the traction-separation law for fiber bridging is assumed in the form of:

$$\sigma_{nn,F}(\delta_n) = \sigma_{Fu}\left(\frac{\delta_n}{\delta_{Fu}}\right)^{\frac{1}{2}} \quad \text{for} \quad \delta_n \le \delta_{Fu} \qquad (2)$$

where σ_{Fu} = the ultimate tensile stress that the fiber bridging can transmit and δ_{Fu} = COD at which this stress is attained. The post-peak part of the traction-separation law, which represents gradual loss of the fiber-bridging capability due to fiber pullout is simplified by assuming linear softening:

$$\sigma_{nn,F}(\delta_n) = \sigma_{Fu}\left[1 - \frac{\delta_n - \delta_{Fu}}{\delta_{F0} - \delta_{Fu}}\right] \quad \text{for} \quad \delta_{Fu} < \delta_n \le \delta_{F0}$$

$$(3)$$

where δ_{F0} = COD at which the fiber bridging is completely disjoined. The total traction-separation law for a crack opening mode is then obtained by summing up Equations (1) and (2) or (3):

$$\sigma_{nn}(\delta_n) = \sigma_{nn,M}(\delta_n) + \sigma_{nn,F}(\delta_n) \qquad (4)$$

3.1.2 Crack sliding mode

Consistently with the model proposed by Kabele (2002), we consider that sliding of a crack is resisted only by protruding parts of bridging fibers, which undergo uniform elastic shearing within the open width of a crack:

$$\tau_{tn}(\delta_n, \delta_t) = \frac{1}{2}V_f G_{fib}\frac{\delta_t}{\delta_n} \qquad (5)$$

where V_f = fiber volume fraction and G_{fib} = fiber shear modulus.

Note that when a crack forms, it is free of shear stress, since it is perpendicular to the maximum principal stress direction. Before the principal stress direction changes, the crack usually opens so much, that the matrix ligament is lost ($\delta_n > \delta_{M0}$). Also, for such small values of δ_n, Equation (5) gives a very high shearing stiffness compared to the stiffness against opening (Eq. (4)). Furthermore, since ECC does not contain aggregate, matrix crack surfaces are rather smooth and matrix interlock usually does not occur. Consequently the contribution of matrix to crack sliding resistance is neglected in the model.

3.2 Finite element implementation and considerations

The traction-separation law (Eq. (4)) is implemented in the finite element method by the crack band approach. That is, the traction vs. COD relation is transformed into stress vs. inelastic strain relationship; however, by using a characteristic length in the process, the result remains objective with respect to the size of finite elements used—see Červenka & Papanikolaou (2008).

The crack sliding model (Eq. (5)) is implemented by means of a variable shear retention factor, which depends on the COD—see Kabele (2002) and Červenka et al. (2008).

Despite the apparent similarity of the two approaches discussed in this paper there are some principal differences between them (see Fig. 6). As opposed to the homogenization-based model, the individual-crack-based model reflects the formation of each crack and the associated stress drops. It also does not a-priori assume simultaneous formation of a set of multiple cracks within an RVE. These features, however, result in certain constraints with regard to construction of the finite element model. First of all, we must realize that each finite element can accommodate just one crack (in a given direction). Thus the element size should be chosen so as to correspond to the minimum crack spacing. The theoretical minimum spacing for saturated multiple cracking in ECC can be very small (order of 10^{-3} m) (Kanda & Li 1998); however, experimental results on structural elements show that such a dense cracking is rarely achieved in practice—instead the specimens analyzed in the present study showed crack spacing on the order of 10^{-2} m. Use of such an element size for modeling structural elements or details is feasible. Secondly, we must note that should the present model properly capture the process of multiple cracking in a *uniform stress field* (e.g. when simulating a uniaxial tension test), the values of relevant material parameters (in particular the cracking stress σ_{Mc}) may not be uniformly assigned to all

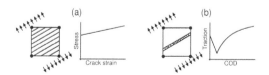

Figure 6. Homogenization-based concept (a) vs. individual-crack-based concept (b).

elements, as it would cause simultaneous cracking of all elements. Instead, the parameters should be prescribed as a stochastic field, which reflects the inherent inhomogeneity of the material—see Example 1 in Kabele (2007). Nevertheless, in real structural elements the stress field is almost never uniform due to their configuration, loading conditions, presence of reinforcement, etc. Then the location of cracks is determined by the extremes of the stress field rather than by the inhomogeneity of the material. Consequently the present modeling concept may be used even with the values of parameters uniformly assigned throughout the finite element mesh.

4 NUMERICAL STUDY

The cracking model outlined in section 3 was employed to numerically simulate experiments on shear beams PVA20-00, PVA20-30 and PVA20-89 described in the earlier work by Kabele & Kanakubo (2007). These beams were produced using ECC with 2% by volume of short PVA fibers (PVA-ECC). The second number indicates the stirrup reinforcement ratio p_w (in hundredths of %). Besides the fracture model, the finite element model used in the present study, in particular the compressive properties of ECC, properties of steel reinforcement, and the finite element mesh, was identical with that employed by Kabele & Kanakubo (2007). Perfect bond between reinforcement and ECC was assumed.

4.1 Effect of crack-normal traction-separation law

At first, the effect of parameters that describe the response of opening cracks on the computed overall behavior of shear beams was analyzed. To this end, four sets of parameters denoted as b1 to b4 and listed in Table 1 were used. The parameter G_{fib} was kept constant, equal to 110 MPa.

Figure 7 compares the analytical and experimental results in terms of shear load vs. translational angle for the beam PVA20-00. It is obvious,

Table 1. Parameters of the crack-normal traction-separation law.

Param. set	σ_{M_t} (MPa)	G_M (N/m)	σ_{Tu} (MPa)	δ_{Tu} (mm)	δ_{T0} (mm)
b1	3.41	20	4.06	0.2	2.27
b2	2.03	20	2.42	0.2	2.27
b3	2.03	20	3.02	0.13	2.27
b4	2.75	20	3.02	0.13	2.27

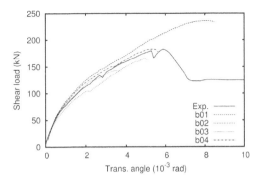

Figure 7. Effect of the used crack-normal traction-separation law on the computed response of beam PVA20-00.

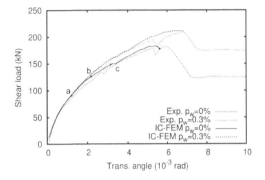

Figure 8. Experimental (Exp.) and analytical (IC-FEM) results of tests on beams with stirrup reinforcement ratios $p_w = 0$ (PVA20-00) and $p_w = 0.3\%$ (PVA20-30).

that the best fit is obtained with parameters set b4. Set b1 leads to a gross overestimation of the peak load, while using sets b2 and b3 results in predicting lower load carrying capacity.

The set b4 was used to predict the behavior of the beam PVA20-30, which was moderately reinforced for shear. As seen in Figure 8 the peak load and translational angle were calculated very accurately. Figure 8 also shows that the present model captures the pre-peak response much better than the homogenization-based approach (compare with Fig. 3).

Figure 9 shows the sequence of cracking computed for beam PVA20-00. It is seen that fracture initially developed in isolated bands, before multiple cracking spread throughout the beam. It should be also noticed that the direction of multiple cracks is consistent with the experiment: the cracks in the central part are less inclined than those on the sides (compare Figs. 4 and 9c). The final failure mode, with localization of fracture into inclined cracks in the side thirds of the depicted area and evolution of a splitting crack along the reinforcement layer, also agrees well with the experiment.

Figure 10 shows the traction-COD/CSD and stress-strain diagrams recorded in three points of the beam marked by the white dots in Figure 9c. The stresses and strains are transformed into coordinate systems aligned with cracks in the respective

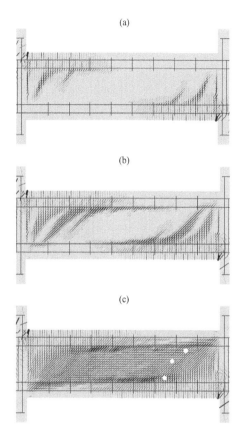

points. From Figure 10c it is obvious that the compressive stress parallel with the cracks remains on a relatively low level, much below the uniaxial compressive strength of 39 MPa. Figures 10a, b show that the cracks undergo significant opening and sliding and that shear stresses transmitted across the cracks may reach similar magnitudes as the normal stresses.

Since the parameters listed in Table 1 do not directly indicate the composite overall behavior under uniaxial tension, we used them to simulate direct tension tests. To this end, the minimum crack spacing was set to 18 mm, which corresponded to the size of elements used in the preceding analyses of shear beams. Each potential crack plane was assigned a different value of σ_{Mc}, with the minimum corresponding to the value of σ_{Mc} listed in Table 1 and the maximum being less than σ_{Fu}. The resulting stress-strain curves are compared with a set of experimental data in Figure 11. Note that the set b1 fits the experimental data in terms of strength but it is on the lower side in terms of strain capacity. Results with all other sets, including the set b4, which provided the best reproduction of experiments on shear beams, fall under the uniaxial test data. This indicates that the actual performance of a composite, when it is exposed to a complex loading history involving sliding of cracks, is worse than that measured in the uniaxial test.

4.2 *Effect of crack sliding resistance*

In the next series of numerical experiments, we examined the influence of the assumed crack-sliding resistance on the computed overall response of shear beams. It is evident from Equation (5), that if the fiber volume fraction V_f is kept constant, the shear resistance of a crack is solely determined by parameter G_{fib} (for a given crack width δ_n). For the crack opening mode, the parameters set b4 was used in all of the forthcoming calculations.

Figure 8 shows that in the beams PVA20-00 and PVA20-30, which were unreinforced or moderately reinforced for shear, the best results were obtained

Figure 9. Evolution of cracking in beam PVA20-00 obtained with the individual-crack-based model. Phases (a) thru (c) correspond to points marked in Figure 8.

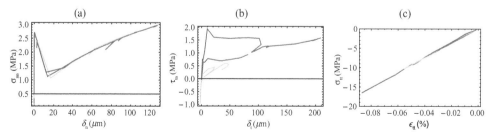

Figure 10. Records of (a) normal traction vs. COD, (b) tangential traction vs. CSD, (c) transversal compressive stress vs. strain calculated at cracks marked by white dots in Figure 9c.

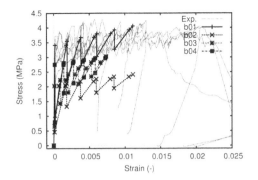

Figure 11. Experimental and simulated uniaxial stress-strain curves of PVA-ECC. Simulation captures only the phase of multiple cracking (up to stress σ_{Fu}).

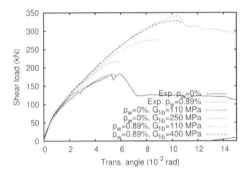

Figure 12. Effect of parameter G_{fib} on the computed response of beams with stirrup reinforcement ratios $p_w = 0$ (PVA20-00) and $p_w = 0.89\%$ (PVA20-89).

with $G_{fib} = 110$ MPa. Figure 12 shows that if G_{fib} is increased to 250 MPa, the load carrying capacity of the unreinforced beam PVA20-00 is significantly overestimated. At the same time, we can see that using $G_{fib} = 110$ MPa for the beam PVA20-89 (with a heavy shear reinforcement) causes underestimation of the peak load and that a much better fit is obtained when G_{fib} is set to 400 MPa.

To explain these observations we can recall the following features of the present model:

– Protruding portions of fibers bridging a crack are assumed to remain elastic when crack sliding occurs.
– Conventional steel reinforcement is modeled by truss elements, which transfer only axial force. Thus they cannot model a dowel effect of reinforcing bars (transfer of shear across a crack provided by shear stiffness of the bar).

As seen in Figure 10, cracks were exposed to significant sliding. This sliding may have caused damage and rupture of bridging fibers. However,

as the model did not explicitly account for fiber damage, it was necessary to reduce the apparent value of fiber shear modulus G_{fib} to 110 MPa. (Note that the fiber axial Young modulus of 40 GPa is quoted by the producer). In a beam with a heavy shear reinforcement (PVA20-89), increasing G_{fib} may have substituted the missing dowel effect of stirrups.

5 CONCLUDING REMARKS

In this paper, two approaches to numerical modeling of cracking in R/SHCC members have been explicated and compared: a widely used homogenization-based approach and a novel individual-crack-based approach. From the presented results it is obvious, that the latter provides more accurate reproduction of experimental results, especially when the SHCC material is exposed to a complex non-proportional loading history. The proposed methodology entails certain constraints on the size of the finite element mesh; however these requirements can be accommodated in finite element models of usual structural elements or details.

Numerical results reveal that in structural members, such as shear beams, stress redistribution occurs after cracking, which causes that cracks are exposed not only to opening, but also shearing. The analyses also indicate that under these conditions the PVA-ECC material exhibits reduced strength and deformational capacity, compared to its performance in a uniaxial tensile test. Since the tested material lacks aggregate that would otherwise provide matrix interlock, shearing of cracks may cause excessive loading of bridging fibers, resulting in their rupture and eventually reduced capacity of the material. It should be noted that similar phenomena have been also observed in a recent experimental work by Suryanto et al. (2008).

ACKNOWLEDGMENTS

The presented research has been supported by the Ministry of Education, Youth and Sports of the Czech Republic under Research Plan MSM6840770003.

REFERENCES

Červenka, J. & Papanikolaou, V.K. 2008. Three dimensional combined fracture-plastic material model for concrete. *International Journal of Plasticity* 24 (12): 2192–2220.
Červenka, V., Jendele, L. & Červenka, J. 2008. *Atena program documentation, part 1, theory*. Prague: Červenka Consulting.

Fukuyama, H. & Suwada, H. 2003. Experimental response of HPFRCC dampers for structural control. *Journal of Advanced Concrete Technology* 1 (3): 317–326.

Fukuyama, H., Suwada, H. & Mukai, T. 2006. Test on high-performance wall elements with HPFRCC. In Fischer, G. & Li, V.C. (eds.), *High Performance Fiber Reinforced Cementitious Composites (HPFRCC) in Structural Applications, Honolulu, Hawaii, USA, 2005*. RILEM Publications S.A.R.L.

Han, T.-S., Feenstra, P. & Billington, S.L. 2003. Simulation of highly ductile fiber-reinforced cement-based composite components under cyclic loading. *ACI Structural Journal* 10 (6): 749–757.

Kabele, P. 2002. Equivalent continuum model of multiple cracking. *Engineering Mechanics (Association for Engineering Mechanics, Czech Republic)* 9 (1/2): 75–90.

Kabele, P. 2007. Multiscale framework for modeling of fracture in high performance fiber reinforced cementitious composites. *Engineering Fracture Mechanics* 74 (1–2): 194–209.

Kabele, P. & Horii, H. 1997. Analytical model for fracture behavior of pseudo strain-hardening cementitious composites. *Concrete Library International (Japan Society of Civil Engineers)* 29: 105–120.

Kabele, P. & Kanakubo, T. 2007. Experimental and numerical investigation of shear behavior of PVA-ECC in structural elements. In Reinhardt, H.W. & Naaman, A.E. (eds.), *Proceedings of the Fifth International Workshop on High Performance Fiber Reinforced Cementitious Composites (HPFRCC5), Mainz, Germany, July 10–13, 2007*. RILEM Publications S.A.R.L.

Kabele, P., Takeuchi, S., Inaba, K. & Horii, H. 1999. Performance of engineered cementitious composites in repair and retrofit: Analytical estimates. In Reinhardt, H.W. & Naaman, A.E. (eds.), *High Performance Fiber Reinforced Cement Composites (HPFRCC 3), Mainz, Germany*.

Kanda, T. 1998. *Design of engineered cementitious composites for ductile seismic resistant elements*. Ph.D. Thesis. University of Michigan.

Kanda, T. & Li, V.C. 1998. Multiple cracking sequence and saturation in fiber reinforced cementitious composites. *Concrete Research and Technology (Japan Concrete Institute)* 9 (2): 1–15.

Lepech, M.D. & Li, V.C. 2009. Application of ECC for bridge deck link slabs. *Materials and Structures* Published Online on 30 July, 2009.

Li, V.C. 2003. On engineered cementitious composites (ECC)—a review of the material and its applications. *Journal of Advanced Concrete Technology* 1 (3): 215–230.

Maruta, M., Kanda, T., Nagai, S. & Yamamoto, Y. 2005. New high-rise RC structure using pre-cast ECC coupling beam. *Concrete Journal (Japan Concrete Institute)* 43 (11): 18–26.

Qian, S. & Li, V.C. 2009. Influence of concrete material ductility on headed anchor pullout performance. *ACI Materials Journal* 106 (1): 72–81.

Suryanto, B., Nagai, K. & Maekawa, K. 2008. Influence of damage on cracking behavior of ductile fiber-reinforced cementitious composites. In Tanabe, T., Sakata, K., Mihashi, H., Sato, R., Maekawa, K. & Nakamura, H. (eds.), *Creep, Shrinkage and Durability Mechanics of Concrete and Concrete Structures: Proceedings of the Concreep 8 Conference, Ise-Shima, Japan, September 30–October 2, 2008*.

Suwada, H. & Fukuyama, H. 2006. Nonlinear finite element analysis of shear failure of structural elements using high performance fiber reinforced cement composite. *Journal of Advanced Concrete Technology* 4 (1): 45–57.

Takeuchi, S. 1999. *Modeling of ECC fracture behavior under shear and analysis of its performance in antiseismic retrofit*. M.Eng. Thesis. University of Tokyo.

Xoxa, V. 2003. *Investigating the shear characteristics of high performance fiber reinforced concrete*. Master of Applied Science Thesis. University of Toronto.

Advances in Cement-Based Materials – van Zijl & Boshoff (eds)
© *2010 Taylor & Francis Group, London, ISBN 978-0-415-87637-7*

Derivation of a multi-scale model for Strain-Hardening Cement-based Composites (SHCC) under monotonic and cyclic tensile loading

P. Jun & V. Mechtcherine
Institute of Construction Materials, Faculty of Civil Engineering, TU Dresden, Germany

P. Kabele
Department of Mechanics, Faculty of Civil Engineering, CTU Prague, Czech Republic

ABSTRACT: Paper presents the derivation of constitutive relations for Strain-Hardening Cement-based Composites (SHCC) under monotonic and cyclic tensile loading. These constitutive relations result from a sound physical model which links different levels of observation from the behaviour of individual fibres to the overall response of the composite material. As an experimental basis for modelling, single fibre pullout tests were used. First, the pullout behaviour as well as de-bonding and failure of single fibre incorporated in the composite are presented and discussed for monotonic tests with regard to different testing parameters. Subsequently, a set of multi-linear descriptions for the fibre-matrix interaction can be defined. Statistical approach is employed in order to deal with the variance of results observed. The stress-displacement relation for one representative crack in SHCC is obtained by adding crack-bridging contributions of all fibres crossing the crack plane. Fibre distribution in the composite as well as the embedded lengths and inclinations of fibres involved in crack bridging are the main parameters considered. Furthermore, the behaviour of single fibre pullout under cyclic loading is investigated. The results obtained are incorporated into the model, so that also unloading and reloading stress-crack opening relations for a single crack in SHCC are derived. Based on these formulas constitutive relations for macro-stress-strain behaviour of SHCC (considering multiple cracking) will be developed in the next step.

1 INTRODUCTION

This paper treats Strain-Hardening Cement-based Composites (SHCC), which display high ductility and strain capacity when subjected to tensile loading. Such specific behaviour results from progressive multiple cracking, achieved by the optimized crack-bridging action of short, thin, well distributed polymeric fibres. The characteristic behaviour of SHCC in tension under monotonic, quasi-static loading has been studied intensively in the last few years; see Mechtcherine (2007). However, in practise, the majority of concrete structures are exposed to more or less severe cyclic loadings such as temperature changes, traffic loads, wind gusts, sea waves, vibrations due to the operation of machinery or, in extreme circumstances, earthquakes.

Mechtcherine & Jun (2007) investigated the macroscopic behaviour of SHCC by using dumbbell-shaped specimens under different loading regimes: deformation controlled monotonic and cyclic as well as load controlled cyclic and creep regimes. Only a moderate effect of the loading regime on

tensile behaviour of SHCC was found. This is in accordance with the results of Fukuyama et al. (2002) but contradicts the findings of Douglas & Billington (2006). However, the SHCC materials investigated in both these references showed generally less ductile behaviour.

Since the stress-strain behaviour observed on the macro-level (where the material can be assumed to be homogenous) depends on a number of micro- and meso-mechanical phenomena, a multi-scale approach is needed to develop a sound physical material model as a basis for a material law. For representative volume elements of the material, Kabele (2007) defined the stress vs. strain relation, which considers micromechanical phenomena and further employs spatial averaging in order to link the scales of observation.

In order to describe micromechanical phenomena, a series of experiments including single fibre tension tests, single fibre pullout tests as well as optical observations were performed by the authors; see Jun & Mechtcherine (2008).

In this paper, a multi-scale modelling approach is introduced, based on which constitutive relations for

SHCC under monotonic and cyclic tensile loading are developed. This approach implies that the macroscopic behaviour of SHCC in the hardening state can be considered as a result of the development and joint action of multiple parallel cracks. The stress-crack opening relation for each of these individual cracks results from a great number of pullout responses by fibres crossing the crack plane. Nevertheless, the focus is first to be directed at characteristic fibre pullout behaviour. Then the action of all fibres crossing a crack plane is described under consideration of the variance of the test results, while the embedded fibre length and inclination serve as the main parameters. The model developed for individual cracks in SHCC is to be upgraded to describe the macroscopic behaviour of the composite in the next step. This description is based on the interaction of the uncracked matrix with gradually developing system of cracks in serial interconnection.

2 MATERIAL UNDER INVESTIGATION

Material composition used in this investigation was developed by Mechtcherine & Schulze (2005). Table 1 gives the mix components and their mass proportions. A mix containing a combination of Portland cement 42.5R HS and fly ash was utilized as the binder. The fine aggregate was uniformly graded quartz sand with particle sizes of 0.06 mm to 0.20 mm. Furthermore, polyvinyl alcohol (PVA) fibres, 2.25% of the total mix by volume, 12 mm in length and with a density of 1300 kg/m³, were used. A super-plasticiser (SP) and a viscosity agent (VA) were added to the mix in order to adjust its rheological properties.

The average characteristics of this material in tension were the first crack stress of 3.6 MPa, tensile strength of 4.7 MPa and strain capacity of 2.5%. These values were derived from 5 displacement controlled tests performed on dumbbell shaped specimens with cross-section of 24 mm × 40 mm and the gauge length of 100 mm; for details see Mechtcherine & Jun (2007). The compressive strength of 34.8 MPa and the strain capacity in compression of 0.75% were derived from 15 displacement controlled tests on cubes with a side length of 100 mm. The displacement rate used for tensile as well as the compressive tests was 0.01 mm/s, for details see Jun & Mechtcherine (2007).

Table 1. Mass proportions of the SHCC components.

Cement	Fly ash	Quartz sand	Water	SP	VA	PVA Fibers
1.00	2.33	1.67	1.04	0.05	0.01	0.09

3 EXPERIMENTAL INVESTIGATIONS ON MICRO LEVEL

3.1 Single fibre tension tests

12 mm long PVA fibres (Kuraray Co., Ltd., Kuralon K-II. REC15) with a diameter of 40 µm were used as reinforcement. To test its mechanical properties, the ends of individual fibre were glued to the adapters mounted on the base plate and cross-head of the testing machine, respectively. Fibre deformations were measured by evaluating the series of images made using a high resolution camera with a macro lens during the tests. The free length of the fibre was 5 mm. Tests were performed by using the loading rate of 0.01 mm/s. Single fibres were investigated under deformation controlled monotonic and cyclic regime as well as under load controlled creep regime. Five tests were performed for each loading regime. Further details may be found in Jun & Mechtcherine (2008).

3.2 Single fibre pullout tests

The test setup used for fibre pullout tests is based on the procedure described by Kabele et al. (2006). The fibre was inserted into a hollow medical cannula with a blunt tip. The position of the fibre was fixed by introducing wax into the cannula when the desired embedded length was attained. Since the bond between wax and fibre was much weaker than the bond between fibre and matrix, the specimens were not damaged when the cannulae were pulled-out.

The future embedded length of the fibre was subsequently estimated by analysing the images made by a high resolution camera. The embedded length corresponded to the part of the fibre exiting the cannula, using the constant diameter of the cannula as a reference. Finally, the frame with a number of cannulae was fixed to the mould and the SHCC matrix without fibres was added until the ends of cannulae lay 1 mm beneath the matrix surface. This arrangement prevented possible influence on the bonding properties by carbonation. Figure 1 shows some details of the production of the specimens.

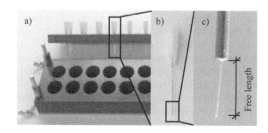

Figure 1. Production of specimens for the fibre pullout tests: a) mould and cannulae frame, b) single cannula, c) detail of a fixed fibre.

Moulds were stored 2 days in a climate room at 20°C and 65% RH and 26 days sealed in foil at room temperature before de-moulding. Matrix cylinders were then clamped into the testing machine and the fibres were glued to the load adapter, as described in the case of single fibre tests. The free length between the embedded and glued ends of the fibre was 5 mm, i.e. equal to the free length in the fibre tension tests.

The experimental program included deformation controlled monotonic and cyclic tests. The loading rate was again 0.01 mm/s. A displacement increment of 0.1 mm was prescribed in cyclic pullout tests for each cycle. The fibres' embedded length varied between 1 and 6 mm with an increment of 0.5 mm. Five tests were performed under a monotonic loading regime for each embedded length, i.e. 55 tests in total.

4 EXPERIMENTAL RESULTS

4.1 *Single fibre tension tests*

In the first step the testing approach chosen was validated by tests on single fibres of 100 mm free length. The results obtained were in accordance with data provided by the fibre producer (i.e. the tensile strength of 1600 MPa, the strain capacity of 6% and the modulus of elasticity of 40 GPa), who used long fibre lengths for testing. However, tests with a free fibre length of 5 mm relevant as a reference for the fibre pullout tests revealed pronounced differences in the values of the modulus of elasticity and the strain capacity. Details and discussion on this subject may be found in Jun & Mechtcherine (2008). Table 2 gives the average fibre characteristics obtained.

In general, the response of single fibres was nearly linear; however, after unloading a considerable irreversible component of the deformation was observed. Thus, the material behaviour can be classified as elastic-plastic.

Additionally, single fibres were tested in load controlled creep regime. The time-dependent deformation components were found to be insignificant and, hence, negligible.

4.2 *Single fibre pullout tests*

Typical stress-deformation response consisted of an elastic part (before full de bonding) and a pullout part, which exhibited softening or hardening behaviour. Fibre failures could be observed in individual tests at any stage. Figure 2 shows the results obtained from pullout tests under monotonic loading. Responses in individual tests vary significantly, even for fibres having similar embedded

Table 2. Measured characteristics of single fibre.

Free length l [mm]	Tensile strength σ [MPa]	Strain capacity ε [%]	Tangent modulus E [GPa]	Non-reversible component δ [%]
5	1620	11	15	41

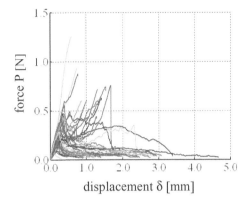

Figure 2. Pullout responses obtained under deformation controlled monotonic loading regime for embedded lengths between 1 and 6 mm.

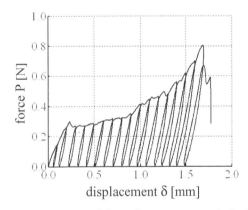

Figure 3. Characteristic pullout response obtained under deformation controlled cyclic loading regime for an embedded length of 3.5 mm.

lengths. In order to understand this phenomenon, optical investigations using Environmental Scanning Electron Microscope (ESEM) were performed, see Section 4.3.

The tests performed under deformation-controlled cyclic loading did not reveal any pronounced effect of the loading regime on the pullout behaviour. Furthermore, these tests showed consistent results with respect to the shape and inclination of unloading and reloading branches. Thus, only one representative result is shown here (Figure 3).

Cyclic loops were not influenced by the scenario of subsequent failure, i.e. fibre breakage before or after hardening, fibre pullout after hardening or in the softening regime. It could be concluded that the unloading and reloading behaviour was only dependent on force reached.

Since the pullout deformation was found to be irreversible, the shape of unloading and reloading branches during pullout testing is primarily related to the deformation of the fibre within its initial free length and the length of region where a partial de-bonding of the fibre from the matrix occurred.

In the region where de-bonding occurred the fibre deformation behaviour is influenced by friction between fibre and matrix. This phenomenon can be also time-dependent; see Boshoff et al. (2009). However, the time dependency of the pullout process is neglected in this investigation because the creep contribution is insignificant due to the relatively short duration of the tests.

Because of the pullout setup arrangement, the results obtained are influenced by the deformation of the fibre free length between the point where it exits the matrix and glued fibre end. However, no considerable free length is involved in cracking of SHCC in the hardening regime. Therefore, this deformation component has to be subtracted from the results in order to obtain a representative description of single fibre pullout behaviour.

4.3 *Optical investigations*

Optical investigations of pullout specimens were performed after mechanical testing. Matrix cylinders were split into two pieces and investigated by ESEM in order to obtain additional information regarding the pullout process.

From Figure 4 the arrangement used in the pullout tests can be clearly recognised. There is a

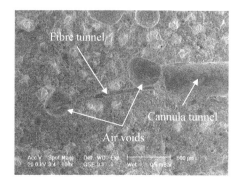

Figure 4. Split pullout specimen; air voids change the effective embedded length.

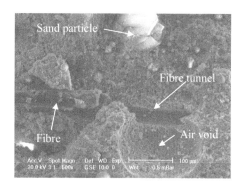

Figure 5. Split pullout specimen; local imperfections and particles' arrangement in the fibre vicinity.

tunnel where the cannula with its embedded fibre was previously. Furthermore, the figure displays a much narrower tunnel left after the fibre pullout. On this example it can be clearly seen that the measured embedded length of the fibre can differ significantly from the length that is really in contact with matrix due to local imperfections of the matrix, such as air voids in fibre vicinity. Since the contact area is a crucial parameter, this difference may play an important role.

Furthermore, the fibre pullout response is influenced by arrangements of aggregates in fibre vicinity, cf. Figures 4 and 5. Considering the scale of observation, the microstructure around the fibre is highly inhomogeneous. Since the sand and fly ash particles are stiffer than the fibre, the contact during pulling-out process will result in fibre damage and subsequently in a decrease in failure load. Both, the deviation of the true contact area from nominal value derived from embedded length and fibre diameter and the damage of fibre by the surroundings particles explain, at least partly, the pronounced variance of pullout results observed.

Apart from the local damage caused by contact with the surrounding particles, a stress increase in the regions where fibre locally lost the contact to the matrix is a possible explanation for the fibre failure during pullout process.

Mechanisms leading to the hardening behaviour in pullout tests are discussed in Jun & Mechtcherine (2008).

5 PREPARATION FOR MODELLING

5.1 *Consideration of the fibre deformation*

As was already mentioned in Section 4.2, fibre free length of 5 mm between its embedded and glued ends contributed to the overall deformation during the pullout test. Since the deformation of the fibre itself is known from tensile tests on single fibres, it can be subtracted from the deformation values measured in the pullout test. The fibre deformation to be subtracted δ_s depends on the force level and can be estimated by Eq. 1 which was derived on the basis of single fibre tests:

$$\delta_s = 0.59 \cdot \delta_t(P) + 0.41 \cdot \delta_t(P_h) \qquad (1)$$

where $\delta_t(P)$ = actual deformation as a function of actual force P, P_h = highest force related to the point where the softening or the unloading starts.

Eq. 1 takes into account that the total deformation of fibre consists of an elastic and a plastic component. The percentage of these two components related to the total deformation was found to be 59% for the elastic and 41% for the plastic component, respectively. When the force increases, the actual force P is equal to the highest force P_h in each point. As result, the total fibre free length deformation (100%) is subtracted as a function of the force reached. In the case of softening or unloading (for cyclic loading), the contribution of the elastic deformation decreases with decreasing actual force, while the plastic deformation component remains constant as defined for the highest force P_h reached before load decrease.

5.2 *Fibre distribution in the composite*

Three dimensional random distribution of fibres in SHCC is considered, no wall effect is taken into account. Based on this assumption, the probability density functions for the distribution of fibres embedded lengths and inclinations can be derived according to Li et al. (1991) using Eq. 2 and 3:

$$p(z) = \frac{2}{L_f} \quad 0 \le z \le \frac{L_f}{2} \qquad (2)$$

where $p(z)$ = probability density function of variable z, z = distance between centre of the bridging fibre and the crack surface, L_f = fibre length;

$$p(\phi) = \sin \phi \quad 0 \le \phi \le \frac{\pi}{2} \qquad (3)$$

where $p(\phi)$ = probability density function of variable ϕ and ϕ = fibre inclination to the direction perpendicular to the crack plane.

Eq. 2 implies that the distribution of the fibres' embedded lengths along the crack is constant, i.e. each particular embedded length from the possible range has the same probability of participation

in crack bridging action as any other. Eq. 3 shows that the distribution of fibres with respect to their inclinations is determined by a sine function. The assumption adopted determines the fibre inclination by one end located in the centre of an imaginary hemisphere base and the second end on the surface of this hemisphere. Therefore, the probability that the fibre is going to be oriented perpendicularly to the crack plane ($\phi = 0$) is the lowest. The probability increases with inclination approaching the parallel orientation of the fibre ($\phi = \pi/2$) to the crack.

Furthermore, according to Li et al. (1991) Eq. 4 can be used for estimating the number of fibres N_f involved in crack bridging:

$$N_f = \frac{A_c V_f}{2 A_f} \qquad (4)$$

where A_c = cross-section area, V_f = fibre volume fraction in the composite and A_f = fibre cross-sectional area.

The estimation using Eq. 4 was confirmed in this study by visual investigation of SHCC fracture surfaces. The number of fibres (either pulled out or broken) counted on one fracture surface was multiplied by two, following the assumption that only the half of the fibres can be seen on each fracture surface. For the SHCC investigated there are approximately 1000 fibres bridging 1 cm² of the crack surface in projection.

5.3 *Defining typical pullout responses*

Four basic modes could be observed in the single fibre pullout tests:

– Fibre fails before the de-bonding from the matrix is completed;
– Fibre fails after de-bonding and, as a consequence of fibre damage caused by friction with the inhomogeneous matrix, during the hardening regime;
– Fibre is pulled out after slip hardening;
– Fibre is pulled out after complete de-bonding of fibre from the matrix; no slip hardening occurs.

In very rare cases fibre failed in the softening regime after foregoing hardening. This scenario was not considered in respect of the modelling.

All measured force displacement curves were divided into four groups according to the failure scenarios described above. The curves were subsequently approximated by linear or multi-linear plots, which were built by connecting characteristic points. The number and position of these points depend on a failure scenario: the first point was always the respective curve's origin

249

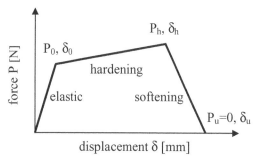

Figure 6. Schematic presentation of force-displacement curve describing fibre pullout behaviour.

Eqs. 5 give a functional description of the single fibre pullout model for all defined failure modes:

$$
p = \begin{cases}
k\delta & 0 \leq \delta \leq \delta_0 \\
P_0 + \left(P_h - P_0\right)\left(\dfrac{\delta - \delta_0}{\delta_h - \delta_0}\right) & \delta_0 \leq \delta \leq \delta_h \\
P_h\left(1 - \dfrac{\delta - \delta_h}{\delta_u - \delta_h}\right) & \delta_h \leq \delta \leq \delta_u \\
0 & \delta_u \leq \delta
\end{cases}
\tag{5}
$$

where k = fibre stiffness = $E * A_f$, E = fibre modulus of elasticity, A_f = fibre cross-sectional area, other characteristics are given in Figure 6.

Based on this description and knowing the number of bridging fibres, the pullout responses can be generated within the given ranges using the constant distribution of probability. The number of responses generated for each specific failure scenario reflects the percentage of corresponding responses in the experimental investigation. Fibre responses are generated only for a representative unit crack plane area of one square centimetre in order to limit the computation time. It was proven that this simplification does not influence the prediction of the material behaviour on the meso-level when larger cracked cross-sections are considered.

(force = 0, displacement = 0). In describing the case of fibre failure before complete de-bonding, only one additional point is needed to indicate the force and deformation at failure. Three points were used to describe the response of fibres broken during hardening or the fibre pulled out after softening. Four points were needed to describe the response of fibres pulled out after hardening, cf. Figure 6.

Upper and lower limits of force and displacement values were defined for each group of characteristic points. Because of the limited experimental data for different embedded lengths and the high variation of the pullout tests' results, it was not possible to determine clearly the statistical distribution of the values. A constant probability distribution within the chosen intervals for force and displacement was therefore used at this stage, which is sufficient for the development of a modelling framework.

6 DERIVATION OF CONSTITUTIVE RELATIONS ON MICRO-LEVEL

6.1 Pullout response

As indicated in section 5.3, pullout responses were described by using a number of linearly interconnected characteristic points. Figure 6 shows a schematic response for the case with a hardening stage and a subsequent complete fibre pullout without fibre failure. This is the most "complex" curve shape of the four modes possible, and it incorporates the characteristic points of other three possible failure modes as well. The case when the fibre is broken before the completion of de-bonding is described by one line only, following the elastic behaviour of de-bonded fibre free length. Fibres broken during pullout hardening or pulled out in the softening regime without showing hardening at all are described by two lines.

6.2 Extension of the pullout results

Because of the limitations of the pullout test setup used, the effect of fibre inclination could not be studied directly. The declination from vertical direction results during fibre pullout in increased friction of the fibre at the edge of the cracked matrix. In order to consider the effect of fibre inclination the approach according to Kanda & Li (1998) was adopted.

The enlargement of the pullout force due to an increased friction at the edges of the cracked matrix can be accounted for by using Eq. 6:

$$
P_\phi = P_{\phi=0} \cdot e^{f\phi}, \quad f = 0.5
\tag{6}
$$

where P_ϕ = enhanced pullout force as a function of fibre inclination ϕ, $P_{\phi=0}$ = pullout force for perpendicular direction to the crack plane, f = parameter determined from pullout experiments under different inclinations by Kanda & Li (1998).

However, when the friction increases locally, the fibre tends to fail earlier because of the additional damage. Therefore, the fibre load carrying capacity is limited to a maximum value according to Eq. 7:

$$P_{fu} = P_{fu}^{n} \cdot e^{-f'\phi}, \quad f' = 0.3 \qquad (7)$$

where P_{fu} = fibre maximum load carrying capacity, P_{fu}^{n} = original load carrying capacity (for fibre pullout perpendicular to the crack plane), ϕ = fibre inclination and f' = parameter determined by Kanda & Li (1998).

For all fibre responses, the inclinations are generated according to Eq. 3. Subsequently, the maximum pullout forces are increased according to Eq. 6 and, simultaneously, the fibre maximum load carrying capacity is corrected, creating upper boundaries for fibre responses to be generated.

The latest extension of the results is related to the double side de-bonding; see Wang et al. (1988). Pullout tests describe only single side fibre de-bonding and pullout behaviour. However, in SHCC fibres can be partially or fully de-bonded from both sides of the crack. According to this phenomenon, the pullout deformation, before the failure is localized to one fibre end, is multiplied by 2. The softening branch is then only moved to the new origin at a higher displacement, while its slope remains the same.

6.3 *Generated pullout responses*

Figure 7 shows a number of the generated responses of fibres fully pulled out from the matrix after showing hardening behaviour. Extensions described in Section 6.2 are introduced here already. These extensions led partly to considerable changes in the failure mode. Due to the introduc-

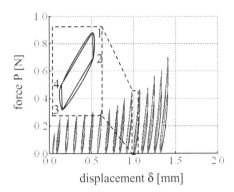

Figure 8. Unloading-reloading loops from cyclic pull-out tests and the model description of the loops by four straight lines.

tion of fibre inclination, and the related increase in the maximum pullout force accompanied by a decrease in fibre maximum load carrying capacity, some fibres fail before softening occurs.

A change in the failure mode due to fibre inclination to the crack plane can also occur for fibres which would otherwise show softening behaviour after de-bonding or which are broken after showing some hardening (see sections 5.3 and 6.1 for the description of the considered failure modes).

6.4 *Cyclic behaviour*

Figure 8 shows unloading-reloading loops extracted from a force-displacement diagram obtained in a fibre pullout test (compare Figure 3). The figure also contains a simplified linear description of the loops. At the beginning of unloading, the pullout response typically shows slight increase in deformation, which might result from pullout creep. At the beginning of the reloading a reverse tendency could be observed, i.e. a slight decrease of the measured displacement. These time-dependent components are, however, insignificant for the test rate used and therefore can be neglected. Loops are described by four points connected by straight lines. Three parameters are used to define a loop. The first defines the descent from maximal force while the displacement remains constant (the descent from point 1 to point 2 in Figure 8). The second parameter gives the displacement decrease due to unloading (from point 2 to point 3 in Figure 8). And finally, the third parameter defines the increase in force in the reloading regime without a change in displacement values (from point 3 to point 4 in Figure 8).

These three parameters are defined for each force level reached and approximated by linear functions by employing the least square method. Based on the linear dependencies observed, the

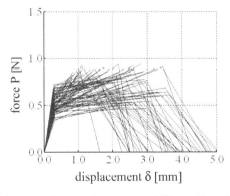

Figure 7. Generated responses of fibre pullout for a reference failure mode 'complete fibre pullout after hardening', the consideration of the fibre inclination to the crack plane led to a premature failure of some fibres; the effect of double side de-bonding is also considered here.

shapes of the loops are defined as a function of the force reached. This procedure is further used to describe the cyclic fibre pullout behaviour.

7 DERIVATION OF CONSTITUTIVE RELATIONS ON MESO-LEVEL

Formation and opening of just one crack of many in SHCC is regarded as representative for the material behaviour at the meso-level. Presented modelling approach assumes that a crack is formed suddenly and through the entire cross section in the direction perpendicular to the loading direction. A corresponding stress-displacement curve giving a characteristic material response on this level of observation can be derived by adding the force contributions of all fibres bridging the crack using Eqs. 8 and 9:

$$P(\delta) = \sum_{i=1}^{n} P_i(\delta_i) \qquad (8)$$

where $P(\delta)$ = crack bridging force, P_i = the carrying force of ith fibre as a function of displacement δ, and:

$$\sigma(\delta) = \frac{P(\delta)}{A} \qquad (9)$$

where $\sigma(\delta)$ = tensile stress, A = the cross-section area of the tensile specimen.

The same approach, i.e. adding the force contributions of a number of fibres involved in crack bridging is used also for determination of unloading and reloading branches. Cyclic loops for individual fibres are created according to desired displacement level, according to force acting on an individual fibre on given displacement level. The sum of all cyclic loops then creates the unloading-reloading response on meso-scale.

Figure 9 shows a stress-displacement relation obtained by the procedure described above. The calculated unloading and reloading loops are presented here for displacement before unloading of 0.25, 0.50, 0.75 and 1.00 mm, respectively.

The stress-displacement relation is within the expected limits with respect to the tensile characteristic of the investigated SHCC, see chapter 2. However, no experimental evidence exists to verify the shape of the curve directly. Therefore, the constitutive relationship for entire tensile specimen has to be derived first and then subsequently compared with experimental results under tensile loading in order to verify the accuracy of the model. This is the subject of an ongoing investigation.

The cyclic loops derived show similar behaviour for different stress levels as well as for the hardening and softening branches of the stress-displacement relation. This can be explained by the cyclic behaviour of single fibre during the pullout test. It can be traced back to the fact that cyclic loops from pullout tests do not change significantly their inclination or shape with increasing displacement. The change in the inclination of loops observed on macro level (i.e. for an un-notched specimen under tensile loading), see Mechtcherine & Jun (2007), results most likely from an increasing number of cracks with increasing stress level. This phenomenon will be further studied.

8 SUMMARY AND OUTLOOK

The paper has described the development of a link between the micro- and meso-levels of observation for a Strain-Hardening Cement-based Composite as a basis for the development of modelling framework for such materials under monotonic and cyclic tensile loading. This approach was based on the testing of single fibre pullout behaviour. Typical responses with different failure modes as obtained from the pullout tests were modelled by multi-linear relations which subsequently were corrected in order to eliminate the effect of the test set-up and extended for considering the effect of fibre inclination as well as of fibre de-bonding on both sides of a crack. As a result, a set of representative force-displacement relations for fibre pullout behaviour, including the unloading and reloading regimes was derived.

By observing the simultaneous action of many fibres in a crack plane, a model for stress-displacement relation of a single crack in SHCC was developed. In the next step the effect of particular arrangements of fibres in crack planes should be investigated, which would enable to derive and

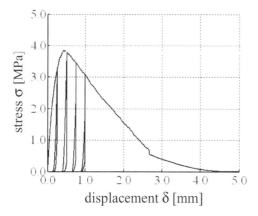

Figure 9. Modelled stress-displacement relation for one crack in SHCC including several unloading and reloading branches.

describe the variations in cracking behaviour. Then the model can be extended to the macro level, at which multiple cracking plays the major role.

After the model verification by comparison of the predicted tensile response with experimental results obtained from uniaxial tension tests on SHCC, the approach can be used, first of all, as an aid in the material design of advanced fibre-reinforced cement based materials like SHCC, but it also can provide a good basis for defining code-like material laws.

REFERENCES

Boshoff, W.P., Mechtcherine, V. & van Zijl. G.P.A.G., 2009, Characterising the Time-Dependent Behaviour on the Single Fibre Level of SHCC—Part 1: Mechanism of Fibre Pull-out Creep. *Cement and Concrete Research* (in press).

Douglas, K.S. & Billington, S.L., 2006. Rate-dependence in high-performance fiber-reinforced cement-based composites for seismic application. *Int. RILEM Workshop on HPFRCC in Structural Appl.*, Honolulu, May 2005, G. Fischer & V.C. Li, RILEM Public., S.A.R.L., PRO 49: pp. 17–26.

Fukuyama, H., Haruhiko, S. & Yang, I., 2002. HPFRCC Damper for Structural Control. *Proceedings of the JCI Int. Workshop on Ductile Fiber Reinforced Cementitious Composites (DFRCC)*, Takayama, Japan, Japan Concrete Institute: p. 219–228.

Jun, P. & Mechtcherine, V., 2007. Experimental investigation on the behaviour of Strain-Hardening Cement-based Composites (SHCC) on different levels of observation. *Int. Workshop NMMF 2007*, Wuppertal, Germany, Aedificatio Publishers, pp. 137–150.

Jun, P. & Mechtcherine, V., 2008. Deformation behaviour of cracked Strain-Hardening Cement-based Composites (SHCC) under sustained and repeated tensile loading. *In: Tanabe, T. et al., Proceedings of 8th Int. Conference on Creep, Shrinkage and durability of Concrete Structure, CONCREEP 8*, Ise-Shima, Japan, Taylor & Francis Group, London, S. pp. 487–493.

Kabele, P., 2007. Multiscale framework for modeling of fracture in high performance fiber reinforced cementitious composites, *Eng. Fracture Mech.*, Vol. 74, pp. 194–209.

Kabele, P., Novák, L., Němeček, J. & Kopecký, L., 2006. Effects of chemical exposure on bond between synthetic fiber and cementitious matrix. *Textile Reinforced Concrete—Proceedings of the 1st Int. RILEM Conference*, Cachan: RILEM Public., ISBN 2-912143-97-7, pp. 91–99.

Kanda, T. & Li, V.C., 1998. Interface Property and Apparent Strength of High-Strength Hydrophilic Fiber in Cement Matrix. *J. of Materials in Civil Eng.*, Vol. 10, No. 1, pp. 5–13.

Li, V.C., Wang, Y. & Backers, S., 1991. A micromechanical model of tension-softening and bridging toughening of short random fiber reinforced brittle matrix composites. *J. Mech. Phys. Solids*, Vol. 39, No. 5, pp. 607–625.

Mechtcherine, V., 2007. Testing Behaviour of Strain Hardening Cement-based Composites in Tension—Summary of recent research. *RILEM-Symposium on High-Performance Fibre Reinforced Cementitious Composites HPFRCC5*, H.-W. Reinhardt and A. Naaman, RILEM PRO 53, pp. 13–22.

Mechtcherine, V. & Jun, P., 2007. Stress-strain behaviour of strain-hardening cement-based composites (SHCC) under repeated tensile loading. *In: Carpinteri, A. et al. (Hrgb.), Fracture Mech. of Concrete Structures*, London: Taylor & Francis, pp. 1441–1448.

Mechtcherine, V. & Schulze, J., 2005. Ultra-ductile concrete—Material design concept and testing. *Ultra-ductile concrete with short fibres—Development, Testing, Applications*, V. Mechtcherine (ed.), ibidem-Verlag, Stuttgart: pp. 11–36.

Wang, Y., Li, V.C. & Backers, S., 1988. Modeling of fiber pull-out from a cement matrix. *Int. J. of Cement Composites and Lightweight Concrete*, Vol. 10, No. 3, pp. 143–149.

Advances in Cement-Based Materials – van Zijl & Boshoff (eds)
© 2010 Taylor & Francis Group, London, ISBN 978-0-415-87637-7

Two-scale modeling of transport processes in heterogeneous materials

Jan Sýkora
CTU in Prague, FCE, Department of Mechanics, Praha, Czech Republic

ABSTRACT: Evaluation of the effective or macroscopic parameters under coupled heat and moisture transfer is presented. The paper gives a detailed summary on the mesotructural solution of coupled transport under steady state conditions. The principles of the method rely on a uncoupled two-scale computational homogenization approach, which was successfully applied to numerical analysis of Charles Bridge in Prague. Computational homogenization is presented here in the framework of a coupled multi-scale analysis in which the material response is obtained from the underlying mesostructure by solving a given boundary value problem defined on a periodic unit cell (PUC). Solving a set of problem equations on meso-scale provides a up-scaled macroscopic equations. They include a number of effective (macroscopic) transport parameters, which are necessary for a detailed analysis of the state of a structure as a whole. A reliable methodology of the prediction of these quantities is one of the goals of our contribution. Special attention is paid to the description of heat and moisture transport in masonry wall, which is strengthened by using engineered cementitious composites (ECC).

1 INTRODUCTION

Moisture damage is one of the most important factors limiting a buildings service life. High moisture content causes corrosion, wood decay and other degradation of structure. Consequently, studying the mechanism of coupled heat and moisture transport in porous building materials is essential to improve their lifetimes.

For decades, many researchers have dealt with the modeling of transport phenomena in the porous media. The first developed models were focused on the analysis of soils. Richards (Richards 1931) established the equation of unsaturated flow in porous materials, which is on the basis of Darcys law. Philip and Vries (Philip and de Vries 1957) and Luikov (Luikov 1975) developed a moisture migration model at inhomogeneous temperature profiles, where moisture migration affected by temperature gradient was taken over. In the building area, the first technique to evaluate moisture in building materials was the Glasers method. Pedersen (Pedersen 1992) and Kiinzel (Kiinzel and Kiessl 1997) have developed more complex models that take into account the liquid and vapor diffusive transport. An extensive historical review and many others models can be found in (Černý and Rovnaníková 2002) and (Lewis and Schrefler 1999).

The objective of the paper is to incorporate the description of heat and moisture transport into the multiscale framework by using homogenization techniques. This step allows us to obtain more precise results of the material behavior.

Computational homogenization is presented here in the coupled framework of a multi-scale analysis in which the material response is obtained from the underlying mesostructure by solving a given boundary value problem defined on a periodic unit cell (PUC), see (Oezdemir, Brekelmans, and Geers 2008). Two distinct levels are considered. The first scale, on the order of decimeters, is mesostructural and serves to provide effective material properties of a heterogeneous material. The second scale (macrostructural), on the order of meters, is identified as a typical scale of structures. Solving a set of problem equations on meso-scale provides us with up-scaled macroscopic equations. They include a particular set of effective (macroscopic) transport parameters, which are necessary for a detailed analysis of the state of a structure as a whole, see (Sýkora, Vorel, Krejčí, Šejnoha, and Šejnoha 2009). A reliable methodology of the prediction of these quantities is one of the goals of our contribution.

The global algorithmic framework is outlined in Fig. 1. The temperature and moisture fields on the macroscopic domain are spatially discretized by finite elements, a proper numerical time integration scheme is introduced to convert the governing differential equations into a fully discrete form. To incorporate the constitutive behavior, a PUC is assigned to each macroscopic integration point which is subdivided in finite elements. At the

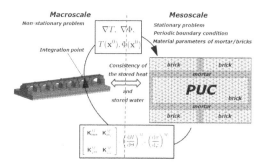

Figure 1. Scheme of coupled multi-scale framework.

mesostructural level, the macroscopic temperature and moisture gradient with the initial temperature and moisture are imposed on the PUC. Upon the solution, the macroscopic conductivity and the macroscopic capacity are extracted and transferred to the macroscopic level (integration point). If the macroscopic transport and capacity coefficients are available, the macroscopic nodal heat and mass balance can be evaluated. If the balance is satisfied the next time increment can be computed. Otherwise, the iterative macroscopic temperature and moisture fields have to be updated using the macroscopic conductivities which are already available in each integration point from the mesostructural analysis.

2 MATERIAL MODEL

There exist a number of material models that allow for the description of non-linear and non-stationary material behavior. We focus on the model by Künzel and Kiessl, see (Künzel and Kiessl 1997; Künzel 1995). It describes all substantial phenomena and the predicted results comply well with experimentally obtained data. It was therefore chosen to perform a case study on a simple masonry mesostructure. In this model, the resulting set of differential equations for the description of heat and moisture transfer, expressed in terms of temperature and relative humidity, assume the following form:

- The energy balance equation

$$\frac{dH}{d\Theta}\frac{d\Theta}{dt} = \nabla[\lambda\nabla\Theta] + h_v\nabla[\delta_p\nabla\{\varphi p_{sat}(\Theta)\}], \quad (1)$$

- The moisture balance equation

$$\frac{dw}{d\varphi}\frac{d\varphi}{dt} = \nabla[D_\varphi\nabla\varphi] + \nabla[\delta_p\nabla\{\varphi p_{sat}(\Theta)\}], \quad (2)$$

where H is the enthalpy of the moist building material, w is the water content of building material, λ is the thermal conductivity, D_φ is the liquid

conduction coefficient, δ_p is the water vapor permeability, h_v is the evaporation enthalpy of the water, p_{sat} is the water vapor saturation pressure, Θ is the temperature and φ is the relative humidity. The storage terms appear on the left hand side of Eqs. 1 and 2. The conductivity heat flux and the enthalpy flux by vapor diffusion with phase changes in Eq. 1 are strongly influenced by the moisture fields. The vapor flux in Eq. 2 is governed by both the temperature and moisture fields due to exponential changes of the saturation vapor pressure with temperature. Vapor and liquid convection caused by total pressure differences or gravity forces as well as the changes of enthalpy by liquid flow are neglected in this model.

For the spatial discretization of the partial differential equations, a finite element method is preferred here to the finite volume technique. Using the weighted residual statement the energy balance equation becomes

$$0 = \int_\Omega \delta w_\Theta \left[-\frac{dH}{d\Theta}\frac{d\Theta}{dt} \right] d\Omega$$
$$+ \int_\Omega \delta w_\Theta \left[\nabla^T \{\lambda\nabla\Theta\} \right] d\Omega$$
$$+ \int_\Omega \delta w_\Theta \left[h_v \nabla^T \{\delta_p\nabla(\varphi p_{sat})\} \right] d\Omega$$
$$+ \int_{\Gamma_\Theta^q} \delta w_\Theta \left[v^T \{\lambda\nabla\Theta\} \right] d\Gamma$$
$$+ \int_{\Gamma_\Theta^q} \delta w_\Theta \left[h_v v^T \{\delta_p\nabla(\varphi p_{sat})\} \right] d\Gamma$$
$$+ \int_{\Gamma_\Theta^q} \delta w_\Theta \bar{q}_\Theta d\Gamma, \quad (3)$$

where δw_Θ is the weighting function such that $\delta w_\Theta = 0$ on Γ_Θ. Applying Green's theorem then yields

$$-\int_\Omega \{\nabla \delta w_\Theta\}^T \left[\{\lambda\nabla\Theta\} + h_v\{\delta_p\nabla(\varphi p_{sat})\} \right] d\Omega$$
$$-\int_\Omega \delta w_\Theta \frac{dH}{d\Theta}\frac{d\Theta}{dt} d\Omega + \int_{\Gamma_\Theta^q} \delta w_\Theta \bar{q}_v d\Gamma = 0, \quad (4)$$

where $\bar{q}_v$ is the prescribed heat flux perpendicular to the boundary Γ_Θ^q. On Γ_Θ the temperature Θ is equal to its prescribed value Θ. Vector v stores the components of the unit outward normal.

The weak formulation for the conservation of mass can be derived analogously

$$0 = \int_\Omega \delta w_\varphi \left[-\frac{dw}{d\varphi}\frac{d\varphi}{dt} \right] d\Omega$$
$$+ \int_\Omega \delta w_\varphi \left[\nabla^T \{D_\varphi\nabla\varphi\} \right] d\Omega$$
$$+ \int_\Omega \delta w_\varphi \left[\nabla^T \{\delta_p\nabla(\varphi p_{sat})\} \right] d\Omega$$
$$+ \int_{\Gamma_\Theta^g} \delta w_\varphi \left[v^T \{D_\varphi\nabla\varphi\} \right] d\Gamma$$
$$+ \int_{\Gamma_\Theta^g} \delta w_\varphi \left[v^T \{\delta_p\nabla(\varphi p_{sat})\} \right] d\Gamma$$
$$+ \int_{\Gamma_\Theta^g} \delta w_\varphi \bar{g}_\varphi d\Gamma \quad (5)$$

256

where δw_φ is the corresponding weighting function such that $\delta w_\varphi = 0$ on Γ_φ. Application of Green's theorem finally gives

$$-\int_\Omega \{\nabla \delta w_\varphi\}^T \Big[\{D_\varphi \nabla \varphi\} + \{\delta_p \nabla(\varphi p_{sat})\} \Big] d\Omega$$
$$-\int_\Omega \delta w_\varphi \frac{dw}{d\varphi}\frac{d\varphi}{dt} d\Omega + \int_{\Gamma_\Theta^{\bar g}} \delta w_\varphi \bar{g}_t d\Gamma = 0. \tag{6}$$

The temperature field Θ and moisture field φ are approximated as:

$$\begin{aligned}
\Theta(\mathbf{x}) &= \mathbf{N}_\Theta(\mathbf{x})\mathbf{r}_\Theta, \\
\varphi(\mathbf{x}) &= \mathbf{N}_\varphi(\mathbf{x})\mathbf{r}_\varphi, \\
\nabla\Theta(\mathbf{x}) &= \nabla\mathbf{N}_\Theta(\mathbf{x})\mathbf{r}_\Theta = B_\Theta(\mathbf{x})\mathbf{r}_\Theta, \\
\nabla\varphi(\mathbf{x}) &= \nabla\mathbf{N}_\varphi(\mathbf{x})\mathbf{r}_\varphi = B_\varphi(\mathbf{x})\mathbf{r}_\varphi,
\end{aligned} \tag{7}$$

where $\mathbf{N}_\Theta(\mathbf{x}) = \mathbf{N}_\varphi(\mathbf{x}) = \mathbf{N}(\mathbf{x})$ is the matrix of shape functions (according to type of used elements) and $\mathbf{r}_\Theta$ and $\mathbf{r}_\varphi$ are the column matrix of nodal values of temperature field Θ and moisture field φ.

Using approximations 7 and Eqs. 4, 6 (considering that p_{sat} is a function of temperature Θ), we obtain a set of equations

$$\mathbf{K}_{\Theta\Theta}\mathbf{r}_\Theta + \mathbf{K}_{\Theta\varphi}\mathbf{r}_\varphi + \mathbf{C}_{\Theta\Theta}\frac{d\mathbf{r}_\Theta}{dt} = \mathbf{q}_{ext}, \tag{8}$$

$$\mathbf{K}_{\varphi\Theta}\mathbf{r}_\Theta + (\mathbf{K}^w_{\varphi\varphi} + \mathbf{K}^V_{\varphi\varphi})\mathbf{r}_\varphi + \mathbf{C}_{\varphi\varphi}\frac{d\mathbf{r}_\varphi}{dt} = \mathbf{g}_{ext}. \tag{9}$$

3 HOMOGENIZATION

In this paragraph, the fundamental steps of computational homogenization is be briefly outlined. Details are available in (Sýkora, Vorel, Krejčí, Šejnoha, and Šejnoha 2009). Let us begin by summarizing the basic equations of homogenization in application to the coupled transport problem. The scale transition between two levels (meso to macro) draws on splitting the local temperature/moisture field into macroscopic and fluctuation parts, respectively, as

$$\Theta(\mathbf{x}) = T(\mathbf{x}) + \Theta^*(\mathbf{x}), \tag{10}$$

$$\varphi(\mathbf{x}) = \Phi(\mathbf{x}) + \varphi^*(\mathbf{x}). \tag{11}$$

Under the assumption that the macroscopic gradients ∇T, $\nabla\Phi$ are constant throughout the material sample, we obtain the well known formula

$$\Theta(\mathbf{x}) = T(\mathbf{x}^0) + \{\mathbf{x} - \mathbf{x}^0\}\nabla T + \Theta^*(\mathbf{x}), \tag{12}$$

$$\varphi(\mathbf{x}) = \Phi(\mathbf{x}^0) + \{\mathbf{x} - \mathbf{x}^0\}\nabla\Phi + \varphi^*(\mathbf{x}). \tag{13}$$

Then the disretized form of energy and moisture balance Eqs. 1 and 2 (note that steady state

conditions are assumed due to negligibly small PUC size) results in

$$\mathbf{K}_{\Theta\Theta}\mathbf{r}^*_\Theta + \mathbf{K}_{\Theta\varphi}\mathbf{r}^*_\varphi = -\mathbf{Q}_{\Theta\Theta}\nabla T - \mathbf{G}_{\Theta\Theta}\nabla\Phi, \tag{14}$$

$$\mathbf{K}_{\varphi\Theta}\mathbf{r}^*_\Theta + (\mathbf{K}^w_{\varphi\varphi} + \mathbf{K}^v_{\varphi\varphi})\mathbf{r}^*_\varphi = -\mathbf{Q}_{\varphi\Theta}\nabla T - \mathbf{G}_{\varphi\varphi}\nabla\Phi. \tag{15}$$

Finally, when the distributions of temperature and moisture fields are known, the effective and macroscopic parameters are calculated from consistency rule and from averaging of fluxes, see (Sýkora, Vorel, Krejčí, Šejnoha, and Šejnoha 2009).

4 APPLICATION

Engineered Cementitious Composite (ECC) is an easily molded and shaped mortar-based composite reinforced with specially selected short random fibers, usually polymer fibers, see (Bruedern, Abeca-sis, and Mechtcherine 2008) and (Li 1993). Though many experimental results have demonstrated the effectiveness of ECC reinforcements for masonry walls (Bruedern, Abecasis, and Mechtcherine 2008), little has been done to understand their impact on the building thermal and moisture envelope. In this section, we show the calculation of effective parameters of masonry wall, which is strengthened by ECC layer.

4.1 Determining the transport coefficients

Before proceeding to the examples, let us start with the determination of the transport coefficients, which appear in the heat and the mass balance Eqs. 1 and 2. The reinforced masonry wall consists of the mortar, bricks and the layer of ECC, which is on the outer side of the wall. A set of experimental measurements was performed to obtain the hygric and the thermal properties of mortar and brick. The material properties of ECC were chosen from the database of the DELPHIN computer code developed at TU Dresden, see (Grunewald 2000). The material parameters of the masonry components, which are used in examples, are listed in Table 1.

The parameters that enter Eqs. 1 and 2 are provided by

- w water content [kgm^{-3}]

$$w = w_f \frac{(b-1)\varphi}{b-\varphi}, \tag{16}$$

where w_f is the free water saturation and b is the approximation factor, which must always be greater than one. It can be determined from the equilibrium water content (w_{80}) at 80% relative humidity by substituting the corresponding numerical values in Eq. 16.

257

Table 1. Material parameters of individual phases.

Parameter	Brick	Mortar	ECC
$w_f\,[\mathrm{kgm^{-3}}]$	311	342	200
$w_{80}\,[\mathrm{kgm^{-3}}]$	46	52	25
$\lambda_0\,[\mathrm{Wm^{-1}K^{-1}}]$	0.46	0.83	1.43
$b\,[-]$	4.31	16.54	1.72
$\rho_s\,[\mathrm{kgm^{-3}}]$	2670	1690	1720
$\mu\,[-]$	19	10	45
$A\,[\mathrm{kgm^{-2}\,s^{-0.5}}]$	1.25	0.22	0.20
$c\,[\mathrm{Jkg^{-1}K^{-1}}]$	840	1000	1000

(w_f—free water saturation, w_{80}—water content at 80% relative humidity, λ_0—thermal conductivity of dry building material, b—thermal conductivity supplement, ρ_s—bulk density of dry building material, μ—water vapor diffusion resistance factor, A—water absorption coefficient, c—specific heat capacity).

- δ_p water vapor permeability $[\mathrm{kgm^{-1}s^{-1}Pa^{-1}}]$

$$\delta_p = \frac{\delta}{\mu}, \qquad (17)$$

where μ is the water vapor diffusion resistance factor and δ is the vapor diffusion coefficient in air $[\mathrm{kgm^{-1}s^{-1}Pa^{-1}}]$ given by, see (Schirmer 1938)

$$\delta = \frac{2.306 \cdot 10^{-5}\,p_a}{R_v \Theta\,p} \left(\frac{\Theta}{273.15}\right)^{1.81}, \qquad (18)$$

with p set equal to atmospheric pressure $p_a = 101325$ Pa and $R_v = R/M_w = 461.5$; R is the gas constant ($8314.41\ \mathrm{Jkmol^{-1}K^{-1}}$) and M_w is the molar mass of water ($18.01528\ \mathrm{kgmol^{-1}}$).

- D_φ liquid conduction coefficient $[\mathrm{kgm^{-1}s^{-1}}]$, see Fig. 2

$$D_\varphi = D_w \frac{dw}{d\varphi}, \qquad (19)$$

where D_w is the capillary transport coefficient and the derivation of the moisture storage function is calculated from

$$\frac{dw}{d\varphi} = \frac{w_f\,b(b-1)}{(b-\varphi)^2}. \qquad (20)$$

- λ thermal conductivity of dry building material $[\mathrm{Wm^{-1}K^{-1}}]$, see Fig. 3

$$\lambda = \lambda_0 \left(1 + \frac{bw}{\rho_s}\right), \qquad (21)$$

where λ_0 is the thermal conductivity of dry building material, ρ_s is the bulk density and b is

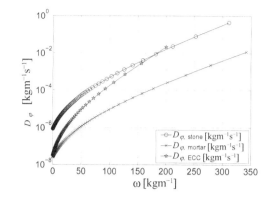

Figure 2. Variation of a phase liquid conduction coefficients D_φ as a function of water content.

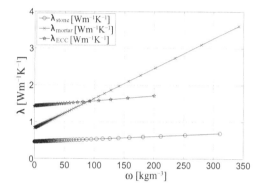

Figure 3. Variation of a phase thermal conductivities λ as a function of water content.

the thermal conductivity supplement. Supplement b indicates how many percent of the thermal conductivity increases per mass percent of the moisture. Its value is determined by the type of building material.

- p_{sat} water vapor saturation pressure [Pa]

$$p_{sat} = 611\exp\left(\frac{a\Theta}{\Theta_0 + \Theta}\right), \qquad (22)$$

where

$$\begin{aligned} a = 22.44 \quad \Theta_0 = 272.44\,^\circ\mathrm{C} \quad \Theta < 0\,^\circ\mathrm{C},\\ a = 17.08 \quad \Theta_0 = 234.18\,^\circ\mathrm{C} \quad \Theta \geq 0\,^\circ\mathrm{C}. \end{aligned} \qquad (23)$$

- h_v evaporation enthalpy of water $[\mathrm{Jkg^{-1}}]$

$$h_v = 2.5008 \cdot 10^6 \left(\frac{273.15}{\Theta}\right)^{(0.167 + 3.67 + 10^{-4}\Theta)}. \qquad (24)$$

4.2 Examples

For simplicity we limit our attention to a mesostructural level (the level of a periodic unit cell), recall Fig. 1 identifying this problem as one particular step in the multi-scale computational scheme. To illustrate the applicability of the proposed method consider the periodic unit cell in Fig. 4. The PUC with the initial conditions ($T(\mathbf{x}^0) = 20$ [°C], $\Phi(\mathbf{x}^0) = 0.5$ [–]) was loaded by the macroscopic temperature gradient $\nabla T = \{\nabla T_x, \nabla T_y, \nabla T_z\} = \{0, 10, 0\}$ and the macroscopic moisture gradient $\nabla \Phi = \{\nabla \Phi_x, \nabla \Phi_y, \nabla \Phi_z\} = \{0, 0.05, 0\}$. The resulting fluctuations of the local temperature and moisture plotted in Figs. 5 and 6 were obtained from the solution of Eqs. 14 and 15. The local temperature/moisture fields are shown in Figs. 5 and 6. The distribution of the local temperature and moisture fields on the PUC allows us to evaluate the effective and macroscopic parameters, which are then utilized in the macroscopic calculation (at this level, masonry wall is represented by the homogeneous material with the effective conductivities and the macroscopic capacities. The real climate boundary and loading conditions are applied). For illustration, the effective submatrix $\mathbf{K}_{\Theta\Theta}^M$ is given

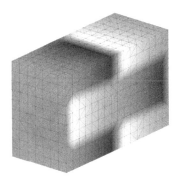

Figure 6.   IM, Fluctuation moisture field (Color bar, relative humidity φ^*—min = black color = –0.0078 [–]; max = white color = 0.106 [–]).

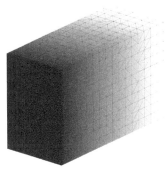

Figure 7.   RM, Local temperature field (Color bar, temperature Θ—min = black color = 20.00 [°C]; max = white color = 23.23 [°C]).

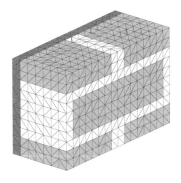

Figure 4.   Periodic unit cell of the reinforced masonry.

by Eq. 26 (note that the non-diagonal elements are obtained due to different spatial structure of the PUC). Verification of the proposed method was carried out on the two-scale model with the same material parameters of the periodic unit cell, see (Sýkora 2008).

$$
\begin{bmatrix} \mathbf{K}_{\Theta\Theta}^M & \mathbf{K}_{\Theta\varphi}^M \\ \mathbf{K}_{\varphi\Theta}^M & \mathbf{K}_{\varphi\varphi}^M \end{bmatrix}_{(6\times6)} \Rightarrow
$$

$$
\mathbf{K}_{\Theta\Theta}^M = \begin{bmatrix} 0.62 & -1.89 \cdot 10^{-4} & 2.33 \cdot 10^{-4} \\ -1.89 \cdot 10^{-4} & 0.64 & 9.08 \cdot 10^{-6} \\ 2.33 \cdot 10^{-4} & 9.08 \cdot 10^{-6} & 0.66 \end{bmatrix}. \tag{25}
$$

To capture the influence of macroscopic loading conditions on effective parameters consider the periodic unit cell displayed in Fig. 4. The PUC was loaded by several different heat and moisture macroscopic gradients. In addition, the macroscopic/ initial temperature and macroscopic/initial relative

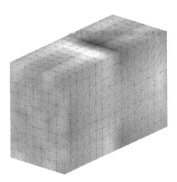

Figure 5.   Fluctuation temperature field (Color bar, temperature Θ^*—min = black color = –0.0078 [°C]; max = white color = 0.106 [°C]).

humidity also varied. The distribution of the [1,1] element of effective conductivity matrix as a function of the macroscopic moisture and temperature and variation of the macroscopic heat storage capacity as a function of the macroscopic moisture and temperature appear in Figs. 9 and 10. As evident from the presented figures, the calculated effective parameters are dependent both on the initial and loading conditions. Therefore, an application of this method in a coupled multi-scale framework is inevitable.

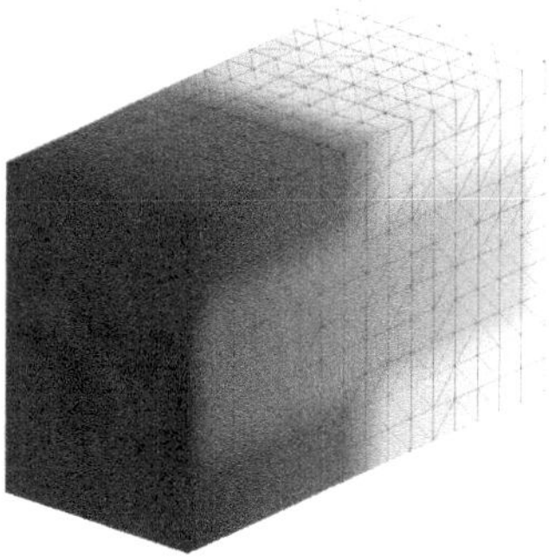

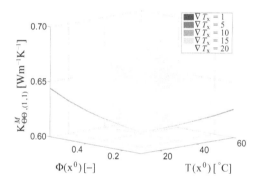

Figure 8. IM, Local moisture field (Color bar, relative humidity φ—min = black color = 0.50 [–]; max = white color = 0.52 [–]).

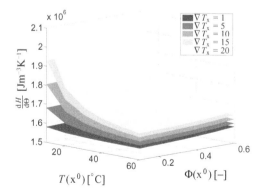

Figure 9. Variation of the [1,1] element of effective conductivity matrix as a function of the macroscopic temperature and moisture.

Figure 10. Variation of the macroscopic heat storage capacity as a function of the macroscopic temperature and moisture.

5 CONCLUSIONS

In this contribution, the reinforced masonry wall was chosen as an example for evaluation of the effective parameters at the mesostructural level. Presented approach offers the possibility of including the mesostructural morphology and mesostructural material behavior in macrolevel, where typical structures are analyzed. The following interconnected topics are subject to investigation:

- The high computational effort can be reduced by parallel computing which can has a simple algorithmic structure. In a parallel computing framework, the macroscopic problem is handled by a single master processor and the micro level problems are distributed to the slave processors.
- Description of the effect of heat and moisture transport in porous materials across interfaces between two porous materials, which manifest itself by the jumps in both the capillary pressures and temperature fields and which in classical models is mostly not dealt with. These phenomena especially gain on importance in the regions situated in the immediate vicinity of surfaces of bulky structures exposed to thermal radiation of the sun.
- Limitation of the proposed method is the missing part of mechanical behavior, which may play a crucial role in transport phenomena and in estimation of the effective parameters, when the cracks appear.

ACKNOWLEDGEMENTS

The financial support provided by the GAČR Grant No. 103/08/1531 and CTU0920211 is gratefully acknowledged.

REFERENCES

Bruedern, A.-E., D. Abecasis, and V. Mechtcherine (2008). Strenghtening of masonry using sprayed strain hardening cement-based composites (shcc). In *Seventh International RILEM Symposium on Fibre Reinforced Concrete: Design and Applications.*

Černý, R. and P. Rovnaníková (2002). *Transport Processes in Concrete.* London: Spon Press.

Grunewald, J. (2000). *DELPHIN 4.1 Documentation, Theoretical fundamentals.* Dresden: TU Dresden.

Künzel, H. (1995). Simultaneous heat and moisture transport in building components. Technical report, Fraunhofer IRB Verlag Stuttgart.

Künzel, H. and K. Kiessl (1997). Calculation of heat and moisture transfer in exposed building components. *Int. J. Heat Mass Tran.* (40), 159–167.

Lewis, R.W. and B.A. Schrefler (1999). *The Finite Element Method in the Static and Dynamic Deformation and Consolidation of Porous Media.* John Wiley&Sons, Chichester, England, 2nd edition.

Li, V.C. (1993). From micromechanics to structural engineering the design of cementitious composites for civil engineering applications. *Journal of Structural Mechanics and Earthquake* 10(2), 37–48.

Luikov, A.V. (1975). System of differential equation of heat and mass transfer in capillary-porous bodies. *Int. J. Heat Mass Tran.* 18(1), 1–14.

Oezdemir, I., W. Brekelmans, and M. Geers (2008). Computational homogenization for heat conduction in heterogeneous solids. *Int. J. Num. Meth. Engng.* (73), 185–204.

Pedersen, C.R. (1992). Prediction of moisture transfer in building constructions. *Building and Environment* 27(3), 387–397.

Philip, J.R. and D.A. de Vries (1957). Moisture movement in porous materials under temperature gradients. *Transactions of The American Geophysical Union* (38), 222–232.

Richards, L.A. (1931). Capillary conduction of liquids through porous media. *Physics* 1(3), 318–333.

Schirmer, R. (1938). ZVDI. *Beiheft Verfahren-stechnik 6*, 170.

Sýkora, J. (2008). Multiscale modeling of transport processes in masonry structures. CTU in Prague.

Sýkora, J., J. Vorel, T. Krejčí, M. Šejnoha, and J. Šejnoha (2009). Analysis of coupled heat and moisture transfer in masonry structures. *Materials and Structures* (0), 0–0. Article in press. Available at http://arxiv.org/abs/0804.3554v1.

Advances in Cement-Based Materials – van Zijl & Boshoff (eds)
© *2010 Taylor & Francis Group, London, ISBN 978-0-415-87637-7*

Computational modelling of Strain-Hardening Cement Composites (SHCC)

G.P.A.G. van Zijl
Stellenbosch University, South Africa

ABSTRACT: Strain-Hardening Cement Composites (SHCC) are designed to combine crack control and robust mechanical resistance in a durable construction material. Constitutive models that capture the dominating mechanisms of mechanical and time-dependent behaviour of this class of construction materials are essential for accurate structural analysis. Such models may be complex in order to capture the complex behaviour of fibre-reinforced cement-based construction materials. However, the value of such models lies in the ability it affords the specialist to predict structural behaviour beyond that measured in physical experiments, which is a cost-effective extension of physical experimental data. With the aid of such complex models, thorough understanding and eventual simpler analytical models can be derived. This paper describes a constitutive model for fibre-reinforced strain-hardening cement composites. In the derivation of the failure criteria and their dependence on strain hardening/softening, reference is made to results of characterising tests, including uniaxial tension and compression tests, shear tests, rate tests, as well as creep loads. The model is based on multi-surface, anisotropic, rate-dependent, computational continuum plasticity. Verification of the model is presented by analysis of two series of laboratory tests, namely SHCC and steel bar reinforced SHCC tested in flexure, and SHCC specimens tested in shear.

1 INTRODUCTION

Strain-Hardening Cement Composites (SHCC) form a recent class of high performance fibre reinforced cement-based composites (HPFRCC). It has highly ductile composite tensile behaviour, with tensile deformability expressed as average strain at the ultimate tensile resistance in the range 3%–5%, for relatively low fibre volume content ($1.5\% < V_f < 2.5\%$) of short, polymeric fibres. This crack width arrest is in fact a mechanism of the large deformability, through formation of multiple cracks, leading to pseudo strain-hardening response in tension. There is evidence that the crack control is maintained also in more general loading conditions, including flexure, shear and combined flexure and shear (van Zijl 2007), both in pure SHCC and with steel bar reinforcement (R/SHCC) (Fischer and Li 2004). The crack control is ensured by balanced properties of the cement matrix, fibres and their interfaces (Li et al. 1995). The conditions for pseudo strain-hardening are well understood and the data base of experimental results towards confirming micro-mechanical composite design models is growing.

To enable sound application of SHCC, design models are required, which should eventually be incorporated in standards for structural design with this construction material. Such models do not yet exist, and the pool of experimental data is limited relative to traditional construction materials like reinforced concrete and structural steel. It is argued that computational models, in combination with physical experiments, are effective in developing insight in the behaviour of structures manufactured of such a new material, thereby assisting in the formulation of simpler design models. Such computational models should capture the main mechanisms of behaviour with reasonable accuracy to be objective and allow prediction of structural behaviour beyond that tested in physical experiments.

While detailed micro-models may prove more accurate by capturing interaction of the heterogeneities in SHCC (fibres, hcp matrix, fibre-matrix interfaces), so-called macro-models are pragmatic and allow viable analysis and prediction of structural behaviour. In this paper a macro-model based on continuum plasticity is proposed and elaborated. Such a model does not distinguish between the various ingredients in SHCC, but considers it to be a homogeneous continuum. The use of such continuum models for concrete or other cement-based materials has become standard in research environments and are even available in commercial finite element packages such as DIANA (2008), which is used in this research.

However, a computational macro-model appropriate for SHCC does not yet exist commercially. A model based on continuum damage was proposed recently by Boshoff and van Zijl (2007), but provided for nonlinear behaviour in only the tensile regime. Also, reported SHCC response to cyclic loading (Boshoff 2007, Jun and Mechtcherine 2009) indicates that elastic unloading, as used in a plasticity approach, is more appropriate for SHCC in the strain-hardening regime, than the secant unloading used in the damage approach. Kabele (2002) also developed a continuum model for SHCC, despite acknowledging that a discrete approach is more appropriate for the localisation. In the models mentioned above, time-dependence has not been incorporated.

In this chapter, a continuum model is proposed for both tensile and compressive behaviour, in terms of a multi-surface continuum plasticity model. Anisotropy is considered in terms of different strengths and hardening-softening responses in orthogonal directions. This has been shown to be relevant for certain manufacturing processes, for instance extrusion, which predominantly orientates fibres in the extrusion direction. Creep is incorporated in the form of visco-elasticity. In tension, cracking rate-dependence is considered, based on experimental evidence of rate-enhanced resistance (Boshoff and van Zijl 2007b).

The model is verified by the analysis of SHCC and R/SHCC beams tested in three-point bending (Visser and van Zijl 2007), as well as Iosipesco shear experiments on SHCC specimens (van Zijl 2007).

2 SHCC BEHAVIOUR

SHCC have been developed and characterised at Stellenbosch University in an ongoing research programme. The ingredients and mass proportions of a particular SHCC mix used extensively in laboratory experiments are given in Table 1. All results reported in this paper are for this mix, or a mix with negligible deviations. Note that chemical additives are used in reasonable quantities to obtain appropriate levels of workability and consistency.

Specimens have been produced by casting as well as extrusion. In this paper reference is made only to cast specimens. Specimens are typically tested at

Table 1. Typical SHCC mix proportions.

Ingredient	Mass	Kg/m³
Cement (CEM I 42.5)	0.5	531
Fly Ash	0.5	531
Water	0.4	425
Sand (φ_s < 0.21 mm)	0.5	531
PVA fibre, L_f = 12 mm, d_f = 0.04 mm	0.024	26

14 days, until which date they are cured in water. Full details are given by Visser and van Zijl (2007).

The tensile results here are for thin dumbbell specimens, shown in Figure 1, with nominal thickness of 16 mm and uniform gauge area of 80 mm long by 30 mm wide.

The ductility of SHCC is also exhibited in flexure. Figure 2 shows the responses of SHCC beams of dimensions 100 mm × 100 mm × 500 mm, spanning 400 mm in three point bending. Also shown are the three-point bending responses of such SHCC beams reinforced in addition with 1% by volume of tensile reinforcing steel bars.

It has been shown that creep and loading rate dependence may be significant for SHCC (Boshoff 2007) rendering it essential to be considered for accurate prediction of structural behaviour (van Zijl and Boshoff 2009). The phenomena of

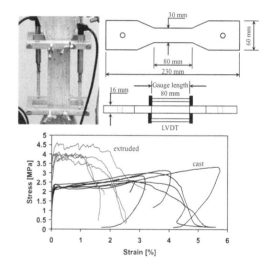

Figure 1. Uniaxial tensile responses of cast (considered here) and extruded SHCC specimens (Visser and van Zijl 2007).

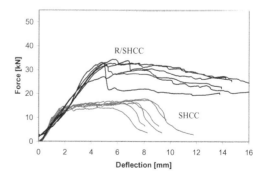

Figure 2. Flexural response of cast SHCC and R/SHCC beams subjected to three-point bending (Visser and van Zijl 2007).

264

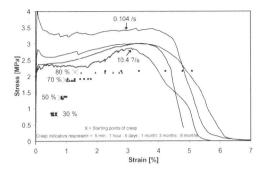

Figure 3. Tensile stress-strain responses to monotonic tests at various loading rates, and creep tests of pre-cracked specimens at various sustained load levels (Boshoff and van Zijl 2007b).

rate-enhanced first cracking and ultimate tensile resistance, as well as creep fracture have been shown experimentally. Creep fracture at high sustained tensile loads was observed also recently by Jun and Mechtcherine (2009). Figure 3 summarises uniaxial tensile rate and creep test results obtained at Stellenbosch University. Full details are given in Boshoff (2007).

3 CONSTITUTIVE MODEL FOR SHCC

To capture the typical mechanical behaviour of SHCC shown in the previous section, a computational model based in computational plasticity has been developed recently. A full, elaborate description of the model can be found in van Zijl (2009). For clarity, a brief outline is given here.

3.1 Anisotropic Rankine continuum plasticity

The model is formulated in plane with stress and strain vectors

$$\sigma^T = \lfloor \sigma_x \ \sigma_y \ \tau \rfloor, \quad \varepsilon^T = \lfloor \varepsilon_s \ \varepsilon_y \ \gamma \rfloor \tag{1}$$

The material law can be expressed as follows

$$\dot{\sigma} = D^{ve}\dot{\varepsilon}^{ve} + \Sigma = D^{ve}\left(\dot{\varepsilon} - \dot{\varepsilon}^p - \dot{\varepsilon}^0\right) + \Sigma \tag{2}$$

where D^{ve} is an equivalent time-dependent stiffness modulus, $\dot{\varepsilon}^{ve}$, $\dot{\varepsilon}^p$ and $\dot{\varepsilon}^0$ are the visco-elastic, plastic and initial strain vectors and Σ is a viscous stress vector, which accounts for the history. It can be shown that the visco-elastic stiffness is computed from the spring stiffnesses (E_i) and dashpot viscosities (η_i) of the elements in the Maxwell chain shown in Figure 4. The stress history term accumulates the stresses in each chain element at the end of the previous time step. Eq. (2) is integrated with

linear time integration. To incorporate the various sources of time dependence, the limit functions are enhanced by consideration of bulk visco-elasticity through the Maxwell-chain, as well as cracking viscosity (η_{cr}), as schematised in Figure 4. Note that the latter is activated only in tension.

The limit stresses are treated in multi-surface plasticity fashion, whereby the plastic strain increment is expressed as follows:

$$\Delta\varepsilon^p = \Delta\lambda \frac{\partial f}{\partial \sigma} \tag{3}$$

with λ the amount of plastic flow/crack strain and f the flow function. For tension, the anisotropic Rankine yield function is chosen, which is a maximum principal stress failure criterion, formulated as

$$f_t = \frac{(\sigma_x - \sigma_{tx}) + (\sigma_y - \sigma_{ty})}{2}$$
$$+ \sqrt{\left(\frac{(\sigma_x - \sigma_{tx}) + (\sigma_y - \sigma_{ty})}{2}\right)^2 + \alpha_t \tau^2} \tag{4}$$

The current admissible stresses in the material x, y directions are denoted by σ_{tx} and σ_{ty} respectively. Simple expressions for the tensile resistance are given in Figure 5. The strain-hardening and subsequent exponential strain-softening tensile response is typical of SHCC. The same functions are adapted in compression, which is a simplified, yet realistic approximation of the observed uniaxial compression response.

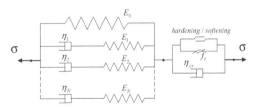

Figure 4. Schematic representation of the model

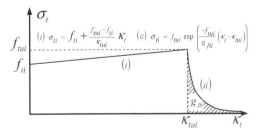

Figure 5. Adopted limit function evolution with equivalent strain.

265

Separate expressions are used in the orthogonal directions, with separate sets of model parameters. The hardening-softening is governed by equivalent tensile strain κ_t, with an appropriate measure for SHCC considered to be the maximum principal plastic strain given by

$$\dot{\kappa}_t = -\dot{\varepsilon}_1^p = \frac{\dot{\varepsilon}_x^p + \dot{\varepsilon}_y^p}{2} + \sqrt{\left(\frac{\dot{\varepsilon}_x^p - \dot{\varepsilon}_y^p}{2}\right)^2 + \left(\frac{\dot{\gamma}}{2}\right)^2} \qquad (5)$$

The equivalent strain increment can be shown to be directly related to the plastic flow increment ($\Delta k_t = \Delta\lambda_t$). The parameter α_t controls the shear contribution to tensile failure.

In compression, the limit function is

$$f_c = -\frac{(\sigma_x - \sigma_{cx}) + (\sigma_y - \sigma_{cy})}{2} \\ + \sqrt{\left(\frac{(\sigma_x - \sigma_{cx}) + (\sigma_y - \sigma_{cy})}{2}\right)^2 + \alpha_c \tau^2} \qquad (6)$$

The equivalent compressive strain is

$$\dot{\kappa}_c = -\dot{\varepsilon}_2^p = -\frac{\dot{\varepsilon}_x^p + \dot{\varepsilon}_y^p}{2} + \sqrt{\left(\frac{\dot{\varepsilon}_x^p - \dot{\varepsilon}_y^p}{2}\right)^2 + \left(\frac{\dot{\gamma}}{2}\right)^2} \qquad (7)$$

As for tension, it can be shown that $\Delta k_c = \Delta\lambda_c$.

The intersection between the two limit surfaces is treated in a consistent way, as proposed by Koiter (1953)

$$\Delta\varepsilon^p = \Delta\lambda_t \frac{\partial f_t}{\partial \sigma} + \Delta\lambda_c \frac{\partial f_c}{\partial \sigma} \qquad (8)$$

The multi-surface limit function is shown in Figure 6. A dashed line, which allows for confinement strengthening is also suggested, but not used here. A research project at Stellenbosch University

is currently exploring the biaxial behaviour experimentally.

3.2 *Cracking rate model*

In tension, a cracking rate-dependence is introduced following Wu and Bažant (1993) and van Zijl et al. (2001) as follows:

$$\Delta w = \Delta w_r \sin h\left[\frac{\sigma - \sigma_t(w)}{c_1\left[\sigma_t(w) + c_2 f_t\right]}\right] \qquad (9)$$

where $\dot{w}_r$ is a constant, reference crack opening rate. The crack width w, which is assumed to be spread uniformly across the fracture process zone of width l_b, can be related to the equivalent strain as follows:

$$\kappa_t = \frac{w}{l_b} \qquad (10)$$

with eqs. (9, 10) the tensile strength equations (Figure 5) can be rewritten as

$$\sigma_{ti} = \left(f_{ti} + \frac{f_{tui} - f_{ti}}{\kappa_{tui}}\kappa_t\right)\left[1 + c_1 \sin h^{-1}\left(\frac{\dot{\kappa}_t}{\dot{\kappa}_r}\right)\right] \\ + c_1 c_2 f_{tui} \sin h^{-1}\left(\frac{\dot{\kappa}_t}{\dot{\kappa}_r}\right) \quad \forall \; \kappa_t \leq \kappa_{tui} \qquad (11)$$

$$\sigma_{ti} = f_{tui} \exp\left(\frac{-f_{tui}}{G_{fti}/l_b}(\kappa_t - \kappa_{tui})\right)\left[1 + c_1 \sin h^{-1}\left(\frac{\dot{\kappa}_t}{\dot{\kappa}_r}\right)\right] \\ + c_1 c_2 f_{tui} \sin h^{-1}\left(\frac{\dot{\kappa}_t}{\dot{\kappa}_r}\right) \quad \forall \; \kappa_t > \kappa_{tui}$$

$$(12)$$

where the index i refers to the materials axes. The coefficient c_2 is a small offset value to prevent singularity of eq. (9). The reference strain rate $\dot{k}_r$ is directly related to $\dot{w}_r$ cf. eq. (10). It should be a sufficiently low cracking rate at which the cracking strengths are rate-independent. In the current model, the rate terms represent all physical processes causing rate dependence, here both cracking viscosity and fibre pull-out rate-dependence. Fibre pull-out under sustained load, as well as at various monotonic loading rates was shown to be an important mechanism of time-dependence (Boshoff et al. 2009).

An alternative simplified rate model, with only one model parameter, can be formulated as follows:

$$\sigma_{ti} = f_{ti} + \frac{(f_{tui} + m_i \dot{\kappa}_t) - f_{ti}}{\kappa_{tui}}\kappa_t \quad \forall \; \kappa_t \leq \kappa_{tui} \qquad (13)$$

$$\sigma_{ti} = \left(1 + \frac{m_i \dot{\kappa}_t}{f_{tui}}\right)f_{tui} \exp\left(\frac{-f_{tui}}{G_{fti}/l_b}(\kappa_t - \kappa_{tui})\right) \quad \forall \; \kappa_t > \kappa_{tui}$$

$$(14)$$

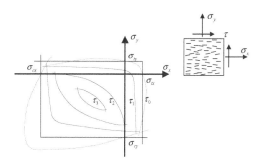

Figure 6. Limit function for biaxial behaviour of SHCC.

where the cracking viscosity m_i may differ in orthogonal material directions. The constitutive model has been implemented as a user material in DIANA (2008), with fully implicit treatment also of a consistent tangent matrix (stiffness) to ensure quadratic convergence in the iterative-incremental nonlinear solution process.

4 BEAM IN THREE POINT BENDING

The model is verified by analysis of three-point bending tests on SHCC and R/SHCC beams reported by Visser and van Zijl (2007), where full details of the tests can be found.

4.1 Model parameter determination

The uniaxial tensile behaviour of the cast SHCC specimens shown in Figure 1 applies. The model parameters are shown in Table 2. Note that the rate and creep parameters were computed by inverse analysis of test results by Boshoff and van Zijl (2007b) on SHCC of similar mix and quasi-static tensile reponse. Methods for parameter estimation are described in more detail by van Zijl (2009). An elastic-perfect plastic model was used for the steel reinforcement. Full bond was considered between SHCC and the steel bars.

Table 2. Model parameters.

Tension		Compression	
$f_{tx} = f_{ty}$ (MPa)	1.8	$f_{cx} = f_{cy}$ (MPa)	20
$f_{tux} = f_{tuy}$ (MPa)	2.0	$f_{cux} = f_{cuy}$ (MPa)	30
$\kappa_{tux} = \kappa_{tuy}$ (−)	0.04	$\kappa_{cux} = \kappa_{cuy}$ (−)	0.02
$g_{ftux} = g_{ftuy}$ (N/mm²)	0.01	$g_{ftux} = g_{ftuy}$ (N/mm²)	5
Creep			
E_0 (MPa)	360	v	0.3
E_1 (MPa)	3380	$\zeta_1 = \eta_1/E_1$ (s)	10^0
E_2 (MPa)	3380	$\zeta_2 = \eta_2/E_2$ (s)	10^1
E_3 (MPa)	3380	$\zeta_3 = \eta_3/E_3$ (s)	10^2
E_4 (MPa)	375	$\zeta_4 = \eta_4/E_4$ (s)	10^3
E_5 (MPa)	375	$\zeta_5 = \eta_5/E_5$ (s)	10^4
E_6 (MPa)	375	$\zeta_6 = \eta_6/E_6$ (s)	10^5
E_7 (MPa)	375	$\zeta_7 = \eta_7/E_7$ (s)	10^6
Rate:		c_1 (−)	0.05
	350; 1200;		
m_x (N.s · mm⁻²)	5000	c_2 (−)	0.001
	350; 1200;		
m_y (N.s · mm⁻²)	5000	$\dot{\kappa}_r$ (/s)	10^{-8}
Steel: Elastic-perfect plastic			
E (GPa)	200	Yield stress (MPa)	450
v	0.3	Steel area (mm²)	101

4.2 Flexural results

The measured load-deflection responses are shown in Figure 2. Recall that these beams were of dimension 100 mm × 100 mm × 500 mm, spanning 400 mm and loaded in three point bending. Note that the Materials Testing Machine cross-head displacement is shown on the horizontal axis, which does not account for local crushing at supports or other deformations in the load application setup. Although the linear elastic part of the responses is short, i.e. roughly up to 3–4 kN when first cracking occurs, the deflection response is exaggerated by this simplification. The displacement rate of the cross-head was controlled to be 3 mm/minute. This displacement rate was also applied in the finite element analysis, executed with Diana (2008), incorporating the constitutive model described in this paper as a user material. The computed responses are shown in Figure 7. Four-node plane stress elements of size 5 mm × 5 mm were used throughout, as shown in the Figure.

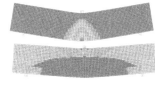

SHCC, showing localization at onset of peak response.

R/SHCC, showing maximum principal strains above f_t.

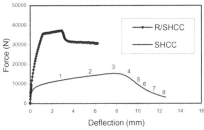

Figure 7. Computed flexural responses of SHCC and R/SHCC beams.

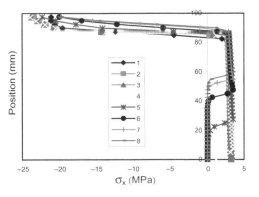

Figure 8. Mid-span horizontal normal stress along the cross-section.

Peak resistance and overall ductility are captured with reasonable accuracy. Note that the shown responses were obtained with the one parameter rate model, using $m = 1200$ N.s $\cdot$ mm^{-2}.

The necessity to consider inelastic behaviour in tension as well as compression is illustrated in Figure 8 by normal stresses along the cross-section at mid-span of the beam, at coinciding stages of the response shown in Figure 7.

4.3 *Regularisation*

To study the objectivity of the computed results with regard to element size, two additional element densities (coarse: 10 mm square, fine: 2.5 mm square) were prepared and used for the analyses. In Figure 9 it is clear that the strain-hardening is objective, but the peak and strain-softening parts are not objective. The shown results are for the three-parameter rate model, with parameter values given in Table 2.

The analyses were repeated with the one-parameter model, see Figure 10.

It appears that a strong rate contribution manages to regularise the computed response, resulting

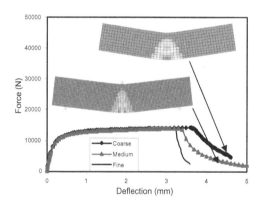

Figure 9. Mesh objectivity study with 3-parameter rate model.

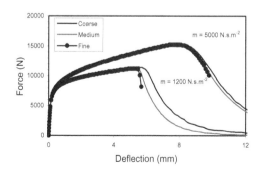

Figure 10. Mesh objectivity study with 1-parameter rate model.

in mesh objective results. This is currently under investigation.

5 IOSIPESCO SHEAR TEST

An attempt was recently made by Boshoff and van Zijl (2007) to analyse the Iosipesco shear tests reported by van Zijl (2007). A model based on isotropic damage was used, and found to under predict the experimentally observed shear resistance with more than 30%. The test was re-analysed with the current model, and the results presented here.

The setup is shown in Figure 11. The SHCC specimen is 280 mm long, 140 mm high, and 20 mm thick. The central notch was optimized for uniform shear stress distribution in the reduced section, and also for shear dominated failure. The reduced sectional area is 36.3 mm x 20 mm. For full details, the reader is referred to van Zijl (2007).

Also seen in Figure 11 are the crack patterns found on non strain-hardening specimens, with sub-critical fibre volume at $V_f = 1\%$, as well as the multiple inclined cracks in the strain-hardening specimens ($V_f = 2\%$).

The model parameters used in the previous section in the analysis of the SHCC beams, are used also for the shear analysis, except for the Elastic modulus and rate parameters. This is because the uniaxial tensile response of the SHCC was similar to that of the cast specimens of Visser and van Zijl (2007), except for a lower E-modulus (E = 8520 MPa). This lower value was used for the shear analysis, and the creep model spring constants were adjusted proportionally.

It must be noted that the strain rate in the shear analysis is higher than the strain rate in the flexural tests. To avoid strong over-estimated resistance, the acting strain-rate regime must be estimated, and the appropriate cracking viscosity for this rate must be selected, in the case of the single-parameter rate model. The rate of 1.5 mm/minute of vertical deflection controlled in the tests (van Zijl 2007) results in a diagonal strain rate of 0.0033 s^{-1}, gauged by the LVDT attached in the notch area as shown in Figure 11. Considering a power law regression of the cracking viscosity variation with strain rate, the value of m = 350 N.s $\cdot$ mm^{-2} was calculated and adopted. A fine element mesh of four-node plane stress elements was used with element side lengths ranging from 1 mm in the notch area to 2.3 mm.

The computed response is shown in Figure 12 to be in good agreement with the experimental responses. Note that the diagonal strain was calculated from FE displacement results at two nodes, simulating the diagonal LVDT measurement in the experiment.

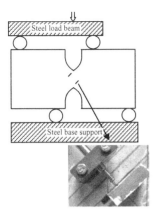
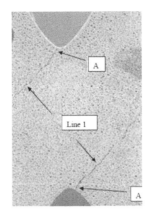
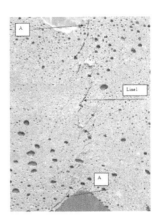

Figure 11. Iosipesco shear test setup. Cracking in non strain-hardening (centre) and strain-hardening specimens (van Zijl 2007).

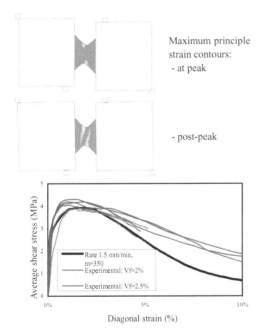

Maximum principle strain contours:

- at peak

- post-peak

Figure 12. Iosipesco shear test computed results, compared with experimental results by van Zijl (2007).

It must be noted that the three-parameter model does not require adjustment of the rate model parameters, but applies to a wide range in loading rates. This is due to the hyperbolic sine functions which, apart from strong gradients at the origin (zero equivalent strain rate), have relatively even gradients. The same set of parameters as stated in Table 2 for the three-parameter model produces a similar response as shown in Figure 12 for the one-parameter models with adjusted parameter value (to 350 N.s · mm^{-2}).

6 CONCLUSIONS

A computational model has been presented for SHCC or R/SHCC, which incorporates

- biaxial, anisotropic failure surfaces,
- strain-hardening and subsequent strain-softening in tension as well as compression,
- rate effects for bulk creep, as well as a stronger rate effect in the cracking zone, to capture rate-dependent cracking.

The model has been verified by comparison of computed results with experimental observations in:

- three-point bending of SHCC and R/SHCC beams,
- Iosipesco shear of SHCC.

The computed results have been shown to be sensitive to the cracking rate terms, especially in the single parameter rate model. The cracking viscosity is computed by inverse analysis of rate tests. However, a single viscosity constant is not valid at various testing rates of a particular material, requiring that the loading rate be considered in the selection of this parameter. In the three-parameter model, this is not the case, and a single set of parameter values characterized from rate test results apply to a wide range in loading rates.

The model has been shown to be objective with respect to finite element mesh density, for strong rate terms, in agreement with earlier findings for concrete (van Zijl et al. 2001). A suitable set of rate model parameters which also regularizes the computation of localization is currently under investigation.

In addition to the rate model improvement, appropriate biaxial failure surfaces are currently investigated through biaxial laboratory experiments on SHCC currently in progress at Stellenbosch

University. The presented model has proven invaluable in the design of the setup and control of the biaxial test, to be reported in the near future.

ACKNOWLEDGEMENT

The support by the Ministry of Trade and Industry through the Technology and Human Resources for Industry Programme (THRIP), as well as the industrial partners of the THRIP project ACM-S, is gratefully acknowledged.

REFERENCES

Boshoff, W.P. 2007. *Time-dependant behaviour of ECC*, PhD dissertation, Stellenbosch University, South Africa.

Boshoff, W.P., Mechtcherine, V. and van Zijl, G.P.A.G. 2009. Characterising the time-dependent behaviour on the single fibre level of SHCC: Part 1: Mechanism of fibre pull-out creep, accepted for publication in *Cement and Concrete Research,* 39 (2009), pp. 779–786.

Boshoff, W.P., Mechtcherine, V. and van Zijl, G.P.A.G. 2009b. Characterising the time-dependent behaviour on the single fibre level of SHCC: Part 2: Rate effects in fibre pull-out tests, *Cement and Concrete Research*, 39 (2009), 787–797.

Boshoff, W.P., and van Zijl, G.P.A.G. 2007. A computational model for strain hardening fibre reinforced cement-based composites, *Journal of the South African Institution of Civil Engineers*, 49(2) 24–31.

Boshoff, W.P. and van Zijl, G.P.A.G. 2007b. Time-dependent response of ECC: Characterisation of creep and rate dependence, *Cement and Concrete Research*, 37, pp. 725–734.

Fischer, G. and Li, V.C. 2004. Effect of fiber reinforcement on the response of structural members, Fracture mechanics of concrete and concrete structures, Vail, USA, April 2004, pp. 831–838.

Jun, P. and Mechtcherine, V. 2009. Deformation behaviour of cracked SHCC under sustained and repeated tensile loading, Proceedings of 8th International Conference on Creep, Shrinkage and Durability of Concrete and Concrete Structures (CONCREEP 8), Oct 2008, Ise-Shima, Japan, pp. 487–493.

Kabele, P. 2002. Equivalent continuum model of multiple cracking, *Engng. Fracture Mechanics*, 9(1/2), pp. 75–90.

Koiter, W.T. 1953. Stress-strain relations, uniqueness and variational theorems for elastic-plastic materials with singular yield surface, Q. Appl. Math. **11**(3), pp. 350–354.

Li, V.C. Mishra, D.K. Wu, H.C. 1995. Matrix design for pseudo strain-hardening FRCC, *Materials and Structures,* 28, pp. 586–595.

Van Zijl G.P.A.G. 2007. Improved mechanical performance: Shear behaviour of strain hardening cement-based composites (SHCC). *Cement and Concrete Research,* 37(8), pp. 1241–1247.

Van Zijl G.P.A.G. 2009. Computational modeling of SHCC, ISE Report ISI2009-20, Stellenbosch University.

Van Zijl G.P.A.G. and Boshoff W.P. 2009. Mechanisms of creep in fibre-reinforced strain-hardening cement composites (SHCC), Proceedings of 8th International Conference on Creep, Shrinkage and Durability of Concrete and Concrete Structures (CONCREEP 8), Oct 2008, Ise-Shima, Japan, pp. 753–759.

Van Zijl G.P.A.G., de Borst R. and Rots J.G., 2001. The role of crack rate dependence in the long-term behaviour of cementitious materials. *Int. J. Solids and Structures,* 38(30–31), 5063–5079.

Visser C.R. and van Zijl G.P.A.G. 2007. Mechanical characteristics of extruded SHCC. *Proc. International RILEM CONFERENCE on High performance fibre reinforced cement composites*, 10–13 July 2007, Mainz, Germany, pp. 165–173.

Wu, Z.S. and Bažant, Z.P. 1993. Finite element modelling of rate effect in concrete fracture with influence of creep, *Creep and Shrinkage of Concrete* (eds. Z.P. Bažant and I. Carol), E.&F.N.Spon, London, pp. 427–432.

Advances in Cement-Based Materials – van Zijl & Boshoff (eds)
© *2010 Taylor & Francis Group, London, ISBN 978-0-415-87637-7*

Numerical modelling of strain hardening fibre-reinforced composites

J. Vorel & W.P. Boshoff

Department of Civil Engineering, Stellenbosch University, South Africa

ABSTRACT: Conventional concrete suffers in most cases from insufficient tensile strength and ductility. Strain Hardening Cement-based Composite is a type of High Performance Concrete (HPC) that was engineered to overcome these weaknesses. In comparison to conventionally reinforced concrete, SHCC demonstrates fine, multiple cracking under tensile loading which can be seen as a noticeable advantage from a durability point of view. The primary objective of the presented research is to develop a constitutive model that can be used to simulate structural components with SHCC under different types of loading. In particular, the constitutive model must be efficient and robust for large-scale simulations while restricted number of material parameters is needed. For the modelling of specific behaviour in tension, the application of classical material models used for quasi-brittle materials is not straightforward. The proposed numerical model is based on a rotating crack assumption to capture specific characteristics of SHCC, i.e. the strain hardening and softening, the multiple cracking and the crack localization. The multiple orthogonal crack patterns are allowed which is in accordance with the observations presented in Suryanto et al. (2008). The accuracy of the developed model is investigated by the comparison of the numerical results with experimental data.

1 INTRODUCTION

Strain Hardening Cement-based Composite (SHCC) is a type of High Performance Concrete (HPC) that was developed to overcome the brittleness of conventional concrete. Even though there is no significant compressive strength increase compared to conventional concrete, it exhibits superior behaviour in tension. It has been shown to reach a tensile strain capacity of more than 4% during a pseudo strain hardening phase (Li & Wang 2001, Boshoff & van Zijl 2007). This pseudo strain hardening is achieved by the formation of fine, closely spaced multiple cracks with crack widths normally not exceeding 100 µm (Li & Wang 2001). These fine cracks, compared to large (larger than 100 µm) localised cracks found in conventional concrete, have the advantage of increased durability. For a further discussion of the mechanical properties of SHCC, the reader is referred to Boshoff et al. (2009a,b).

Several scholars have simulated SHCC mechanical behaviour with the Finite Element Method (FEM). Kabele (2000) formulated a model to simulate the mechanical behaviour of SHCC using a smeared cracking approach. Despite acknowledging that a discrete cracking model would be best for the final localising crack, Kabele decided to use a smeared cracking approach for the localisation.

This is due to the uncertainty of the position of the final localising crack. Another model was proposed by Han et al. (2003). This model was created to simulate the behaviour of SHCC under cyclic loading to test the improvement of structural response if SHCC elements are used to dissipate energy during earthquake loadings. Computational modelling of SHCC was also performed by Simone et al. (2003) who used an embedded discontinuity approach for the final material softening. This method would have the same kinematic characterisation as one obtained with interface elements for discrete cracking, but does not require remeshing procedures. Their conclusion was that it did not simulate the experimental results of SHCC satisfactorily due to the simplicity of the model.

Boshoff (2007) created a simple damage mechanics based model for the tensile behaviour of SHCC. This was implemented numerically using the FEM. Even though numerous shortcomings still exist, the model showed relative good results. Remaining issues include an unresolved mesh dependence and the under prediction of the deformation when analysing a structure with a strain gradient.

The primary objective of the presented research is to develop a constitutive model that can be used to simulate structural components with SHCC under different types of loading conditions. In particular, the constitutive model must be efficient and robust for large-scale simulations while restricted number of material parameters is needed. The proposed model is outlined and the results of the preliminary implementation are shown.

2 MODEL DEFINITION

For the modelling of specific behaviour of SHCC in tension, the application of classical constitutive material models used for quasi-brittle materials is not straightforward. The proposed numerical model is based on a rotating crack assumption to capture specific characteristics of SHCC, i.e. the strain hardening and softening, the multiple cracking and the crack localization. Multiple orthogonal crack patterns are allowed which is in accordance with the observations presented in Suryanto et al. (2008). A schematic representation of orthogonal cracking using the rotating crack model is shown using global and local axes in Figure 1. A complete description of the rotating crack model can be found in Rots (1988).

The presented model is implemented in a commercially available software package, DIANA (TNO DIANA BV. 2008), for a 2D case using a coaxial rotating crack model (RCM) with two orthogonal cracks as described in Han et al. (2003). This numerical approach is classified as the smeared cracking approach.

When implementing the model into a nonlinear fine element code, the incremental-iterative procedure based on a strain increment is assumed. Therefore, the strain vector $\varepsilon = \{\varepsilon_{11}, \varepsilon_{22}, \gamma_{22}\}^T$ reads

$$\varepsilon^{(i)} = \varepsilon^{(i-1)} + \Delta\varepsilon, \tag{1}$$

where i stands for an increment number and $\Delta\varepsilon$ is a strain increment. The rotating crack model evaluates a given strain state and generates the inelastic strain in the principal directions of the strain. Therefore, it is inevitably required to introduce a transformation tensor $(\mathbf{T}_\varepsilon, \mathbf{T}_\sigma)$ interconnecting global and a principal strain $e = \{e_1, e_2, 0\}^T$ or stress $s = \{s_1, s_2, 0\}^T$, respectively

$$e = \mathbf{T}_\varepsilon\varepsilon, \quad s = \mathbf{T}_\sigma\sigma \tag{2}$$

Using the standard transformation rule the tensors are

$$\mathbf{T}_\varepsilon = \begin{bmatrix} n_{11}^2 & n_{12}^2 & 2n_{11}n_{12} \\ n_{21}^2 & n_{22}^2 & 2n_{21}n_{22} \\ n_{11}n_{12} & n_{21}n_{22} & n_{11}n_{22}+n_{12}n_{21} \end{bmatrix}, \tag{3}$$

$$n = \begin{bmatrix} \cos\alpha & \sin\alpha \\ -\sin\alpha & \cos\alpha \end{bmatrix}, \tag{4}$$

with the relations between $(\mathbf{T}_\varepsilon, \mathbf{T}_\sigma)$

$$\mathbf{T}_\sigma^T = \left(\mathbf{T}_\varepsilon^T\right)^{-1} \text{ and } \mathbf{T}_\varepsilon^T = \left(\mathbf{T}_\sigma^T\right)^{-1}. \tag{5}$$

The rotation angle α can be obtained by means of a standard relation

$$\alpha = \frac{1}{2}\tan^{-1}\frac{\gamma_{12}}{\varepsilon_{11}-\varepsilon_{22}}.$$

The incremental stress-strain law (in the crack orientation) reads

$$\Delta s = \tilde{\mathbf{D}}\Delta e. \tag{6}$$

$$\tilde{\mathbf{D}} = \begin{bmatrix} \dfrac{ds_1}{de_1} & \dfrac{ds_1}{de_2} & 0 \\ \dfrac{ds_2}{de_1} & \dfrac{ds_2}{de_2} & 0 \\ 0 & 0 & \dfrac{s_1-s_2}{2(e_1-e_2)} \end{bmatrix}, \tag{7}$$

where $\tilde{\mathbf{D}}$ is the tangent material stiffness matrix. The derivation can be found in Jirásek & Zimmermann (1998). The stiffness matrix is transformed to the global coordinates using the standard transformation rule

$$\mathbf{D} = \mathbf{T}_\varepsilon^T\tilde{\mathbf{D}}\mathbf{T}_\varepsilon. \tag{8}$$

2.1 Poisson's ratio effect and equivalent principal stresses

It has to be mentioned that the rotating crack approach does not automatically include the effect of Poisson's ratio as the stress is evaluated on the base of the individual principal strains. In Han et al. (2003) the definition of equivalent strain is used to take this effect into account. This approach is reliable when a model formulation does not

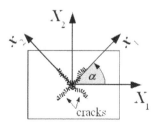

Figure 1. Coordinates and transformation angle.

permit residual deformations by cyclic loading, i.e. by changing state (tension to compression and vice versa). However, in the model presented in this paper permanent (residual) deformations are allowed. Therefore, a new approach was employed to treat the effect of Poisson's ratio. The principal strain (e) is used to determine the equivalent stress ($\hat{s}$) from the simplified uniaxial stress-strain diagram (see Sections 2.2 and 2.3). The final stresses are consequently evaluated as

$$\begin{Bmatrix} s_1 \\ s_2 \end{Bmatrix} = \frac{1}{1-v_{12}v_{21}} \begin{bmatrix} 1 & v_{12} \\ v_{21} & 1 \end{bmatrix} \begin{Bmatrix} \hat{s}_1 \\ \hat{s}_2 \end{Bmatrix}, \tag{9}$$

$$v_{12} = v_0 E_1/E_0, \; v_{21} = v_0 E_2/E_0 \tag{10}$$

where E_0 and v_0 stand for Young's modulus and Poisson's ratio of the undamaged material respectively. The parameters E_1, E_2, v_{12} and v_{21} represent the characteristics of the damaged material in a given direction and are defined in Section 2.2. The isotropic elastic material is represented in the state without cracks ($E_1 = E_2 = E_0$, $v_{12} = v_{21} = v_0$) and the orthotropic when the crushing or cracking starts.

$$\begin{Bmatrix} \hat{s}_1 \\ \hat{s}_2 \end{Bmatrix} = \begin{Bmatrix} E_1 e_1^{el} \\ E_2 e_1^{el} \end{Bmatrix} \tag{11}$$

Equation 6 satisfies the condition of symmetry for orthotropic materials. Combining Equations 9 and 11 further gives:

$$\begin{Bmatrix} s_1 \\ s_2 \end{Bmatrix} = \frac{1}{1-v_{12}v_{21}} \begin{bmatrix} E_1 & v_{12}E_2 \\ v_{21}E_1 & E_2 \end{bmatrix} \begin{Bmatrix} e_1^{el} \\ e_2^{el} \end{Bmatrix}, \tag{12}$$

where $v_{12} E_2 = v_{21} E_1$ and superscript $\cdot el$ represents the elastic part.

2.2 Equivalent stress

The equivalent stress state in principal direction is determined by the stress function σ as a function of the current principal strain and associated history parameters.

The stress function is based on the uniaxial strain-stress diagrams in compression and tension. The experimental data are idealised to obtain a suitable mathematical representation of this constitutive model.

2.2.1 Tension
The material response for virgin loading in tension (Figure 2) is described for each individual part by Equation 14. The model parameters are depicted in Figure 2. The elastic part is assumed to be linear

whereas the hardening and the softening sections are defined by means of Hermit functions.

The unloading and reloading scheme shown in Figure 3 is based on the experiments presented in Mechtcherine & Jůn (2007). The unloading curve is based on the polynomial function and the reloading is assumed to be linear (Eq. 15). The partial unloading and reloading is incorporated by using:

$$\begin{aligned} \varepsilon_{T\max}^* &= \min\left(\varepsilon_{T\max}, \varepsilon_{Tprl}\right), \\ \varepsilon_{Tul}^* &= \max\left(\varepsilon_{Tul}, \varepsilon_{Tpul}\right), \end{aligned} \tag{13}$$

where $\sigma_{T\max}^*$, σ_{Tul}^* are associated stresses and $\varepsilon_{T\max}$ is maximum strain experienced in previous steps with stress $\sigma_{T\max}$. The evolution of inelastic strain is assumed to be linearly dependent on $\varepsilon_{T\max}$ (Equation 14). This simplification correlates well with recent, unpublished cyclic tensile results done at Stellenbosch University, see Figure 4.

$$\varepsilon_{Tul} = \max\left[0, b_T\left(\varepsilon_{T\max} - \varepsilon_{T0}\right)\right], \tag{14}$$

The parameter a_T governs the unloading trajectory and must be determined from the experimental tests as well as the material characteristic b_T.

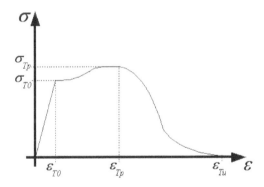

Figure 2. Schematic tensile response.

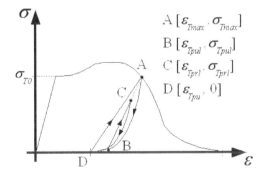

Figure 3. Loading and unloading diagram for tension.

273

$$\sigma_T\left(\varepsilon \geq \varepsilon_{T\max}\right) = \begin{cases} E_0\varepsilon, & 0 \leq \varepsilon \leq \varepsilon_{T0} \\ \sigma_{T0} + \left(\sigma_{Tp} - \sigma_{T0}\right)\left[-2\left(\dfrac{\varepsilon - \varepsilon_{T0}}{\varepsilon_{Tp} - \varepsilon_{T0}}\right)^3 + 3\left(\dfrac{\varepsilon - \varepsilon_{T0}}{\varepsilon_{Tp} - \varepsilon_{T0}}\right)^2\right], & \varepsilon_{T0} < \varepsilon \leq \varepsilon_{Tp} \\ \sigma_{Tp}\left[2\left(\dfrac{\varepsilon - \varepsilon_{Tp}}{\varepsilon_{Tu} - \varepsilon_{Tp}}\right)^3 - 3\left(\dfrac{\varepsilon - \varepsilon_{Tp}}{\varepsilon_{Tu} - \varepsilon_{Tp}}\right)^2 + 1\right], & \varepsilon_{Tp} < \varepsilon < \varepsilon_{Tu} \\ 0, & \varepsilon_{Tu} \leq \varepsilon \end{cases} \tag{15}$$

$$\sigma_T\left(\varepsilon < \varepsilon_{T\max}\right) = \begin{cases} E_0\varepsilon, & 0 \leq \varepsilon_{T\max} \leq \varepsilon_{T0} \\ \sigma_{T\max}^*\left(\dfrac{\varepsilon - \varepsilon_{Tul}}{\varepsilon_{T\max}^* - \varepsilon_{Tul}}\right)^{a_T}, & \varepsilon_{T0} < \varepsilon_{T\max} \leq \varepsilon_{Tu}, \quad \dot{\varepsilon} < 0 \\ \sigma_{Tul}^* + \left(\sigma_{T\max} - \sigma_{Tul}^*\right)\left(\dfrac{\varepsilon - \varepsilon_{Tul}}{\varepsilon_{T\max} - \varepsilon_{Tul}^*}\right), & \varepsilon_{T0} < \varepsilon_{T\max} < \varepsilon_{Tu}, \quad \dot{\varepsilon} \geq 0 \\ 0, & \varepsilon_{Tu} \leq \varepsilon_{T\max} \end{cases} \tag{16}$$

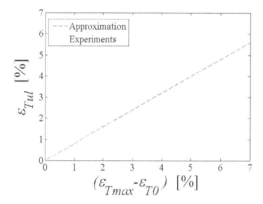

Figure 4. Evolution of inelastic strain ε_{Tul}

To ensure proper energy dissipation during localising, the crack band approach is used which relates the strain ε_{Tu} to the crack opening for the complete force transfer loss (w_d) and element size (h) (Equation 17). The crack opening can be considered as a half of the fibre length (Boshoff 2007). In the present paper the equivalent crack band width is evaluated by projecting the element into the direction normal to the crack at its initiation (h). This is done for each cracking direction separately

$$\varepsilon_{Tu} = \varepsilon_{Tp} + w_d/h. \tag{17}$$

As seen in Equation 14, the damage and cracking strains are driven by a single material parameter, namely b_T. By considering the standard definition of the damage parameter ω

$$E_T = (1 - \omega) E_0, \tag{18}$$

where E_T denotes actual elastic modulus, the damage variable can be determined by introducing Equation 14 into 18 as:

$$\omega = 1 - E_T / E_0 = 1 - \frac{\sigma_{T\max}}{\left(\varepsilon_{T\max} - \varepsilon_{Tul}\right)E_0}. \tag{19}$$

The transverse strain ratio in Equation 10 can be then evaluated as well as

$$v_{ij} = -v_0\left(1 - \omega\right). \tag{20}$$

This definition assures the decreasing influence of Poisson's ratio while the material cracks.

2.2.2 Compression
The virgin compression loading response is shown in Figure 5 and is defined mathematically in Equation 23.

The unloading and reloading scheme (Eq. 24) is depicted in Figure 6 and is based on a similar assumptions as for tension:

$$\begin{aligned} \varepsilon_{C\min}^* &= \max\left(\varepsilon_{C\min}, \varepsilon_{Cprl}\right), \\ \varepsilon_{Cul}^* &= \max\left(\varepsilon_{Cul}, \varepsilon_{Cpul}\right), \end{aligned} \tag{21}$$

where $\sigma_{C\min}^*$, σ_{Cul}^* are associated stresses and $\varepsilon_{C\min}$ is minimum strain reached in previous steps with stress $\sigma_{C\min}$. The evolution of inelastic strain is again assumed to be linearly dependent on $\varepsilon_{C\min}$

$$\varepsilon_{Cul} = \min\left[0, b_C\left(\varepsilon_{C\min} - \varepsilon_{C0}\right)\right]. \tag{22}$$

The material parameters a_T and b_T have to be determined from experimental test results.

$$\sigma_C\left(\varepsilon \le \varepsilon_{Cmin}\right) = \begin{cases} E_0\varepsilon, & 0 > \varepsilon \ge \varepsilon_{C0} \\[2mm] \sigma_{Cp} - \left(\sigma_{Cp} - \sigma_{C0}\right)\left(\dfrac{\varepsilon_{Cp} - \varepsilon}{\varepsilon_{Cp} - \varepsilon_{C0}}\right)^{E\frac{\varepsilon_{Cp} - \varepsilon_{C0}}{\sigma_{Cp} - \sigma_{C0}}}, & \varepsilon_{Co} > \varepsilon \ge \varepsilon_{Cp} \\[4mm] \sigma_{Cp}\left[2\left(\dfrac{\varepsilon - \varepsilon_{Cp}}{\varepsilon_{Cu} - \varepsilon_{Cp}}\right)^3 - 3\left(\dfrac{\varepsilon - \varepsilon_{Cp}}{\varepsilon_{Cu} - \varepsilon_{Cp}}\right)^2 + 1\right], & \varepsilon_{Cp} > \varepsilon > \varepsilon_{Cu} \\[3mm] 0, & \varepsilon_{Cu} \ge \varepsilon \end{cases} \tag{23}$$

$$\sigma_C\left(\varepsilon > \varepsilon_{Cmin}\right) = \begin{cases} E_0\varepsilon, & 0 > \varepsilon_{Cmin} \ge \varepsilon_{C0} \\[2mm] \sigma_{Cmin}^*\left(\dfrac{\varepsilon - \varepsilon_{Cul}}{\varepsilon_{Cmax}^* - \varepsilon_{Cul}}\right)^{a_C}, & \varepsilon_{C0} > \varepsilon_{Cmin} \ge \varepsilon_{Cu}, \ \dot{\varepsilon} > 0 \\[3mm] \sigma_{Cul}^* + \left(\sigma_{Cmin} - \sigma_{Cul}^*\right)\left(\dfrac{\varepsilon - \varepsilon_{Cul}^*}{\varepsilon_{Cmin} - \varepsilon_{Cul}^*}\right), & \varepsilon_{C0} > \varepsilon_{Cmin} > \varepsilon_{Cu}, \ \dot{\varepsilon} \le 0 \\[3mm] 0, & \varepsilon_{Cu} \ge \varepsilon_{Cmin} \end{cases} \tag{24}$$

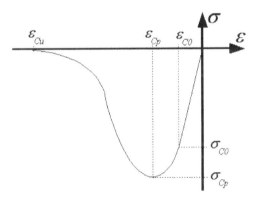

Figure 5.   Schematic compressive response.

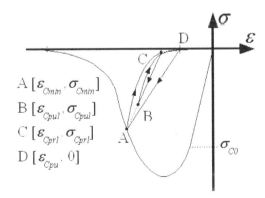

Figure 6.   Loading and unloading for compression.

The dissipated energy during the crushing should also be mesh-independent as for tensile cracking. Therefore, the strain ε_{Cu} is defined with respect to the mesh size as

$$\varepsilon_{Cu} = \varepsilon_{Cp} + d_c\left(\varepsilon_{Cu}^{\text{test}} - \varepsilon_{Cp}\right)/h, \tag{25}$$

where d_c stands for the localisation band in real material and h represents the equivalent crack band (element size). The product $d_c(\varepsilon_{Cu}^{\text{test}} - \varepsilon_{Cp})$ can be seen as the displacement needed for releasing correct energy during material softening.

The damage parameter is determined, in a similar fashion as for tensile, Equation 19, as:

$$\omega = 1 - E_C/E_0 = 1 - \frac{\sigma_{Cmin}}{\left(\varepsilon_{Cmin} - \varepsilon_{Cul}\right)E_0} \tag{26}$$

2.3   Biaxial behaviour

To demonstrate the complex behaviour of the proposed approach the failure envelope in space of principal stresses is shown in Figure 7. The boundaries are influenced by the transverse strain ratio of cracked and crushed material which is expected when the failure criterion is based on principal strains. This disadvantage of the presented model can be solved by a defining dependence between tensile and compressive strength. Nevertheless, the real shape of failure envelope for SHCC will only be included at a later stage as the bi-axial behaviour is currently under investigation at Stellenbosch University.

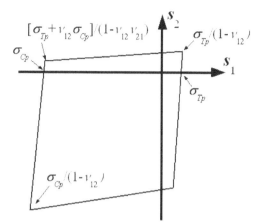

Figure 7. Failure envelope in the principal stress space.

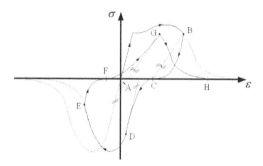

Figure 8. Schematic cyclic behaviour.

2.4 Cyclic loading

The above described model is adjusted for cyclic loading when the orientation of the principle stresses change. The residual deformations are assumed to be dependant on the inelastic strain. Therefore, a simple linear definition is employed and the permissible closing (opening) strain is evaluated as

$$\varepsilon_{T(C)}^{cl} = b_{T(C)}^{cl} \varepsilon_{Tul(Cul)} \tag{27}$$

where b_T^{cl} and b_C^{cl} are material parameters and can therefore be calculated from reverse cyclic loading tests. The trajectories of reloading after the stress state change are in good agreement with experimental results presented in Billington (2004).

To demonstrate the model response, a loading change from tension to compression to tension (A–G) is shown in Figure 8:

- A-B: initial virgin loading, Equation 15,
- B-C: unloading, Equation 16,
- C-D: cracks closing and compressive loading,

- D-E: virgin loading, Equation 23,
- E-F: unloading, Equation 24,
- F-G: cracks reopening and tensile loading,
- G-H: virgin loading, Equation 15.

Note that if the loading follows the stress state change, the loading path has the tangent equal to the actual modulus. The unloading from this stage is assumed to be the same as defined in Equations 16 and 24, respectively. The residual strains are determined under consideration of the actual modulus.

For the sake of simplicity the functions are not presented here, but will be presented in a forthcoming publication.

2.5 Summary of model parameters

The presented model is relatively simple and all parameters can be set up by means of uniaxial tensile and compressive test results. For the simulation of cyclic loading the relevant experimental data are also required.

The failure envelopes are described by the parameters limiting the elastic, hardening or softening part of the uniaxial diagrams:

$$\varepsilon_{T0(C0)}, \varepsilon_{Tp(Cp)}, \sigma_{Tp(Cp)}, w_d, \varepsilon_{Cu}, d_c.$$

The unloading and reloading paths as well as loading after stress state change are driven by

$$a_{T(C)}, b_{T(C)}, b_{T(C)}^{cl}.$$

3 IMPLEMENTATION AND APPLICATION

As mentioned in the previous section, the constitutive model is implemented in the commercial available finite element code DIANA version 9.3 using the "User supplied subroutine" option to demonstrate its suitability for SHCC. Newton-Raphson iterative procedure is used for the solution of nonlinear equations.

Finite element analyses of the flexural tests reported in Boshoff (2007) are performed to verify the constitutive model analyses. The three-point bending test is introduced using parameters presented in Boshoff (2007), see Table 1. The obtained results are compared with experimental data as well as a smeared crack model based on a damage mechanics formulation done by Boshoff (2007).

3.1 Three-point bending test

The experimental setup is depicted in Figure 9. It is noteworthy that the beam is relatively thin

Table 1. Model parameters (Boshoff 2007).

Prop.	Value	Prop.	Value
E	9200 MPa		
ν	0.35	ε_{C0}	–4.89E-3
ε_{T0}	2.42E-4	ε_{Cp}	–5.89E-3
ε_{Tp}	3.92E-2	σ_{Cp}	–50.0 MPa
σ_{Tp}	2.79 MPa	ε_{Cu}	–2.00E-1
w_d	6.0 mm	d_C	1.3 mm
a_T	3	a_C	3
b_T	0.8	b_C	0.8
b_T^{cl}	0.6	b_C^{cl}	0.8

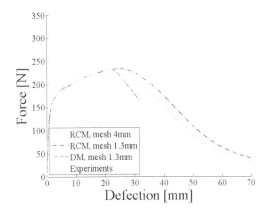

Figure 10. Three-point bending test.

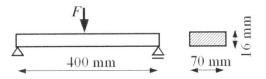

Figure 9. Three-point bending test setup.

which complicated the numerical modelling. As mentioned in Boshoff (2007), this beam demonstrates strong alignment of fibres close to the surface which results in a stronger, more ductile response. Therefore the results of the model are expected to underpredict the flexural behaviour. This issue is however not addressed in this paper.

3.1.1 Model description

The numerical model is based on experimental data obtained over the past 5 years by the Institute of Structural Engineering based at the Department of Civil Engineering, Stellenbosch University. Due to the lack of a reverse cyclic loading some parameters are set up using the engineering judgement of the authors as this will not have a significant influence on the presented results. All the parameters used are listed in Table 1.

To examine the feasibility of the proposed numerical approach, two different meshes are used. The flexural test is modelled using linearly interpolated, four node plane stress elements. The finite element mesh is refined towards the middle of the beam with the size of the elements in the expected softening and localisation zone 1.33 mm × 1.33 mm and 4.0 × 4.0 mm. The former dimension of elements in the middle of the beam is chosen in accordance with the theory introduced in Boshoff (2007) to deal with the crack spacing. The boundary conditions of the model are shown in Figure 9 as well as the beam dimensions. The other mesh size is chosen to study the mesh sensitivity.

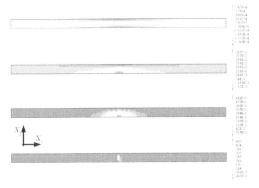

Figure 11. Evolution of ε_{11} during three-point flexural test simulation.

3.1.2 Results

The presented crack rotating model (RCM) is used to obtain the force-deflection diagrams. These results are plotted in Figure 10 together with the experimental data and response produced by the model based on a damage mechanics formulation (DM) by Boshoff (2007). As can be seen, the numerical models demonstrate good agreement with experimental data in the elastic as well as hardening part. The discrepancy is detected for the softening part. This is probably caused by the fibres alignment close to the surface which is not taken into account for generally used numerical models and the interested readers are referred to Boshoff (2007). The mesh dependency is observed by comparison of the two different mesh sizes.

To provide entire overview about the model the evolution of strain in the longitudinal direction is shown in Figure 11. After the first elastic stage the strain starts to localise into the elements which undergo hardening and then softening. Note that more than one element can soften before the strain is finally localized into one single crack. This is believed to be the reason for the shown mesh dependence (Boshoff 2007).

4 CONCLUSION AND FUTURE WORK

In this paper a two-dimensional numerical model for Strain Hardening Cement-based Composites was introduced. This approach is based on a rotating crack model implemented in the commercially available software package DIANA. The presented model takes into account:

- strain hardening and softening in tension as well as in compression,
- nonlinear unloading,
- nonlinear loading after stress state change—crack closing,
- the effect of Poisson's ratio.

The accuracy of the introduced approach was demonstrated by means of a three-point flexural test. Nevertheless, the model suffers some mesh dependency as is shown in Figure 10. This must be treated before the proposed approach will be used for larger structural components under different loading conditions. The authors also intend to incorporate the time-dependent behaviour in the model.

REFERENCES

Billington, S. (2004). Damage-tolerant cement-based materials for performance-based erthquake engineering design: Research needs. *Fracture Mechanics of Concrete Structures* , pp. 53–60.

Boshoff, W. (2007). *Time-Dependant Behaviour of Engineered Cement-Based Composites*. Ph.D. thesis, University of Stellenbosch.

Boshoff, W. & van Zijl, G. (2007). Time-dependant response of ECC: Characterisation of creep and rate dependence. *Cement and Concrete Research*, 37, pp. 725–734.

Boshoff, W., Mechtcherine, V. & van Zijl, G. (2009a). Characterising the time-dependant behaviour on the single fibre level of SHCC: Part 1: Mechanism of fibre pull-out creep. *Cement and Concrete Research*, 39, pp. 779–786.

Boshoff, W., Mechtcherine, V. & van Zijl, G. (2009b). Characterising the time-dependant behaviour on the single fibre level of SHCC: Part 2: The rate effects on fibre pull-out tests. *Cement and Concrete Research*, 39, pp. 787–797.

Han, T.-S., Feenstra, P. & Billington, S. (2003). Simulation of Highly Ductile Fiber-Reinforced Cement-Based Composite Components Under Cyclic Loading. *ACI Structural Journal*, 100 (6), pp. 749–757.

Jirásek, M. & Zimmermann, T. (1998). Analysis of Rotating Crack Model. *Journal of Engineering Mechanics,* 124 (8), pp. 842–851.

Kabele, P. (2000). *Assessment of Structural Performance of Engineered Cemetitious Composites by Computer Simulation*. A habilitation thesis, Czech Technical University in Prague.

Li, V. & Wang, S. (2001). Tensile Strain-hardening Behavior of PVA-ECC. *ACI Materials Journal*, 98 (6), pp. 483–492.

Mechtcherine, V. & Jůn, P. (2007). Stress-strain behaviour of strain-hardening cement-based composites (SHCC) under repeated tensile loading. *Fracture Mechanics of Concrete Structures*, (pp. 1441–1448).

Rots, J. (1988). *Computational modeling of concrete fracture*. Ph.D. thesis, Delft University of Technology.

Simone, A., Sluys, L. & Kabele, P. (2003). Combined continuous/discontinuous failure of cementitious composites. *Proceedings for EURO-C 2003* (pp. 133–137). Computational Modelling of Concrete Structures.

Suryanto, B., Nagai, K. & Maekawa, K. (2008). Influence of damage on cracking behavior of ductile fibre-reinforced cementitious composite. *Proceedings of 8th International Conference on Creep, Shrinkage and Durability of Concrete and Concrete Structures*, pp. 495–500.

Tno Diana BV. (2008). DIANA Finite Element Analysis.

Advances in Cement-Based Materials – van Zijl & Boshoff (eds)
© 2010 Taylor & Francis Group, London, ISBN 978-0-415-87637-7

Some pertinent observation on the behaviour of high performance concrete-steel plate interface

Sekhar K. Chakrabarti

School of Civil Engineering, Surveying and Construction, University of Kwazulu-Natal,
Durban, South Africa

ABSTRACT: Due consideration to account for the presence of concrete-steel plate interfaces should be given in the analysis and design of structures in which high performance concrete (HPC)-steel plate composite construction is adopted for main structural members, for connections between different members, and for providing different means to facilitate attachment of components on the body of the structure. Such a situation has been typically existed in the primary reactor containment dome of the present generation Indian Nuclear Power Plant Reactor Building. Typically, such primary reactor containment domes are built as post-tensioned prestressed concrete structures using HPC. The steel embedded parts which are made up of steel liner plates, and located along the edges of the steam generator openings in the dome, facilitate attachment of the dished head type steel sealing arrangement. HPC is normally used for such a construction in order to take of the situations arising out of the pre-stressing and other associated details for anchorages and the steel embedded plates at the openings. The issues related to the possible separation that may occur at the concrete-steel plate interfaces, are important in respect of leakages in the event of such separation. The present paper deals with some pertinent observation on the behaviour of HPC-steel plate interface based on the findings of a three-dimensional finite-element study of interface specimens using relevant experimental data.

1 INTRODUCTION

High performance concrete-steel plate composite construction is generally adopted for main structural members, for connections between different members, and for providing different means to facilitate attachment of component on the body of the main structure. Such a situation has been typically existed in the primary reactor containment dome of the present generation Indian Nuclear Power Plant Reactor Building (Warudkar 1997). Typically, such primary reactor containment domes are built as post tensioned prestressed concrete structures using high performance concrete. The steel embedded parts which are made up of steel linear plates and are located along the edges of the steam-generator openings in the dome, facilitate attachment of the dish head type steel sealing arrangement (Figure 1).

The anchors between the steel plates and the adjacent concrete are provided to achieve a composite action of the structure. But, such a provision for achieving a composite action may not cater for the desired no-separation condition at the interface between the steel plate and concrete at all the locations. Such separation may be crital in the

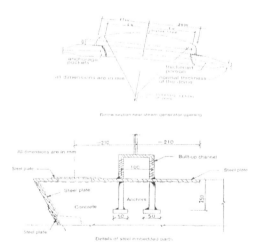

Figure 1. Steel embedded parts around the steam generator openings of an inner containment dome.

context of permeability requirement (e.g., leakage) of the reactor.

The issue of leakage has been addressed by Chakrabarti et al. (1994, 1997), specifically for the case of steel-concrete interfaces at penetration assemblies of the steam-generator openings in the dome of a typical Indian Nuclear Containment Building.

The authors considered the strength development mechanism at and around the steel-concrete interfaces, along with issue of permeability of the interfaces. The reduction of the tensile strength of the steel-concrete interface compared to the tensile strength of concrete was observed experimentally by Basu and Rajagopalan (1997). The reason for this reduction of tensile strength was assigned to the absence of any friction or interlocking phenomenon at the interface. The authors observed that the adhesive force between the steel and concrete, which is the only force resisting the tension, is too weak compared to the aggregate interlocking and cohesive bond in concrete matrix. It was also observed that the pressure gradients through the interface zones due the presence of air in voids, can cause additional pressure force.

Chakrabarti and Basu (1999) has dealt with the issues related to the performance of the steel-concrete interfaces with respect to the separation that may occur in the prestressed containment dome due to situation arising out of prestressing cable and other associated details for the prestressing anchorages, steel embedded parts at the openings, and passive reinforcements. The authors concluded that examination of leakage possibilities through the steel-concrete interface zones at and around the steam-generator openings in the dome, should be regarded as one needed safety requirement to be complied with in addition to the strength and stability requirements. It was also felt that the knowledge about the local steel-concrete interface behaviour will be of help in designing the interface adequately against possible separation. Chakrabarti and Basu (2001) emphasised on the important issues related to the properties of interface with respect to separation including the findings of an experimental study on the behaviour of steel plate-concrete interface for different loading conditions.

Experimental and analytical studies were conducted on the profiled steel-deck concrete flooring system considering the presence of shear connectors (Schuster 1976, Porter & Ekberg 1976 & Porter et al. 1976). Though some design recommendations have been made on such systems without shear connectors, nothing in particular has been observed regarding the behaviour of the steel-concrete interface zone in such cases. Ong & Mansur (1986) carried out experimental studies on the shear-bond capacity of similar composite slabs with and without anchors. For a similar test set-up, the shear bond failure of a specimen without shear connectors has been observed to occur at a lesser load compared to that for a specimen with shear connectors; but the load-deflection behaviours have been almost the same.

Menrath et al. (1998) proposed a friction model for the behaviour of steel-concrete interface layer along with shear transferring devices, in a steel-girder concrete-deck composite system. The structure has been modelled for two-dimensional finite-element analysis using line elements for modelling the interface layer with shear connectors, and two-dimensional planar elements for modelling the steel and concrete portions.

The present paper deals with some pertinent observations on the behaviour of high performance concrete-steel plate interface, base on a three-dimensional finite element study of the interface specimens using relevant experimental data.

2 THE FINITE ELEMENT STUDY

The linear elastic behaviour of high performance concrete-steel plate interface has been studied through three dimensional finite element analysis of interface specimens using the relevant experimental data (Chakrabarti & Basu 2001). The sample specimens used were modelled using solid elements for each of the load cases (direct tension, direct compression, direct shear, combined tension and shear, and combined compression and shear) to study the stress and deformation behaviour at and around the interface. Three dimensional isoparametric solid tetrahedron elements of isotropic materials were used for modelling the different solid components of each specimen and the test floor, where as, one-dimensional gap elements were used for capturing situations in which contact between parts of the structures is dependant upon the deformations resulting from applied loads. Convergence studies have been made on the direct tension, direct compression and direct shear specimens; based the observations on these convergence studies, one analysis has been carried out for each of the remaining two specimens (combined tension and shear, combined compression and shear). The finite element models along with the geometry, loading and constraints for each of the specimens are presented in Figures 2–6. The responses of the high performance concrete-steel plate interfaces of each specimen were observed to be within the linear elastic range for the adopted loading in the experiments (Chakrabarti & Basu 2001). It is to be noted here that based on the findings of the experimental investigation (Chakrabarti & Basu 2001), the thickness of the interface zone has been assumed to be expected up to uniform depth of 4 mm below the bottom surface of the steel plate in each specimen.

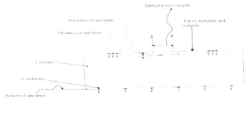

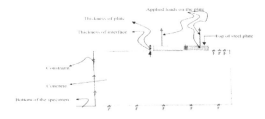

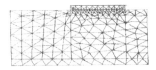

Figure 2. Finite element model for direct tension test.

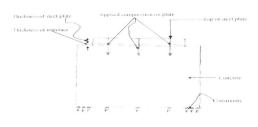

Figure 5. Finite element model for tension and shear test.

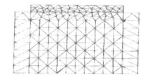

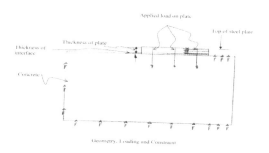

Figure 3. Finite element model for direct compression test.

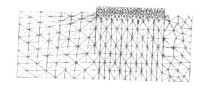

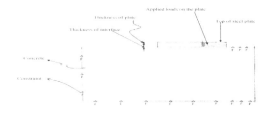

Figure 6. Finite element model for compression and shear test.

3 PERTINENT OBSERVATIONS

Results of the finite element analyses of the high performance concrete-steel interface specimens were obtained in the form of deformation data, stress data (normal stresses, shear stresses, principle stresses and von Mises stresses), strain data (normal strains, shear strains, principal strains and von Mises strains) to compare with those obtained in the experiments. Following pertinent observations have been made based on these results.

Figure 4. Finite element model for direct shear test.

3.1 Observation on direct tension specimen

- Both the experimental and analytical results show predominantly vertical (upward in the direction of applied tension) displacement of the steel plate.
- For the interface zone the magnitudes of normal stresses in the vertical direction have been observed to be significantly high compared to other stress components. The distribution of this stress component has not been uniform; from a very high value of tensile stress at four locations right below the load application points at the top of the steel plate, it has gradually decreased towards the centre of the interface layer and at two parallel edges away from the load application points. The distribution of maximum principal stresses (all tensile) has been almost the same as that of the normal stresses (vertical) with maximum stress of the same magnitude occurring at the same locations. The distribution of von Mises stresses also has been similar.
- The specimen as well as the interface has been predominantly under vertical tensile strains with strain concentration occurring mainly at the interface zone. The distribution patterns and magnitudes of the principal strains and vertical normal strains have been almost identical. Similar strain distribution has also been observed for the von Mises strains.
- At the interface, the average experimental strain has been found to be about 27% higher than the analytical stain.

3.2 Observations on direct compression specimen

- Both the experimental and the analytical results show predominantly vertical (downward in the direction of the applied compression) displacement of the steel plate.
- For the interface layer, the magnitudes of normal stresses in the vertical direction have been observed to be significantly high compared to the other stress components. The distribution of these normal stresses has been almost uniform over the entire area of the interface. The distribution of principal stresses shows occurrence of no tensile principal stress anywhere. The von Mises stress distribution has also been of the same nature.
- The specimen as well as the interface has been subjected to vertical compressive strains with high values occurring in the interface zone.
- At the interface, the average experimental strain has been observed to be about 9% higher than the analytical strain.

3.3 Observations on direct shear specimens

- Both the experimental and analytical results show predominant shear deformation at the interface zone in conformity with the applied load.
- Distributions of shear stresses (on horizontal planes) and von Mises stresses have indicated the occurrence of stress concentration at the interface.
- The shear strain in the vertical plane (containing the applied load) has been very high in the interface relative to steel and concrete.
- At the interface, the average experimental strain has been observed to be about 50% higher than the analytical strain.

3.4 Observations on combined tension and shear specimen

- Both the experimental and analytical results indicate predominant shear deformation at or near the interface.
- Distribution of the shear stresses (on the horizontal planes) and von Mises stresses have indicated the occurrence of stress concentrations at the interface.
- The horizontal shear strain has been very high in the interface. The magnitude of the maximum vertical tensile strain in the interface has been observed to be about one-sevenths of the maximum horizontal shear strain.
- At the interface, the average experimental strain has been observed to be about 31% higher than the analytical strain.

3.5 Observation on combined compression and shear specimen

- Predominant shear deformation at/near the interface has been observed.
- Occurrence of predominant shear stresses (on the horizontal plane) and vertical normal compressive stresses of comparable magnitude in the interface has been observed.
- The interface zone has been subjected to very high horizontal shear strains compared to the vertical compressive normal strains; the maximum magnitude of the vertical compressive normal strains have been found to be about 20% of the maximum horizontal shear strain.
- At the interface, the average experimental strain has been observed to be about 37.5% higher than the analytical strain.

4 CONCLUSION

- The interface has been observed to be the zone of stresses and strain concentration for all the specimens, causing significant relative deformation between the steel plate and the main concrete body, in conformity with the mode of load application.

- The occurrence of high strain values contributing to the relative displacement of the steel plate with respect to the body concrete signifies the importance of incorporating a separate interface layer in modelling.

ACKNOWLEDGEMENT

Observations made in the paper is an outcome of the research project, "Investigation on Performance of Steel-Concrete Interface at Penetration Assemblies in the Inner Containment Dome of a Reactor Building," sponsored by the Atomic Energy Regulatory Board, India.

REFERENCES

Basu, P.C. and Rajagopalan, N. "Behaviour of concrete with reinforcement under split tension," Transactions of the 14th International Conference On Structural Mechanics In Reactor Technology, Div.H, August 1997, pp. 65–72.

Chakrabarti, S.K. and Basu, P.C. "Importance of the Properties of Steel Plate-Concrete Interface in the Local Behaviour at Large Openings in the IC Dome of a Reactor Building," Transactions of the 16th International Conference On Structural Mechanics In Reactor Technology, Washington, DC, USA, August, 2001, http://www.engr.ncsu.edu/SMiRT_16/1761.pdf.

Chakrabarti, S.K. and Basu, P.C. "Issues Related to Design and Construction of Prestressed Concrete Inner Containment Dome Of Reactor Buildings With Large Openings," Transactions of the 15th International Conference On Structural Mechanics In Reactor Technology, Vol. VI, paper H07/1, August 1999, pp. VI-283-VI-289.

Chakrabarti, S.K., Kumar, A. and Basu, P.C. "Leakage through concrete-steel interface at and around containment assemblies: Some New Observations," proceedings Second International Conference on Advanced Reactor Safety, Orlando, Florida, USA., 1997.

Chakrabarti, S.K., Sai, A.S.R., and Basu, P.C. "Examination of leakage aspect through concrete steel interface at and around the containment penetration assemblies," proceedings, Third International Conference on Containment Design and Operation, Toronto, Canada, 1994.

Menrath, H., Haufe, A. and Ramm, E. "A model for steel-concrete structure," Proceedings, Euro-C 1998 Conference on Computational Modelling of Concrete and Concrete Structures, Vol. 1, April 1998, pp. 33–42.

Ong, K.C.G. and Mansur, M.A. "Shear-bond capacity of composite slabs made with profiled sheetings," International Journal of Cement Composites and Lightweight Concrete, Vol. 8, No. 4, 1986, pp. 231–237.

Porter, M.L. and Ekberg, C.E. "Design recommendation for steel-deck floor slabs," Journal of the Structural Division, Proceedings, American Society of Civil Engineers, Vol. 102, No. ST11, 1976, pp. 2121–2136.

Porter, M.L. and Ekberg, C.E. Greimann, L.F and Elleby, H.A.,"Shear-bond analysis of steel-deck reinforced slabs," Journal of the Structural Division, Proceedings, American Society of Civil Engineers, Vol. 102, No. ST12, 1976, pp. 2255–2268.

Schuster, R.M. "Composite steel-deck concrete floor systems," Journal of the Structural Division, Proceedings, American Society of Civil Engineers, Vol. 102, No. ST5, 1976, pp. 889–917.

Warudkar, A.S., "Nuclear Power Plant Containment Design: Evolution And Indian Experience," Joint WNO/OECD-NEA Workshop, Prestress Loss in NPP Containment, Civavx NPP, Poitiers, France, 1997.

Specifications, design guidelines and standards

Advances in Cement-Based Materials – van Zijl & Boshoff (eds)
© 2010 Taylor & Francis Group, London, ISBN 978-0-415-87637-7

Enabling the effective use of High Performance Fibre Reinforced Concrete in infrastructure

E.P. Kearsley & H.F. Mostert

Department of Civil Engineering, University of Pretoria, Pretoria, South Africa

ABSTRACT: The construction industry can reduce the volume of material required for a given structure by using High Performance Concrete (HPC) efficiently. It is a well established fact that an increase in the compressive strength of concrete results in a more brittle failure, which could result in sudden, catastrophic failure under unexpected loads or dynamic forces. Unexpected brittle failure can be prevented by the addition of suitable fibre reinforcing, resulting in High Performance Fibre Reinforced Concrete (HPFRC). Although researchers have successfully manufactured HPFRC in laboratory conditions for years, the implementation of this material on a large scale has been limited by both the lack of standard design procedures and standard material testing methods. This paper focuses on test methods and experimental results will be used to justify the need for developing test methods that will yield results that can be used in the design of HPFRC structures.

1 INTRODUCTION

Concrete is the most widely used construction material in the world, resulting in the cement and concrete industry being identified as one of the industries contributing significantly towards the environmental impact of human activity. Researchers have spent years finding ways of reducing the impact by optimizing the properties and use of materials. Sustainable growth can only be achieved if the materials manufactured and used, as well as the structures designed and built are cost-effective, environmentally friendly and give durable and trouble free service performance for their specified design life. In the recent past significant attention has been paid to durability, but it is important to realize that durability should not only include the resistance to material degradation, but also resistance to structural damage (Swamy, 2008). It is a well established fact that an increase in the compressive strength of concrete results in a more brittle failure, which could result in sudden, catastrophic failure under unexpected loads or dynamic forces.

The construction industry can reduce the volume of material required for a given structure by using High Performance Concrete (HPC) efficiently. Unexpected brittle failure can be prevented by the addition of suitable fibre reinforcing, resulting in High Performance Fibre Reinforced Concrete (HPFRC). Although researchers have successfully manufactured HPFRC in laboratory conditions for years, the implementation of this material on a large scale has been limited by both the lack of standard design procedures and standard material testing methods. This paper focuses on test methods and experimental results will be used to justify the need for developing test methods that will yield results that can be used in the design of HPFRC structures.

The South African National Roads Agency Limited (SANRAL) recently embarked on major investment in road infrastructure by upgrading highways. Sections of Ultra Thin Continuously Reinforced Concrete Pavement (UTCRCP) or Ultra Thin Heavy Reinforced High Performance Concrete (UTHRHPC) with thicknesses as little as 40 mm have been constructed on an experimental basis. The relevance of both the design parameters and the test methods for quality control that have been used for un-reinforced concrete pavements in the past still need to be confirmed for UTCRCP. In this paper the material properties and composite behaviour will be discussed and concerns about traditional test methods will be raised.

2 BACKGROUND

2.1 Standard test methods

Standardized procedures for the manufacture and testing of test specimens are required to accurately asses the quality of concrete, as several factors such as rate of loading, specimen size, direction of casting, moisture content, curing conditions and type of testing machine can have a significant influence on the test results recorded.

The compressive strength of concrete is determined from standard test cubes or cylinders that have been cured in temperature controlled water baths and tested 28 days after casting. Normally 150 mm test cubes are used, but the size can be reduced depending on the maximum size off the aggregate (the cube should be at least 4 times the maximum aggregate size). The cubes are tested at a rate of 0.2–0.4 MPa/s perpendicular to the direction of casting and the strength is normally recorded to the nearest 0.5 MPa. The characteristic strength of concrete is normally defined as the strength that must be exceeded by at least 95% of the test results where tests are carried out using standard test methods. Standard cube presses are normally set to record only the maximum load applied to the sample (peak-hold). The standard compressive strength test gives no indication of either the stiffness of the material or the brittleness of the failure (how much warning of the imminent failure occurred).

Since it is very difficult to apply uniaxial tension to a concrete specimen the tensile strength of concrete is determined indirectly and both flexural and splitting tests are conducted to determine the indirect tensile strength of concrete. In the flexural test, the theoretical maximum tensile stress (f_{bt}) is reached in the bottom fibre of the test beam and this is also known as the Modulus of Rupture (MOR), which is used for the design of pavements. The standard size for MOR beams is $150 \times 150 \times 750$ mm but for maximum aggregate sizes of less than 25 mm $100 \times 100 \times 500$ mm beams may be used. The beams are tested in a four point bending test (loaded at third points resulting in a constant bending moment over the middle one-third of the beam) with a span length equal to three times the depth of the beam. The test is conducted by increasing the load at a rate of up to 0.1 MPa/s and the MOR (f_{bt}) is recorded to the nearest 0.1 MPa using ordinary elastic theory as indicated in equation 1:

$$f_{bt} = \frac{PL}{bh^2} \tag{1}$$

where P = maximum total load, L = span, b = width of the beam, and h = depth of the beam.

The flexural testing is mostly conducted with the same equipment that is used for crushing cubes and only the maximum load applied during the test is recorded. When large volumes of fibres are added to concrete mixes, no indication is given whether the MOR value was obtained when the concrete started cracking or whether the highest load was applied long after the concrete cracked and the steel fibres resisted the load. Furthermore

the expression for MOR is based on the elastic beam theory, in which the stress-strain relation is assumed to be linear, so that the tensile stress in the beam is assumed to be proportional to the distance from the neutral axis. There is however a gradual increase in strain with increase in stress above about 50% of the strength. Consequently the shape of the actual stress block near failure load is parabolic and not triangular. The MOR thus overestimates the tensile strength of the concrete and the actual tensile strength is between 30% and 75% of the MOR. Although the MOR is used in the design of pavements, it has historically not been used for quality control purposes during construction as samples are heavy and easily damaged, variability in results is large and the strength is strongly affected by the moisture content during testing. It has been convenient to establish a relation between the MOR and the compressive strength experimentally and then use compressive strength testing for quality control purposes of concrete pavements.

In the splitting test, a concrete cylinder is placed with its axis horizontal, between the platens of a testing machine and the load is increased until failure occurs. Narrow strips of packing material can be placed between the specimens and the platens to prevent local failure at the contact point. The load is applied at a rate up to 0.04 MPa/s and the tensile splitting stress (f_{st}) is calculated to the nearest 0.1 MPa using equation 2:

$$f_{st} = \frac{2P}{\pi Ld} \tag{2}$$

where P = maximum load, L = length of specimen, and d = diameter of specimen.

The split cylinder test is normally conducted in a cube press, which means that only a maximum load value is recorded and the brittleness of the failure is not taken into account at all. As for the MOR beams, the specimen is also tested perpendicular to the direction of casting, which affects the strength especially if fibre reinforcing is not equally dispersed throughout the sample. There is a linear relation between the MOR and the splitting strength, and cores cut from existing pavements can be split to determine the tensile strength of the concrete. The strength from splitting tests is believed to be close to the direct tensile strength of concrete.

The size effect of concrete strength has been well documented and for both compressive and tensile strength smaller specimen yield significantly higher strengths. The size effect is known to be more pronounced for rich mixes, which means that the strength recorded for HPC could be dependant on the size of the specimen tested. When fibre reinforcing is used in relatively small concrete samples, the fibres align themselves, thus affecting the strength by increasing the contribution of the fibres in one plane while decreasing the contribution in other planes. The test method as well as the size of the specimen tested should therefore be carefully selected to ensure useable test results. In this paper the test method sensitivity of HPFRC will be investigated.

2.2 *Pavement specification*

For the UTCRCP currently under construction SANRAL requires concrete with a characteristic compressive strength of 90 MPa and a characteristic MOR of 10 MPa. For quality control purposes the standard cube tests will be used as calibrated against standard MOR test results. During the initial phases of the contract trial mixes should be tested and an additional test is required where round discs are tested to determine the energy absorption of the pavement. This test is conducted according to ASTM1550 but the disc thickness is reduced to 55 mm (which is similar to the UTCRCP thickness) and the disc diameter is reduced to 600 mm. The reinforcing that will be placed in the road is placed in the discs and the discs are cured in water and tested in the direction of casting after 28 days. The discs are supported on ball bearings at 3 points, resulting in a span diameter of 550 mm. The test setup can be seen in Figure 1. A load is applied at centre point and the test is conducted in deflection control (0.2 mm/min) using a closed loop material testing system. The midpoint deflection and the load are recorded and the area under the load deflection graph is used to calculate the energy absorption in Joules. The SANRAL specification requires total energy absorption of 1000 Joules for a deflection of 25 mm. The mix composition specified in the SANRAL contract documentation is indicated in Table 1.

3 EXPERIMENTAL SETUP

A mix composition as indicated in Table 2 was cast repeatedly, while only varying the fibre content as indicated in the table. To ensure constant volume, the content of the mixes was decreased proportionally with increased fibre content. The mix composition as indicated in the table is the composition used for 80 kg/m³ fibres. A water/cement ratio of 0.31 was used for all mixes. The fibres used were 30 mm long hooked ended hard drawn wire fibres with a diameter of 0.5 mm and a tensile strength of at least 1100 MPa.

For each fibre content 100 mm cubes were cast for compressive strength testing 7 days and 28 days after casting. The strength of the mixture was calculated as the average of at least 3 cubes tested. The modulus of elasticity was measured on two 150 mm diameter cylinders cast from each mixture. Both flexural strength (on 100 mm beams) and split cylinder strength were determined for each mixture.

Figure 1. Round disc test.

Table 1. Mix composition for UTCRCP.

Material	kg/m³
Cement (Cem I 42.5R)	480
Pulverised Fuel Ash (PFA)	87
Condenced Silica Fume (CSF)	72
Water	170
6.7 mm stone	972
Sand	689
HRWRA	6.4*
Steel fibre	100
Polypropylene fibre	2

* HRWRA content is in litres.

Table 2. Experimental mix composition.

Material	kg/m³
Cement (Cem I 42.5R)	416
Pulverised Fuel Ash (PFA)	70
Condensed Silica fume (CSF)	35
Water	161
6.7 mm dolomite	860
Dolomite super sand	971
HRWRA	9.2
Steel fibre	0, 80, 115 or 150
Polypropylene fibre	2

* HRWRA content is in litres.

4 EXPERIMENTAL RESULTS

4.1 *Compressive strength*

The results obtained for the compressive strength testing can be seen in Table 3. These results clearly indicate that the inclusion of relatively large steel fibre contents does not result in an increase in compressive strength. At very high fibre contents (150 kg/m³) the slight reduction in compressive strength could be caused by the the fact that it was nearly impossible to fully compact the concrete with such a high fibre dosage. The 7-day strength of all mixes is between 70% and 80% of the 28-day strength. Assuming the fibre content does not affect the compressive strength, the characteristic strength of the concrete can be calculated using

$$f_{cu} = f_m - 1.64\,s \qquad (3)$$

where f_{cu} is the characteristic cube strength, f_m the average strength and s the standard deviation.

If all the 28-day cubes are taken into account the average strength can be calculated as 122.0 MPa and the standard deviation as 8.16 MPa resulting in a characteristic strength of 108.6 MPa.

The modulus of elasticity of the mixtures as indicated in Table 3 also does not increase with increased steel fibre content. The European standard for the design of concrete structures indicate that the E-value of concrete with a characteristic strength of 90 MPa can be assumed to be 44 GPa. The measured E-values are significantly higher than this, but the use of dolomite as aggregate is known to result in high concrete stiffness.

Table 3. Compressive strength.

Fibre content (kg/m³)	7-day strength (MPa)	28-day strength (MPa)	E-value (GPa)
0	92.5	126.5	60.5
80	103.7	130.0	59.1
115	92.0	119.0	58.8
150	86.0	113.5	59.1

Table 4. Tensile strength.

Fibre content (kg/m³)	f_{cu} (MPa)	f_{bt} (MPa)	f_{st} (MPa)
0	126.5	13.5	5.98
80	130.0	13.5	5.75
115	119.0	15.3	7.25
150	113.5	19.1	8.61

4.2 *Tensile strength*

The MOR and splitting strength of the concrete mixtures are indicated in Table 4. These results indicate that the tensile strength of the concrete is not increased by adding as much as 80 kg/m³ of steel fibres to the mixtures, but higher fibre contents does increase the recorded values. Although the compressive strength at higher fibre contents decreased, the measured tensile strengths still increased. The values recorded in Table 4 are the maximum values recorded during testing and these values give no indication of whether the failure was brittle or ductile and whether the concrete cracked at stresses lower than the values indicated in this table.

4.3 *Flexural behaviour*

To get a better idea of the actual failure mode of the fibre reinforced concrete MOR beams were tested in a closed loop Materials Testing System (MTS) in displacement control as shown in the photo in Figure 2 (Elsaigh et al, 2004). The $100 \times 100 \times 500$ mm beams were supported to span 450 mm and loaded at third points. The load and centre point deflection were recorded at a rate of 100 Hz. The loads were used to calculate the bending moments and stresses in the beams.

Typical test results obtained for beams with different fibre contents can be seen in Figure 3. The beams containing no fibre clearly fail in a brittle manner. The load deflection behaviour is linear up to the point of complete failure, indicating that there is no warning of imminent failure. The results in this graph clearly indicate that the actual cracking strength of the concrete is not affected by the fibre content of the beams and the increased flexural strength indicated in Table 4 is a result of increased post-cracked strength where more fibres are bridging the cracks in the concrete. The post-cracked

Figure 2. Flexural beam test setup.

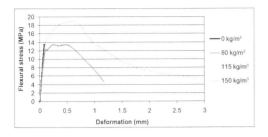

Figure 3. Stress-deformation behaviour of concrete beams.

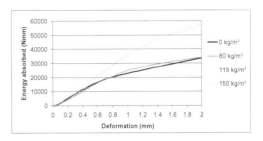

Figure 4. Energy absorbed by beams.

strength of the beam containing 80 kg/m³ fibre is less than the concrete cracking strength and only at a fibre content of 115 kg/m³ does the post-cracked strength of the beam reach the concrete cracking strength. These results correlate with previous unpublished work that indicates that the minimum fibre content that would result in an increase in flexural strength is a function of the compressive strength of the concrete. While a steel fibre content of 60 kg/m³ would be sufficient for 60 MPa concrete, a fibre content of 115 kg/m³ was required in concrete with a compressive strength of 119 MPa to result in a post-cracked strength approximately equal to the concrete cracking strength.

From Figure 3 it can also be seen that a fibre content of 150 kg/m³ results in a post-cracking strength significantly higher than the concrete cracking strength.

One should question whether the MOR test method, where only the peak load is recorded, is a suitable test method for concrete with high fibre contents. With only one value recorded it is impossible to determine whether the maximum stress was recorded when the concrete cracked or whether the fibres contributed to the extent that the post-cracked strength exceeded the concrete cracking strength.

The ability of concrete elements to resist cyclic loading is a function of the energy absorption capacity of the element. The energy absorbed by the beams before failure can be calculated by determining the area under the load-deflection graphs and the energy absorption for different fibre contents can be seen in Figure 4. Reinforced concrete design codes normally limit the deflection of structural members to span/250 (European Committee for Standardization, 2004) and for the 450 mm beam span that would be a deflection of 1.8 mm. Figure 4 clearly indicates that in the deflection range that would normally be experienced by concrete elements the beam containing 80 kg/m³ fibres does not absorb significantly more energy than the beam containing no fibres. Previous research indicated that as little as 30 kg/m³ of the same type of

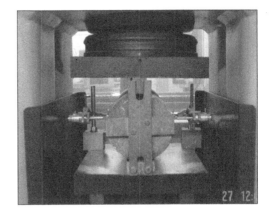

Figure 5. Split cylinder test setup.

fibre has a significant effect on the energy absorbed by 30 MPa concrete beams. These results confirm that the fibre content used in concrete should be a function of the concrete characteristic strength. The higher fibre contents result in significant increases in the energy absorption.

4.4 Splitting behaviour

Linear Voltage Displacement Transducers (LVDT's) were used on the split cylinder test samples to record the increase in cylinder diameter perpendicular to the direction of loading. The test setup can be seen in the photo in Figure 5. The test was conducted in deflection control and the load deformation behaviour was recorded for each cylinder tested. The splitting stress as a function of deformation is plotted in Figure 6 for cylinders with different fibre contents. Although the stresses are much lower than that observed for the flexural strength testing, the behavioural trend is similar to that observed for the beams.

The inclusion of fibres does not prevent the concrete from cracking, but high fibre volumes result in increased post-cracking strengths. The cylinder splitting test is relatively easy and currently

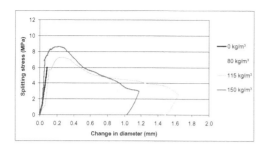

Figure 6. Split cylinder behaviour.

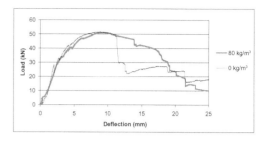

Figure 7. Load deflection behaviour of round discs.

investigations are underway to determine whether the load deformation behaviour of split cylinders can be used to model the post-cracked behaviour of concrete elements.

4.5 Round disc tests

The UTCRCP contains large volumes of steel reinforcing and a steel mesh with 5.6 mm bars at 50 mm spacing has been used in preliminary test sections. The pavement acts as a composite material containing concrete, steel fibres and steel mesh, and the components of the composite can affect the properties of the composite. If for instance the fibres are too long or the bars in the mesh are too closely spaced, it will be impossible to thoroughly place and compact the concrete, resulting in a pavement made up of laminated layers that can de-laminate under load.

The low thickness of the pavement (at 55 mm) will force the fibres to align in mainly the two dimensional plane of the pavement and the flexural strength of the thin pavement should be significantly higher than that predicted using flexural beams. It is therefore important to establish the load deflection behaviour of samples with compositions similar to that of the pavement.

To determine the load deflection behaviour of the UTCRCP 600 mm diameter round discs were cast containing reinforcing bars similar in position and spacing to the bars used in the pavement. The 55 mm thickness of the samples is similar to the pavement thickness and the moulds are filled with FRHPC as specified on site.

The load deflection behaviour for 600 mm discs containing $50 \times 50 \times 5.6$ mm mesh can be seen in Figure 7. One of the samples contained no steel fibres and it can clearly be seen that both these discs support similar maximum loads, but the disc without fibres fails suddenly with the concrete punching. During failure large shards of concrete broke away from the sample, leaving the reinforcing bars completely exposed, as can be seen with the right hand side disc in the photo in Figure 8.

Figure 8. Round discs after failure.

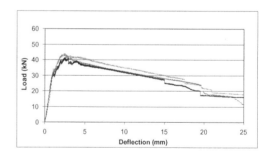

Figure 9. Repeatability of round disc test results.

The left hand side disc contained steel fibres and as can be seen from the graph in Figure 7, the samples containing fibres have a more gradual failure mode, with the load carrying capacity reducing gradually as the deflection increases.

It is desirable that the pavement should not fail in the mode as indicated on the right hand side, and the heavy vehicle simulator test results reported by Kannemeyer et al. (2007) did have a similar failure mode when ineffective fibres were used.

Dupont & Vandewalle (2004) stated that one of the advantages of this test method is the repeatability of the results obtained. To prove this point, 3 identical discs were manufactured using $50 \times 100 \times 5.6$ mm mesh and 80 kg/m³ steel fibres. The load deflection behaviour of these discs can be seen in Figure 9 and it is clear that these 3 discs were manufactured using the same material.

This type of repeatability is not often seen in post-cracked testing. The test setup is relatively cheap and setting up the tests is relatively easy. It is therefore recommended that this type of test should be used for quality control on site.

5 EFFECT OF REINFORCING COMBINATION

The $50 \times 50 \times 5.6$ mm mesh is not a standard mesh size and steel mesh manufacturers find it difficult to manufacture such a tight grid. If the reinforcing bar spacing could be increased, not only the manufacturing of the mesh but also the placing of the concrete would be easier. It has been experimentally proven for normal concrete with low steel content that mesh and steel fibres behave similarly if an equal volume of steel per plan area is used. The round disc test was used to determine the effect of reinforcing combination on the load-deflection behaviour of FRHPC.

In the first phase of the investigation a fibre content of 80 kg/m^3 was maintained while the mesh was changed from $50 \times 50 \times 5.6$ mm to $50 \times 100 \times 5.6$ mm and $100 \times 100 \times 5.6$ mm. The effect of this increase in bar spacing can be seen in the graph in Figure 10. There is no difference in behaviour before the concrete cracks, but once the concrete has cracked there is a clear difference in the behaviour with the higher steel content resulting in the highest post-cracked strength.

In most standard prescribed test methods, it is difficult to distinguish between the results as indicated in Figure 10 and therefore the energy absorption of the samples, as calculated from the area under the load-deflection graph, will again be used as criterion. The energy absorption of the discs can be seen in the graph in Figure 11. The contribution of the steel fibres can now clearly be seen with the energy absorption of the discs significantly reduced as a result of reduced steel content. Not one of the discs tested meets the UTCRCP requirement of

1000 Joules energy absorption and the only way to increase the energy absorption is clearly to increase the steel reinforcing content. The possibility of increasing the energy absorption by increasing the fibre content should be investigated.

The volume of concrete that goes into a 55 mm thick disc with a diameter of 600 mm is 15.55 liters. If the fibre content of the concrete is 80 kg/m^3, the fibre content of each panel should be 1.244 kg. A steel mesh of $50 \times 50 \times 5.6$ mm results in a steel content of 7.734 kg/m^2 and therefore 2.187 kg steel per disc. The total steel content (mesh and fibres) of the disc would be 3.431 kg. The mesh spacing can be changed to $50 \times 100 \times 5.6$ mm or $100 \times 100 \times 5.6$ mm and to keep the total steel content per disc constant the fibre content can be increased proportionally as indicated in Table 5. The fibre content was rounded up to sensible numbers.

The load deflection behaviour for these discs can be seen in Figure 12 and the energy absorption in

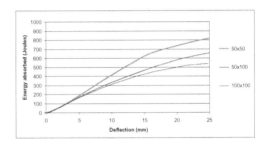

Figure 11.   Effect of mesh on energy absorption.

Table 5.   Equivalent steel contents.

Mesh spacing (mm × mm)	Mesh content (kg)	Extra steel required (kg)	Fibre content (kg/m³)
50×50	2.187	1.24	80
50×100	1.640	1.79	115
100×100	1.093	2.34	150

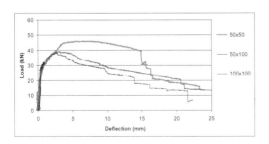

Figure 10.   Effect of mesh on load-deflection behaviour.

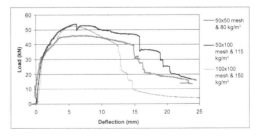

Figure 12.   Effect of steel combination on load-deflection behaviour.

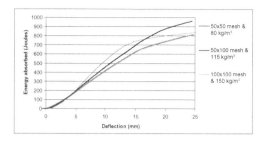

Figure 13. Effect of steel combination on energy absorption.

Figure 13. The energy absorption of the discs does vary significantly with the disc containing $50 \times 100 \times 5.6$ mm mesh and 115 kg/m³ fibre absorbing significantly more energy than the other two discs at large deflections. At lower deflections the $100 \times 100 \times 5.6$ mm mesh with 150 kg/m³ mesh perform better and these results indicate that in future it will be essential to establish suitable deflection limits for these tests.

6 PRACTICALITIES

6.1 Mixing and casting

There is often a tendency on construction sites to mix concrete as rapidly as possibly and therefore it is important to know what the minimum mixing time is that would be required to produce concrete with a uniform composition. The optimum mixing time of concrete depends on the type and size of the mixer, the speed of rotation and the extent of pre-blending of materials that takes place before mixing. Recommended mixing times normally vary from 1 minute to 5 minutes. As a result of both the relatively high fibre content and the very high HRWRA content FRHPC requires significantly longer mixing times than normal concrete. For the majority of large scale mixing facilities mixing times varying from 5 to 10 minutes will be required to ensure uniform concrete composition.

Many suppliers recommend fibres that are glued into strips for better distribution into concrete mixes, but as a result of the low water content of the FRHPC there is not enough water available to dissolve a relatively large volume of glue and the fibres can thus not disperse through the mixture. For low fibre volumes it is possible to get good dispersion of fibres by adding dissolvable bags of fibres to the back of ready mix trucks on site but with the FRHPC the fibres can not be mixed in using this method. The best dispersion of the relatively large volume of fibres is achieved when discrete fibres are added individually to the mixer over a period of time during the mixing process. Internationally automatic systems are available for continuously feeding in fibres.

6.2 Materials

The properties of the FRHPC are highly dependent on the properties of the materials used in the mixtures and relatively small variations, which would be totally acceptable for normal concrete, have significant consequences for FRHPC. The magnitude of the sensitivity can best be explained by using an example to demonstrate the effect of different batches of cement from the same factory.

The mix composition as indicated in Table 2 was used to cast 4 mixes and the results listed before were obtained when fresh bags of cement were used. The same mix composition (with 80 kg/m³ fibre) and materials was then used, but with cement that had been in the laboratory for at least 3 months. No visual signs of deterioration were present, the bags were not wet and no cement clots could be seen in the bags. The mixture containing the old cement was cast on two consecutive days and the results can be seen in Table 6. These results clearly indicate that the strength development of the concrete has been significantly affected by the cement and if FRHPC is to be used on large scale, contractors will have to be aware of the fact that the cement should not be allowed to age and each batch of cement will have to be tested to ensure that the cement can be used to make concrete with the required strength.

6.3 Quality control

During the casting of concrete standard workability tests are normally performed as part of the quality assurance process. The high fibre content of FRHPC makes it impossible to use the standard workability test methods as the mixture has no workability. If the fibres are omitted from the mixture, a collapse slump would be recorded, but the addition of the fibres results in a zero slump being recorded. When energy is applied to the mixture

Table 6. Compressive strength.

Mixture	7-day strength (MPa)	28-day strength (MPa)
New cement	103.7	130.0
Old cement mix 1	59.0	79.0
Old cement mix 2	70.5	86.5

through vibration, the mixture becomes highly workable and flows into moulds.

The cube strength normally used for quality control of concrete can still be used to determine whether the other materials remain constant and were batched correctly, but the cube strength gives no indication of the fibre behaviour and higher strengths would be recorded if the contractor reduced the fibre content. For FRHPC additional quality control tests would be required, to ensure that both the quality and the quantity of the fibres used on a daily basis remain the same than that used in the approved mix design.

7 CONCLUSIONS

Researchers have proven that FRHPC can be used to effectively construct infrastructure. Lack of design codes and construction experience is hampering the large scale use of FRHPC. The standard test methods used for quality control of concrete are not necessarily suitable test methods for FRHPC and the relevance of test methods should be carefully considered before the methods are used to control the production of FRHPC.

ACKNOWLEDGEMENTS

The authors would like to express appreciation for the research grants made by the Duraset business unit of Aveng and the THRIP-program of the South African National Research Foundation that made this research possible.

REFERENCES

Bergh, A.O., Semmelink, C.J., McKay, A.H. & Steyn, A.H. Roodekrans thin concrete experiment sections 4, 5 and 6: Continuously reinforced thin concrete pavements. *Proceedings of the 24th Southern African Transport Conference (SATC2005), Pretoria, 11–13 July 2005*: 180–190.

Dupont, D. & Vandewalle, L. Comparison between the round plate test and the RILEM 3-point bending test. *6th RILEM Symposium on Fibre-Reinforced Concretes (FRC)—BEFIB 2004, 20–22 September 2004, Varenna, Italy*: 101–110.

Elsaigh, W.A., Kearsley, E.P. & Robberts, J.M. Steel fibre reinforced concrete for road paveetn applications. *Proceedings of the 24th Southern African Transport Conference (SATC2005), Pretoria, 11–13 July 2005*: 191–201.

European Committee for Standarization. 2004. *EN 1992-1-1: Eurocode 2: Design of concrete structures—Part 1–1: General rules and rules for building.* Brussels.

Kannemeyer, L., Perrie, B.D., Strauss, P.J. & du Plessis, L. Ultra thin continuously reinforced concrete pavement research in South Africa. *International Conference on Concrete Roads (ICCR), Johannesburg, South Africa, 16–17 August 2007*: 27.

Kearsley, E.P. & Mostert, H.F. Optimizing mix composition for steel fibre reinforced concrete with ultra high flexural strength. *6th RILEM Symposium on Fibre-Reinforced Concretes (FRC)—BEFIB 2004, 20–22 September 2004, Varenna, Italy*: 505–514.

Neville, A.M. 1995. *Properties of concrete.* Essex: Longman group Limited.

Neville, A.M. & Brooks, J.J. 1987. *Concrete Technology.* Essex: Longman group Limited.

Swamy, R.N. 2008. High performance cement based materials and holistic design for sustainability in construction (Part I). *The Indian Concrete Journal*, February 2008: 7–18.

Advances in Cement-Based Materials – van Zijl & Boshoff (eds)
© *2010 Taylor & Francis Group, London, ISBN 978-0-415-87637-7*

Towards a reliability based development program for SHCC design procedures

J.S. Dymond & J.V. Retief
Stellenbosch University, South Africa

ABSTRACT: Principles of structural reliability are used mainly to provide a rational basis for improving the safety and economy of design practice with an extensive experience base. Execution of a development program for design procedures for innovative materials and applications on the basis of structural reliability ensures that the outcome readily conforms to existing practice. Other advantages are that reliability techniques could be applied to identify important sources of uncertainty on which subsequent testing and modelling investigations could focus. The use of Bayesian updating techniques allows optimal utilisation of existing information. This could compensate for the lack of an established experience base and extensive information. The development of a reliability based design model and procedure can highlight the advantages of the innovation. The objective of the paper is to outline the way in which the development of the application of an advanced concrete material such as SHCC should be approached.

1 INTRODUCTION

Design procedures for structures using common, well established materials, such as reinforced concrete or structural steel, are based on structural engineering theory as well as a broad experience base on the behaviour of the material. Satisfactory safety levels are achieved by means of reliability theory combined with acceptable past practice. Great advances are being made in the development of new materials with better structural properties and in order to take advantage of these innovations it is necessary to develop design procedures that will result in an adequate level of reliability without the advantage of a broad experience base or extensive test results. The focus of this paper is to outline a reliability based development program for design procedures for innovative materials with reference to strain hardening cementitious compounds (SHCC).

SHCC is a form of fibre reinforced concrete which displays a number of advantages over conventional concrete. SHCC is able to withstand greater tension forces before cracking than conventional concrete and displays a strain hardening type of behaviour after cracking initiates. When cracking occurs in SHCC, a number of small cracks form which remain only approximately 0,1 mm wide until crack saturation is reached. Crack saturation signals the end of the strain hardening behaviour after which crack localisation occurs, where one

crack opens up and this is followed by failure of the element.

The advantages over conventional concrete are that SHCC displays a ductile type of failure unlike conventional concrete which fails in a brittle manner; some tension resistance of the SHCC concrete matrix may be utilised in design and the smaller width cracks offer greater durability. With these advantages it is desirable to develop design procedures so that this material may be utilised to its full potential by structural engineers while still ensuring adequate safety. Conversely reliability techniques can be used to expose and quantify the advantages of SHCC.

The four basic steps involved in the development of reliable design procedures for a new, innovative material are:

1 Establish the material behaviour
2 Develop an analysis model
3 Verify the analysis model
4 Ensure adequate reliability

These four steps will be discussed in the following sections. Reference will be made primarily to the Eurocode—Basis of Structural Design (EN 1990), specifically Annex C and D which are based on ISO 2394:1998 . The Eurocodes form the basis of the revision of the South African loading code and the basis of structural design in SANS 10160-1 is based on EN 1990. The methods presented here are therefore in harmony with the South African suite of structural design codes.

2 DESCRIPTION OF MATERIAL BEHAVIOR

The various aspects of material behavior include the response to tension, compression and shear (described qualitatively and quantitatively), the relation of sample strength to in situ strength and the relevant statistical characteristics of the above-mentioned parameters.

2.1 Material response to loading

Standard test procedures should be followed as laid out in the relevant codes of practice. A qualitative description of the material behavior and failure modes should be made. This encompasses the development of cracks, crack widths, distribution and direction of cracks and mode of final failure. A typical quantitative description is in the form of stress-strain curves with the defining parameters. The qualitative description should be correlated with the quantitative description to ensure the best understanding of the material behavior.

Figure 1 shows a typical annotated stress-strain curve for SHCC (Visser & van Zijl 2007).

The critical parameters which define the behavior of SHCC are:

1 σ_e = limit of the elastic tensile stress
2 ε_e = limit of the elastic tensile strain
3 σ_u = ultimate tensile stress
4 ε_u = ultimate tensile strain

2.2 Statistical properties of material strength

The statistical characteristics of the material properties are required in order to ensure acceptable levels of structural reliability. Adequate reliability is obtained by determining characteristic values of material strength combined with an appropriate partial factor or by determining the design value of the material strength directly.

Ideally sufficient tests should be performed to obtain stable values of the mean, standard deviation and skewness of the material strength as it will then be possible to assign the most appropriate probability distribution to each parameter. For practical reasons it is often necessary to define the material strength based on smaller samples.

The statistical properties of the material strength will be used later to determine characteristic and design values as well as an appropriate partial material factor.

2.2.1 Bayesian updating

A part of the experimental verification is testing of materials from which the test specimens were made thereby providing further material strength data which could be incorporated into the prior available information by means of Bayesian updating (Holický 2009). When only a mean and standard deviation are known as prior information, the new information may be combined with the prior information in the following manner:

New information: m = sample mean; s = sample standard deviation; n = number of samples; v = degrees of freedom = $n-1$ and prior information m', s', n' (unknown), v' (generally $\neq n'-1$) respectively.

The prior number of samples and degrees of freedom are determined using the coefficients of variation of the population mean and standard deviation, $\delta(\mu)$ and $\delta(\sigma)$ respectively, based on previous experience on the degree of uncertainty of the mean and standard deviation:

$$n' = \left[\frac{s'}{m'\delta(\mu)}\right]^2 ; \quad v' = \frac{1}{2\delta(\sigma)^2} \quad (1)$$

Updated number of samples: $n'' = n + n'$
Updated degrees of freedom: $v'' = v + v' - 1$
$$\text{for} \quad n' \geq 1$$
$$v'' = v + v'$$
$$\text{for} \quad n' = 0$$

Updated mean m'' and standard deviation s'' can then be used to determine characteristic and design values, as discussed in Section 5:

$$m'' = \frac{mn + m''n''}{n''} \quad (2)$$

$$s'' = \sqrt{\frac{vs^2 + v's'^2 + nm^2 + n'm'^2 - n''m''^2}{v''}} \quad (3)$$

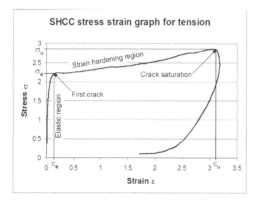

Figure 1. SHCC stress-strain diagram.

2.2.2 Conversion factor

Concrete cubes that have been cast, compacted and cured under controlled, ideal conditions display greater strength than in situ concrete for which the conditions are not as ideal. The conversion factor η represents the ratio of in situ strength to the cube (or cylinder) strength. The conversion factor is also a random variable and its uncertainty should be accounted for either in the determination of the material strength design value or the resistance model design value.

Direct comparisons between cube tests and core samples have been carried out for conventional concrete over a long time period so that the conversion factor has been sufficiently accurately determined. This information is obviously not available in the case of an innovative material such as SHCC.

As a first estimate "in situ" testing could be carried out on specimens that have been cast for laboratory tests. Taking cognisance of the fact that these specimens have also been cast, compacted and cured in more controlled conditions than those on a building site, this provides an upper bound of performance.

In the absence of further tests, engineering reasoning and judgement should be employed to estimate a reasonable value for the conversion factor. Reference could be made to the value of the conversion factor for conventional concrete, where there is information available, taking into account the following:

- The relation between cube test results and laboratory test specimen samples in the innovative material compared to conventional concrete.
- The important aspects of casting, compacting and curing that have an influence on the innovative material and how these differ from conventional concrete.

The variability of the conversion factor should include the uncertainty inherent in the estimation process.

As with all innovations, practical implementation cannot be delayed until there is complete knowledge of every parameter. It should rather be considered as a continuous development process where the initial implementation is based on best estimates and conservative values and the properties are updated and refined as new information becomes available.

3 DEVELOPMENT OF AN ANALYSIS MODEL

An analysis model should be developed to determine the resistance of a structural element constructed using the innovative material. The model should be based on structural analysis theory as well as experience. The resistance of the element will be calculated as:

$$r_t = g_{rt}(\underline{X}) \tag{4}$$

where: X = vector of parameters (basic variables) that affect the resistance of the element for the relevant failure mode.

Further discussion of the analysis models is outside the scope of this paper. Details of the models can be found in Victor & van Zijl (2009).

4 VERIFICATION OF ANALYSIS MODEL

The analysis model should be verified by means of experimental laboratory testing. A typical analysis model includes many parameters and the testing of each of these parameters individually as well as in combination with all the other parameters will lead to a prohibitively large number of test cases. Methods must therefore be employed to reduce the number of test cases necessary to provide an adequate evaluation of the accuracy of the model.

4.1 Identify critical parameters

The performance function for a structural element may be written as:

$$g(\underline{X}) = R - E \tag{5}$$

where R = structural resistance; E = load effect on structure.

The vector of basic variables X for the resistance side of the equation typically consists of various parameters defining the material strength and a number of geometric parameters. In order to effectively design a testing program, the parameters to which the performance function is the most sensitive should be identified and made the focus of the experiments.

There are two contributors to the sensitivity of a model to a given parameter, the deterministic and statistical sensitivity. The deterministic sensitivity of the function $g(X)$ to variable X_i is given by:

$$\alpha_{d,X_i} = \frac{\partial g}{\partial X_i} \tag{6}$$

The statistical sensitivity for variable X_i is given by the standard deviation of the parameter:

$$\alpha_{s,Xi} = \sigma_{Xi} \tag{7}$$

The product of the two contributors gives the sensitivity of the reliability of the structural element to the given parameter:

$$\alpha_{X_i} = \alpha_{d,X_i} \alpha_{s,X_i} = \frac{\partial g}{\partial X_i} \sigma_{X_i} \qquad (8)$$

Both contributors should be considered when designing the experimental test program. If no information is available on the standard deviation of a parameter, a reasonable upper limit should be used based on experience with similar situations.

The critical parameters are those for which the sensitivity is the greatest. These parameters should be the focus of the test program. Engineering reasoning and judgment should be employed when considering the cut-off level of sensitivity for inclusion in the test program.

The parameter sensitivity serves the further purpose of highlighting parameters that require further investigation themselves, by considering both the level of information on a parameter and the sensitivity of the function $g(X)$ to its contribution. These effects should be considered in planning test programs.

4.2 Development of test program

The purpose of the test program is to verify the analysis model by assessing the bias and coefficient of variation over the range of critical parameters likely to be encountered in practice.

4.2.1 Initial model verification

If there is uncertainty as to the validity of the analysis model or the likely failure modes, use of any prior test information can be made or a small pilot test series can be carried out.

The analysis model for reinforced SHCC beams in flexure allows for both tension and compression reinforcement as well as stirrups. Prior information is available in the form of a small sample of beam tests carried out with tension reinforcement only. These test results, being the best available information, should be compared with the analysis model predictions for this situation as a first comparison while incurring little or no cost. If the results compare well, a small pilot study may be designed to include compression reinforcement and stirrups. If the validity of the model is confirmed by the pilot tests, a full test program can then be planned. The approach to be taken should the test results not correspond well with the theoretical predictions will be discussed later.

4.2.2 Design of experimental program

Traditional approaches to the design of an experimental test program tend to result in very large numbers of tests for a relatively small number of parameters.

The One Factor at a Time (OFAT) method consists of setting all test parameters to a given value and varying one parameter over its likely range. The disadvantage of this method is that it does not demonstrate interactions between parameters and it may not achieve the maximum response from the model.

The Full Factorial method consists of combining each parameter with every other parameter at all possible values. Typically each parameter will be considered at two levels denoted "+" and "−". For k parameters, 2^k number of experiments results, quickly becoming a very large number of experiments.

The method of Design of Experiments (O'Connor 1991) has been developed in order to reduce the number of tests required to still obtain an understanding of the main effects (single parameters) and second order interactions (two parameters). It has been shown that generally higher order interactions are not significant. Parameters to which the model is relatively insensitive may be omitted from the test program or may be investigated in a small number of pilot tests to confirm their insignificance. Detailed discussion of Design of Experiments is outside the scope of this paper and reference should be made to specialized literature.

The tests should cover the critical parameters over the range likely to be encountered in practice. The practical lower and upper limits for each parameter could be established by means of surveys or based on expert knowledge.

4.3 Minimize sources of uncertainty

The purpose of the test program is to verify the analysis model by assessing the bias and coefficient of variation in comparison to the test results. It is therefore necessary to minimize the sources of uncertainty in the test results so that the difference between the experimental and theoretical results represents only uncertainty in the model rather than incorporating uncertainties and variations in the constituent parameters. Two methods may be employed to minimize uncertainties, namely quality control during fabrication of test specimens and taking measurements. Taking measurements of the test specimen eliminates the most uncertainty and many parameters may be measured, for example: geometry (height, width, length) and preferably material strength. Strict quality control should be enforced during fabrication to control e.g. reinforcement geometry.

The uncertainties for all variables can not necessarily be quantified but they should at least be identified.

One of the most critical parameters to be measured is material strength. Cube samples should be taken, however, as was discussed previously, material strength in structural elements generally differs from cube strengths and is accounted for by the conversion factor. If the material strength of the test specimen is considered as the cube strength, a conversion factor will be included in the modeling uncertainty. This conversion factor is still not representative of that from the cube strength to the in situ building strength due to laboratory curing conditions for the test specimen. An additional conversion factor representing the ratio between the in situ strength and test specimen strength should be included in the final design calculation method.

It may be argued that the test specimen strength will be close to the cube strength as to be negligibly different because of the ideal casting, compacting and curing conditions in the laboratory. Any small reduction in strength included in the modeling uncertainty will then be merely a slight in-built conservatism. The conversion factor included in the final design calculation will then be the ratio between in situ and cube strengths.

4.4 *Theoretical predictions*

Predictions for the behavior and strength of each test specimen should be made from the analysis model as given in Equation 9:

$$r_{ti} = g_{rt}\left(\underline{X}_i\right) \qquad (9)$$

where r_{ti} = structural element resistance for test specimen i; $\underline{X}_i$ = vector of parameters measured for test specimen i.

A set of r_{ti} values will be obtained for all the test specimens.

4.5 *Interpretation of results*

Results from the test program should be assessed firstly on the basis of engineering judgment and reasoning and secondly statistically. This assessment should be carried out on any available prior information or pilot test runs to eliminate gross discrepancies between the predicted values and test results before the main test series is performed.

Engineering judgment and reasoning should be used to determine if the results look reasonable and if the analysis model acceptably predicts

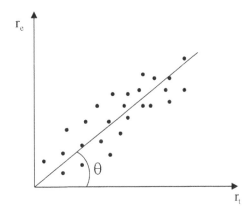

Figure 2. Scatter plot of theoretical and experimental results.

the outcome of the experiment. A scatter plot of the theoretical predictions r_t and experimental results r_e provides a useful tool for this assessment. An example scatter plot is shown in Figure 2.

If the analysis model described the structural behavior exactly and completely then all the points would lie on the line $\theta = 45°$. In practice some scatter can be expected. A systematic deviation should be investigated further to determine whether there were errors in the test procedure or in the analysis model, for example due to neglecting the effect of a certain parameter. The further investigation may consist of another test series or a modification of the analysis model, resulting in updated theoretical predicted values of resistance.

A statistical evaluation of the results may be carried out once the test results are deemed to satisfactorily correlate to the predicted results based on the above assessment. The aim of the statistical evaluation is to determine the bias and uncertainty inherent in the theoretical model. The method presented in EN 1990 will be described.

The probabilistic model of the resistance is set up as $r = br_t\delta$, where r = real structural resistance; b = correction factor (model bias); δ = error term (model uncertainty).

The correction factor is a measure of the tendency of the model to systematically over or under estimate the resistance of the element. The bias is defined as the "least squares" best fit to the slope as given in the scatter plot (Equation 10).

$$b = \frac{\sum r_e r_t}{\sum r_t^2} \qquad (10)$$

The error term is a measure of the relative scatter of the results. The uncertainty is expressed in terms of the coefficient of variation of the error term. The error term for each test specimen is defined as:

$$\delta_i = \frac{r_{e_i}}{b r_{t_i}} \qquad (11)$$

An estimate of the coefficient of variation of the error term V_δ is found by defining Δ such that:

$$\Delta_i = \ln(\delta_i) \qquad (12)$$

The estimate for the expected value is given as:

$$\overline{\Delta} = \frac{1}{n} \sum_{i=1}^{n} \Delta_i \qquad (13)$$

The estimate for the variance is given as:

$$s_\Delta^2 = \frac{1}{n-1} \sum_{i=1}^{n} \left(\Delta_i - \overline{\Delta} \right)^2 \qquad (14)$$

The coefficient of variation is then estimated as:

$$V_\delta = \sqrt{\exp\left(s_\Delta^2\right) - 1} \qquad (15)$$

The accuracy of the model may be assessed by the parameters b and V_δ. A bias value b < 1 indicates an unconservative model that overestimates the resistance of the structure and conversely a bias b > 1 indicates a conservative model that underestimates the strength of the structural member. If the bias value is too far from unity, the analysis model should be reassessed to determine if the resistance is correctly modeled or if the effect of certain parameters was neglected.

It is imperative when applying a correction factor to a model that the entire range of practical situations has been considered. If a model exhibits conservatism in the test range it must be shown that it will exhibit the same degree of conservatism in the practical range which falls outside the test range so as to not over correct the results into the area of unconservatism. In the other case where a model displays unconservative behavior, it should be shown that the most unconservative case has been considered for the determination of the correction factor.

The coefficient of variation of the error term V_δ is a measure of the scatter of the results which is an indication of the ability of the model to accurately predict strength over a range of situations. If the coefficient of variation is very large it may

be reassessed by dividing the test series into subsets, comprising similar structural conditions, and determining b and V_δ for each sub-set. The disadvantage of this is that the number of tests in each sub-set may become very small.

Deliberately vague descriptions of unacceptable model uncertainties have been made in "b too far from unity" and "coefficient of variation is very large". The limiting values should be based on the engineer's discretion, judgment and experience. It is often the case that analysis models which predict structural behavior very accurately can be developed at the expense, however, of simplicity. The optimal point should be found between model accuracy and practical levels of complexity for the design of structures. The position of the optimal point will depend on the type of structure, the cost of materials, the safety implications, the levels of experience of the designer etc.

Once the model has been verified and satisfactory values of b and V_δ have been obtained, the reliability of the structural design may be assessed.

5 RELIABILITY

The codification of design methods includes measures for ensuring adequate structural safety in the form of partial factors. These partial factors have in the past been calibrated to existing practice. New code development depends on the use of reliability calibration methods to determine appropriate values for levels of safety and associated partial factors.

A measure of reliability is the reliability index β which is related to the probability of failure p_f by $p_f = \Phi(-\beta)$, where Φ = cumulative distribution function for the standard Normal distribution.

The Eurocodes are calibrated so that reliability levels for representative structures are as close as possible to the target reliability $\beta_T = 3.8$ (EN 1990). South African codes are calibrated to a minimum target reliability $\beta_T = 3.0$ for ductile failure modes. This was the level of reliability inherent in current practice when limit states design was introduced (Kemp et al. 1987, Milford 1988).

The reliability level of the resistance term of the performance function is ensured by the use of a partial resistance factor γ_M with two constituents γ_m and γ_{Rd} to take into account the uncertainty in the material and the analysis model respectively.

A limit state denotes the border between safe and unsafe states in the structure. With reference to the performance function (Equation 5), the safe state is g > 0 (R > E) the unsafe (failure) state is g < 0 (R < E) and the limit state is g = 0.

The design value method employed by codes consists of determining a design value for each basic variable by means of partial factors. The design is then deemed adequate if the limit states are not reached when the design values are used in the analysis models. A satisfactory design can be expressed as $E_d < R_d$, where E_d = design value of load effects; R_d = design value of resistance.

The design resistance can be expressed as:

$$R_d = R\{X_{d1}, X_{d2}, \ldots; a_{d1}, a_{d2}, \ldots; \theta_{d1}, \theta_{d2}, \ldots\} \quad (16)$$

where X_{di} = design value of material property; a_{di} = design value of geometric property, typically considered as the nominal value; θ_{di} = design value of model uncertainty.

Design values in codes are generally expressed in terms of representative or characteristic values and partial factors as:

$$R_d = R\{\eta X_k / \gamma_m; a_d\} / \gamma_{Rd} \quad (17)$$

where X_k = characteristic value of material strength.

The design values are defined such that there is a given probability of their exceedance which is related to the target reliability index in the following way:

$$P(R \le R_d) = \Phi(-\alpha_R \beta_T) \quad (18)$$

where α_R = reliability sensitivity factor for resistance.

As an approximation made to allow separate calibration of loads and resistance, EN 1990 states that α_R may be taken to be 0.8 provided that $0.16 < \sigma_E / \sigma_R < 7.6$ which is likely for most loads and resistances. For the South African situation of a target reliability index $\beta_T = 3.0$:

$$P(R \le R_d) = \Phi(-0.8 \times 3.0) = \Phi(-2.4) = 0.0082 \quad (19)$$

The design value of the resistance is therefore the 0.0082 fractile of the resistance distribution. The Eurocode target reliability of $\beta_T = 3.8$ corresponds to the 0.0012 fractile.

The partial factor for the material as well as the model uncertainty may be found by determining the design value and dividing it by the characteristic or representative value.

5.1 Material strength

EN 1990 gives guidance on calculating characteristic and design values of material strengths when only a small number of tests have been carried out. The following issues should be considered:

1 Scatter of test data
2 Statistical uncertainty associated with the number of tests
3 Prior statistical knowledge

If the development process for the innovative material is at the stage of deriving design methods, it is likely that the material strength has previously been studied. Some prior knowledge on the material properties will therefore be assumed. In the case of SHCC a number of test series on material strengths have been carried out for which raw data is available.

The characteristic value as calculated according to EN 1990 is the 5th percentile (i.e. 0.05 fractile) of the material strength. The characteristic value so calculated would be multiplied by the conversion factor and divided by the partial material factor to determine the design value of the material strength. In order to use the method of calculating the characteristic value, a suitably determined partial material factor should be available.

Calculation of the design value directly should take into account the conversion factor as well as the required level of reliability. The partial material factor may then be determined by dividing the characteristic or representative value by the design value of the material strength.

If there is sufficient sample information to fully describe a probability distribution for the material strength, the fractile may be calculated by:

$$x_p \text{ such that } P(X \le x_p) = \Phi(x_p) = p \quad (20)$$

where x_p = pth fractile value of variable X (material strength); Φ = cumulative probability distribution function of variable X; p = required fractile value (0.05 for characteristic value, 0.0082 for design value).

There is often insufficient data to fully describe the probability distribution function due to the cost of test programs. It is therefore necessary to estimate the fractile values based on small samples where the additional uncertainty due to the sample size should be taken into account. The method given in EN 1990 is based on the prediction method of fractile estimation (Gulvanessian et al. 2002). The prediction method estimates the pth fractile (x_p) by the prediction limit ($x_{p,pred}$) such that a value of x_i randomly drawn from the population will be less than $x_{p,pred}$ with probability p, i.e. $P(x_i \le x_{p,pred}) = p$. With increasing number of samples, $x_{p,pred}$ asymptotically approaches x_p, the true fractile value.

The prediction method corresponds approximately to the coverage method with a confidence level of 75%. Thus, the probability that $x_{p,pred} < x_p \approx 75\%$. It is conservative for material strengths for the estimated fractile to be less than the true fractile.

Two situations are presented in EN 1990 Annex D, that where the coefficient of variation of the population is known from prior tests and that

303

where it is unknown and the fractile estimation is based on the sample coefficient of variation. The case where the coefficient of variation is known provides less over-conservative fractiles and is therefore preferable. EN 1990 recommends the use of this case together with an upper estimate of the coefficient of variation. The estimate of the coefficient of variation could be based on experience with similar situations but should not be taken as less than 0.10.

The characteristic value of material strength X_k is then calculated as:

$$X_k = m_X(1 - k_n V_X) \qquad (21)$$

where m_X = sample mean of X obtained from test data; V_X = coefficient of variation of X; k_n = fractile estimator.

For the case of V_X unknown, the coefficient of variation is estimated from the sample as:

$$V_X = \frac{1}{m_X}\sqrt{\frac{\sum(x_i - m_X)^2}{n-1}} \qquad (22)$$

The design value is determined from the characteristic value as:

$$X_d = \eta_d \frac{X_k}{\gamma_m} \qquad (23)$$

where η_d = design value of the conversion factor; γ_m = partial material factor.

If the design value is to be estimated directly from the sample, it is calculated as:

$$X_d = \eta_d m_X(1 - k_{d,n} V_X) \qquad (24)$$

where: $k_{d,n}$ = design fractile estimator.

Tables are given in EN 1990 Annex D for k_n and $k_{d,n}$ based on the prediction method for a normal distribution for various sample sizes. It is also shown how to determine the fractiles for a log normal distribution. Log normal distributions are commonly used for resistance variables as they have the advantage that no negative values can occur.

The design fractile factor $k_{d,n}$ given in the Eurocodes is based on the target reliability $\beta_T = 3.8$. The

Table 1. Design fractile factors for $\beta_T = 3.0$.

n	4	5	6	8	10	20	30	∞
V_X known	2.68	2.62	2.59	2.54	2.51	2.45	2.43	2.40
V_X unknown	5.46	4.36	3.83	3.33	3.09	2.70	2.59	2.40

design fractile factors for the South African target reliability of $\beta_T = 3.0$ have been calculated and are given in Table 1.

In the case of SHCC, raw data is available for a number of test series. The usual statistical methods should be employed to ensure that outliers do not unduly influence the results and to confirm that each sample does indeed represent the same population. The number of tests in this case is greater than 30 which is the upper sample size value in the tables. The sample mean and standard deviation are therefore taken as representative of the population and the fractile factors, k_n or $k_{d,n}$ for $n = \infty$ may be used.

5.2 Modeling uncertainty

The method of finding the design value of the resistance is described below, assuming a general resistance function that is not simply a product of basic variables.

The resistance of the structural element is defined as:

$$r = b r_t \delta = b g_{rt}(X_1, \ldots, X_j)\delta \qquad (25)$$

The mean of the resistance is calculated as:

$$E(r) = b g_{rt}\left(E(X_1), \ldots, E(X_j)\right) = b g_{rt}(\underline{X}_m) \qquad (26)$$

The coefficient of variation of the resistance theoretical model is determined by:

$$V_{rt}^2 = \left(\frac{VAR\left[g_{rt}(\underline{X})\right]}{g_{rt}^2(\underline{X}_m)}\right) \cong \frac{1}{g_{rt}^2(\underline{X}_m)} \times \sum_{i=1}^{n}\left(\frac{\partial g_{rt}}{\partial X_i}\sigma_i\right)^2 \qquad (27)$$

The coefficient of variation of the actual resistance is found as:

$$V_r = \sqrt{\left(V_\delta^2 + 1\right)\left(V_{rt}^2 + 1\right) - 1} \qquad (28)$$

where V_δ = coefficient of variation of the error term as found previously.

If the number of tests is small (n < 100) allowance should be made for statistical uncertainties in Δ by considering the distribution of the error term as a t distribution with parameters Δ, V_Δ and n. The design value of resistance may be obtained from:

$$r_d = b g_{rt}(\underline{X}_m)\exp\left(-k_{d,\infty}\alpha_{rt}Q_{rt} - k_{d,n}\alpha_\delta Q_\delta - 0.5Q^2\right) \qquad (29)$$

where:

$$Q_{rt} = \sigma_{\ln(rt)} = \sqrt{\ln\left(V_{rt}^2 + 1\right)}$$
$$Q_\delta = \sigma_{\ln(\delta)} = \sqrt{\ln\left(V_\delta^2 + 1\right)}$$
$$Q = \sigma_{\ln(r)} = \sqrt{\ln\left(V_r^2 + 1\right)} \qquad (30)$$
$$\alpha_{rt} = \frac{Q_{rt}}{Q}; \quad \alpha_\delta = \frac{Q_\delta}{Q}$$

If a large number of tests (n > 100) is available the design resistance may be found by:

$$r_d = bg_{rt}\left(\underline{X}_m\right)\exp\left(-k_{d,\infty}Q - 0.5Q^2\right) \qquad (31)$$

The characteristic value may be found by using the factors k_n and k_∞ in the above expressions.

The partial factor for the resistance model γ_{Rd} may be found by dividing the characteristic value by the design value. The characteristic value is defined in EN 1990 as the value of the material or product property having a prescribed probability of being exceeded (generally the 5% fractile) but may also be a nominal value in some circumstances. It does not make sense when deriving design rules to set the characteristic value of a calculation method to a specified fractile if this requires the use of a partial factor. In this case the characteristic value would be calculated as:

$$r_k = R\{\eta X_k/\gamma_m; a_k\}/\gamma_{Rk} \qquad (32)$$

where r_k = characteristic value of resistance; γ_{Rk} = factor applied to nominal resistance to calculate the characteristic value at a specified fractile level.

In this case it is more reasonable to consider the nominal resistance as the characteristic value and to determine the design partial resistance factor using this value.

$$r_k = R\{\eta X_k/\gamma_m; a_k\} \qquad (33)$$

$$\gamma_{Rd} = \frac{r_k}{r_d} \qquad (34)$$

The design resistance is then calculated by the engineer as:

$$r_d = g_{rt}(X_d; a_d)/\gamma_{Rd} \qquad (35)$$

where $X_d = \eta X_k/\gamma_m$

5.3 Interpretation of results

Two levels of analysis may be considered when determining partial resistance factors. As a first assessment all the tests for all the different design situations may be considered together. In practice for codified design, it is preferable for a particular limit state to have only one partial factor for application across the entire range of practical design situations. This may lead in some cases to over conservative designs because of one particular situation governing the value of the partial resistance factor. If there is concern that this is the case, or to gain a better understanding of the influence of certain parameters, the second level of assessment may be considered.

The second level of assessment comprises dividing the test results into sets that represent different design situations. A disadvantage of this approach is that it results in a smaller number of tests per set and incurs additional statistical uncertainty. The analysis of the correction factor and error term as well as calculation of the design value should be carried out for each set.

A further consideration is the treatment of prior information in the evaluation of partial factors. In the case of R/SHCC, test results are available for a small number of beams with tension reinforcement only. Generally a beam will be designed with tension and compression reinforcement as well as stirrups. As such, these beams fall to the extreme of the practical range, and possibly even outside the practical range, of design situations. It is therefore necessary to ensure that they do not unduly influence the partial resistance factor. It may be warranted in this case to assess these results separately to determine where they lie in the spectrum of partial factor results. Engineering judgment and experience should be used to determine whether these results should be included for the determination of the design values, especially if they are shown to dominate.

6 CONCLUSIONS

A reliability based development program for design procedures for innovative materials has been presented. The steps in the development program were describing the material behavior, developing an analysis model, verification of the analysis model and ensuring adequate reliability.

The material behavior should be described both qualitatively and quantitatively as well as statistically and should be based on the results of laboratory tests. The conversion factor to describe the difference between laboratory samples and in situ strength was highlighted as an important and potentially problematic parameter to determine. Methods from EN 1990 were presented for the estimation of appropriate fractiles from limited samples for characteristic and design values of material strength. The fractiles were modified to be appropriate to the South African situation. Inclusion of additional information from new tests by means of Bayesian updating was discussed.

The development of the analysis model should be based on structural theory and experience. Detailed discussion of this topic is outside the scope of this paper. More information is given by Victor & van Zijl (2009).

The analysis model should be verified by means of an experimental test program. Methods were described ensure that the test program focuses on the critical parameters and that the most useful results are obtained from the minimum number of tests. The method of design of experiments was recommended for the development of the test program. The test program should cover the practical range of parameters.

The sources of uncertainty in the tests should be minimized by measuring as many parameters as possible and ensuring strict quality control for the remaining parameters. Predictions of test results should be made using the theoretical analysis model using the measured values of parameters for each test specimen.

The test results should be compared with the theoretical results to assess the bias and uncertainty which are a measure of accuracy of the model. It should be confirmed that the bias and uncertainty are acceptable over the practical range of parameters.

Reliability of the final design procedure is ensured by determining appropriate partial factors for the material and resistance model. The factors are calculated from the design and characteristic values. It was shown how the design value of the resistance model is calculated based on the target reliability level of $\beta_T = 3.0$ for ductile failures in South Africa.

By following the procedures presented in this paper the development and use of SHCC can be done effectively. Critical sources of uncertainty are identified and treated integrally. Progress with development is tracked quantitatively. The material is qualified for construction practice and its advantages can be clarified and quantified. The end result is the development of design procedures which are in agreement with modern structural design standards.

REFERENCES

EN 1990. 2002. *Eurocode—Basis of structural design.* European Committee for Standardization.

Gulvanessian, H., Calgaro, J.-A. and Holický, M. 2002. *Designers' guide to EN 1990 Eurocode: Basis of structural design.*. London: Thomas Telford.

Holický, M. 2009. *Reliability analysis for structural design.* Stellenbosch: Sun Media. In publication.

ISO 2394:1998. *International Standard: General principles on reliability for structures.* International Organization for Standardization. Geneva, Switzerland.

Kemp, A.R., Milford, R.V. and Laurie, J.A.P. 1987. Proposals for a comprehensive limit states formulation for South African structural codes. *The Civil Engineer in South Africa.* 29(9): 351–360.

Milford, R.V. 1988. Target safety and SABS 0160 load factors. *The Civil Engineer in South Africa.* 30(10): 475–481.

O'Connor, P.D.T. 1991. *Practical reliability engineering.* Chichester: John Wiley & Sons.

SANS 10160-1 (CD) 2009. *Basis of structural design and actions for buildings and industrial structures* Part 1 *Basis of design.* Committee Draft, SABS.

Victor, S.A. and van Zijl, G.P.A.G. 2009. Towards a flexural design model for steel bar reinforced fibre-reinforced strain-hardening cement composites (R/SHCC). In *Advanced concrete materials.* Proc. Int. Conf. Stellenbosch. 17–19 November 2009. Rotterdam: Balkema.

Visser CR and van Zijl GPAG 2007. Mechanical characteristics of extruded SHCC. *Proc. International RILEM CONFERENCE on High performance fibre reinforced cement composites.* 10–13 July 2007. Mainz, Germany. 165–173.

Advances in Cement-Based Materials – van Zijl & Boshoff (eds)
© *2010 Taylor & Francis Group, London, ISBN 978-0-415-87637-7*

A model for building a design tool for ductile fibre reinforced materials

E. Schlangen, H. Prabowo, M.G. Sierra-Beltran & Z. Qian
Delft University of Technology, CiTG, Microlab, Delft, The Netherlands

ABSTRACT: This paper presents a model to simulate fracture of fibre reinforced cement based materials. The model is based on a lattice-type fracture model. Fibres are explicitly implemented as separate elements connected to the cement matrix via special interface elements. With the model multiple cracking and ductile global behaviour are simulated of the composite material. Variables in the model are the fibre dimensions and properties, the fibre volume in the composite, the bond behaviour of fibres and matrix and the cement matrix properties. Especially the fibre bond behaviour seems to have a large influence on the fracture and deformation behaviour. These properties are obtained by testing. The model is used as a design tool for creating fibre cement based composites with any desired mechanical behaviour.

1 INTRODUCTION

Fibre cement based materials have been developed during the past decades for many applications and with many different properties (Naaman 2008). Fibres of different dimension and of different materials have been used, e.g. steel, PVA, PE, glass or several kind of natural fibres. The fibres are added to the cement composites to enhance material toughness, strength, fatigue life, impact resistance and/or cracking resistance. A special type of fibre reinforced cement based composites are the materials that have a very high strain capacity of sometimes more than 6%. An example of these materials is ECC (Engineered Cementitious Composites) as described in (Li 2003). These materials show upon loading a distributed crack pattern with fine cracks and a high strain capacity which make them especially suitable for application where imposed deformation is the loading mechanism. The design of these composites is based on an analytical micromechanical approach (Li 2003) which is recently also applied for wood fibre reinforced concrete (Sierra Beltran & Schlangen 2008).

The parameters that determine the strain capacity in fibre cement based composite can be summarized as follows:

– Fibre parameters: fibre length, fibre diameter, fibre stiffness, fibre strength, fibre volume fraction and fibre shape.
– Matrix parameters: matrix stiffness, matrix strength and matrix fracture energy.
– Fibre-matrix interaction parameters: interfacial frictional bond strength, stiffness and fracture energy.

All these parameters can be derived from measured data obtained in experiments. In (Schlangen et al. 2009) a short overview is given of the tests that are being conducted at the Microlab of Delft University to obtain different properties for various types of fibres and cement matrices.

Experiments, however, are time-consuming, especially while certain time has to be allowed for the hydration of the cement and development of the mechanical properties. Therefore it was decided to develop a 3D numerical model that can be used to predict the strain capacity of fibre reinforced materials. By using the model the number of tests needed to design a new material might be reduced to a large extent. Inputs in the model are all the different parameters for fibres, matrix and interface as discussed above. The model is described in section 2 and 3 of this paper and some examples are given in section 4. Section 5 discusses the influence of a more ductile bond between fibres and cement matrix. The paper ends with a summary of the conclusion and discussion of future work.

2 3D FRACTURE MODEL

In the 3D lattice model used in this research, the two-noded beam element in 3D configuration is adopted as shown in figure 1. Shear effect is taken into account because both slender and non-slender elements can be presented in the mesh used in the simulation.

The complete form of element stiffness matrix is in the dimension of 12×12, which is given below.

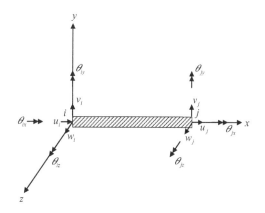

Figure 1. Two-noded beam element in 3D configuration.

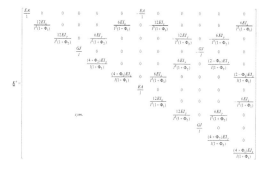

where,

$$\Phi_1 = \frac{12EI_z}{GA_s l^2}$$

$$\Phi_2 = \frac{12EI_y}{GA_s l^2}$$

$$A_s = \frac{A}{\kappa}$$

E = Young's modulus
G = shear modulus
I_z = moment of inertial about z-axis
I_y = moment of inertial about y-axis
J = polar moment of inertial about x-axis
A = cross-sectional area
A_s = shear cross-sectional area
k = shear correction factor

With the transformation matrix for each element the contribution to the global stiffness matrix is constructed. Next the load vector is assembled and the boundary conditions are imposed. With this the system of linear algebraic equations can be solved. In the model developed by the authors an iterative conjugate gradient solver with Jacobi preconditioning is used. The system is solved by minimizing the energy of the system of equations, element by element. In this way memory storage is limited, since there is no need to set up the global stiffness matrix. To simulate fracture sequential linear elastic steps are performed and in each step the element with the highest stress/strength ratio is removed from the lattice to simulate crack growth. In the analysis always the resulting displacement field of the previous step is used as initial guess in the iterative procedure of the conjugate gradient solver. This results in faster convergence, because the total number of equations required is reduced. More details are given in (Qian 2008).

3 MODELLING FIBRE MATERIAL

Simulation of fiber concrete with lattice models is applied by (Bolander 2004). The application of their models is mainly reduction of shrinkage cracks. The model presented in this paper is based on the principle of embedding discrete fibers in a random lattice representing the matrix. Nice examples of continuum models for simulating fracture in fiber concretes can be found in (Kabele 2007 & Radtke et al. 2008).

In the model presented in this paper the (cement) matrix is represented by a random lattice [see also Schlangen & van Mier 1994, Schlangen & Garboczi 1997]. The fibers are discrete beam elements connected to the lattice elements by interface beams. The procedure to generate the network is as follows:

– A cubical grid is chosen. In figure 2 a schematic representation is shown in 2D. The real mesh is 3D, but more difficult to clearly visualize.
– In each cell of the square (cubical for 3D) lattice, a random location for a lattice node is generated.
– Then always the 3 nodes (4 nodes for 3D) which are closest to each other are connected by beam elements.
– Next step is to generate fiber elements. First the number of fibers is calculated that have to be placed in a certain volume based on the length, diameter and volume percentage of fibers. Then the location of the first node of the fiber is randomly chosen in the volume. Next the x, y and z direction of the fiber is chosen randomly which determines the location of the second node. If the second node falls outside the volume the fiber is cut off at the boundary. The fiber-volume of the cut off part of the fiber is subtracted from the already placed fiber-volume in order to ensure enough fibers at the end of the procedure.

– Extra nodes inside the fibers are generated at each location where the fiber crosses the square (cubical for 3D) grid (see figure 2).
– Then interface elements (bond beams) are generated between fiber nodes and the lattice node in the neighboring cell. Also the end nodes of the fibers are connected with an interface element to the cell-node in which the end node is located.
– All the elements in the network are beam elements (with normal force, shear forces, bending moments and torsion moment), which have a local brittle behavior. The beam elements fail only in tension (except for the interface elements, which can also fail in compression) when the stress of the element exceeds its strength. For the fracture criteria only the normal force is taken into account to determine the stress in the beams. The fracture modeling in the lattice model is only a sequence of linear elastic steps, which makes it fast and without numerical difficulties, because no iterations are needed for implementing non-linear material behaviour.

The beam elements in the random lattice representing the cement matrix are all given the same properties. This means that some disorder is build into the lattice. It is also possible to assign different properties to each of the beams (corresponding to the size of the voronoi cell they represent) in the lattice in such a way that the lattice represents a homogeneous material, see (Bolander 2004). The elements in the fiber and of the interface have different properties. The properties of all the components can be determined directly from tests as described in (Schlangen et al. 2009 or Redon et al. 2001] or from fitting results from tests on the composite. The interface properties can for instance being fitted from a simulation of the pull-out of the single fiber. An example of a mesh for a pull-out simulation is shown in figure 3 and 4, as well as a typical pull-out curve that can be obtained by setting the right parameters for the interface beams (parameters in table 1 are used). In the next section

Figure 3. 3D model for simulation of fiber pull-out test.

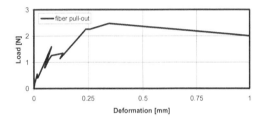

Figure 4. Typical simulated result of a fiber pull out simulation.

Table 1. Input values for the beam elements in the simulations.

	Radius (mm)	E GPa	ν, Poisson's ratio	G GPa	ft MPa	fc MPa
Matrix	0.5	30	0.2	12.50	5	–
Interface	0.1	5	0.333	18.75	100	–300
Fiber	0.022	41.1	0.35	15.22	1640	–

examples of modeling tensile tests with different fiber contents are shown. The parameters represent the pull-out of a PVA-fibre out of a cement matrix. The interface properties are obtained from back-calculation. Note that these properties might not look realistic and also might not be unique. However, modeling an interface, which has a zero-thickness in reality with beam elements having a certain dimension always leads to parameters which should be considered as model-parameters and not realistic values.

4 RESULTS OF SIMULATIONS

To test if the model is able to predict ductile behaviour in fibre cement based material a simulation of a small prismatic specimen loaded in uni-axial

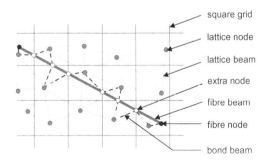

square grid

lattice node

lattice beam

extra node

fibre beam

fibre node

bond beam

Figure 2. Schematic 2D representation of generation of fiber-lattice.

tension is chosen. The dimension of the specimen is $8 \times 5 \times 30$ mm. Two different specimens are created with 1.0 and 2% fibers (by volume) respectively. In figure 5 the 3D mesh of the random lattice is shown as well as the mesh with the 1% fibres used in the simulations. The length of the fibers is 8 mm and the diameter is 44 μm. It represents PVA fibres as used in (Li 2003). For the properties stiffness (E), tensile strength (ft) and compressive strength (fc) values as shown in table 1 are used.

An in-depth study is ongoing to determine realistic parameters for all the properties and to compare them with experiments.

In the simulation the specimens are loaded in uni-axial tension. The crack patterns and load displacement curves in the mesh are shown in figures 6 and 7. In these figures only the beam elements are plotted that break during the simulation. From the figures it can be seen that for both fibre properties a ductile behaviour and multiple cracking is found. In (Schlangen 2009) different fibre volumes and also for larger specimens simulations are shown that lead to brittle single fracture for low volumes and ductile multiple cracking for larger volumes of

fibres. Comparing the load deformation curves for the 1% and 2% case, see figure 8, shows that increasing the volume leads to higher loads, but does not give more ductility. Most probably because in both cases already multiple cracking occurs. More strain

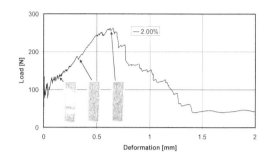

Figure 7. Load deformation curve of simulated tensile test with 2% fibres and crack patterns at different strain.

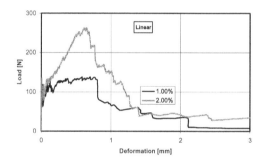

Figure 8. Load deformation curves for both simulated tensile test with 1% and 2% fibres.

Figure 9. Load deformation curves for 4 point bending test with 2% fibres.

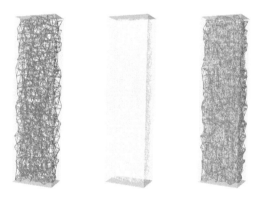

Figure 5. 3D lattice model: a) cement matrix elements, b) 1% fibre elements and c) cement matrix, fibre and interface elements.

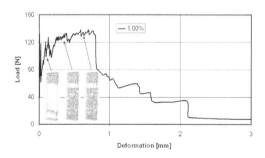

Figure 6. Load deformation curve of simulated tensile test with 1% fibres and crack patterns at different strain.

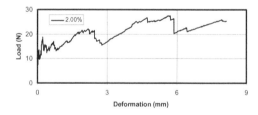

Figure 10. Typical crack pattern in simulated 4 point bending test with 2% fibres.

can be obtained if the pull-out behaviour is more ductile. This will be shown in the next paragraph.

The same properties as used in the tensile tests are adopted to simulate 4-point-bending tests, except for the interface element. After the tensile simulations were performed it was found that actually the interface elements were not stiff enough and to strong. It was found that the stiffness of the interface should be increased to 50 GPa and the strength to 5 and –15 MPa respectively for tensile and compressive strength. In figure 9 is a typical load deformation curve and figure 10 gives a typical crack pattern obtained in a simulation with 2% fibres in the material. Multiple cracking is found between the 2 loading points and a ductile load displacement response. With the properties used now the results resemble the curves on ECC with PVA reported in (Li 1998). However an other way to improve the interface properties is discussed in the next paragraph.

5 OPTIMIZING PULL-OUT BEHAVIOUR

To increase the ductility of the material further it was decided to give different properties to the elements forming the interface or bond between the fibres and the cement matrix. In the simulations in this paragraph these elements are given a sort of plastic behaviour to have more pull-out. To realize this the elements are not removed once the strength of the element is reached in the simulation. But the stiffness of the element is reduced and the strength is kept equal. This procedure is similar as described in [Arslan et al. 1995]. The decrease of the stiffness is done in 3 steps. Two cases are simulated, named NL1 and NL2. The properties used for the interface elements are given in table 2. All the other properties are equal to the properties given in table 1.

The results obtained in the simulation of the tensile tests with the non linear (plastic) interface elements is shown in figures 11 and 12 for the 1% and 2% fibre volumes. A much more ductile behaviour is obtained. It is believed by the authors that these plastic interfaces are more close to the real behaviour observed in the material. Future work will focus on comparing single fibre pull-out tests with simulated behaviour.

Table 2. Stiffness values for the interface elements in the simulations with plastic interfaces..

	E1 GPa	E2 GPa	E3 GPa
Interface Linear	5	–	–
Interface NL1	5	3	1
Interface NL2	5	1	0.15

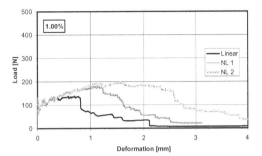

Figure 11. Load deformation curve of simulated tensile test with 1% fibres and non-linear plastic interfaces.

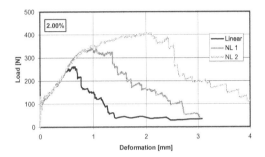

Figure 12. Load deformation curve of simulated tensile test with 2% fibres and non-linear plastic interfaces.

6 CONCLUSIONS AND FUTURE OUTLOOK

A 3D lattice model is developed for simulation of fracture mechanisms in fiber reinforced composites. The model uses discrete beam elements to simulate the fibers. The fiber beams are connected to the matrix beam nodes with interface beams. In principle all the elements in the lattice network have a local brittle behavior. However a global ductile response can be obtained with multiple cracking. The parameters in the model that can be varied and that determine the fracture mechanism and ductility of the material are:

– Fibre parameters: fibre length, fibre diameter, fibre stiffness, fibre strength, fibre volume fraction.
– Matrix parameters: matrix stiffness, matrix strength.
– Fiber-matrix interaction parameters: bond strength and stiffness.

The fibre and matrix parameters are all straightforward because they are dimensional properties or easy to determine properties from tests. These can be directly implemented into the model. The crital parameters, which determine also the behaviour of the material are the fibre-matrix interaction. These parameters can only be obtained by back calculation and fitting from a simulation of a fibre-pull-out experiment. By back calculation the obtained parameters might not be unique and might look unrealistic. However, they should be considered as model-parameters.

In this article also the pull-out behaviour in the model is simulated in a different way, by not giving the bond properties a brittle behaviour but having a local plastic behaviour. By doing this it is shown that the ductility of the composite can be increased to a large extent.

The future work is to determine all the parameters in the model with tests as described in (Schlangen et al. 2009). Furthermore a parameter study will be performed to show the influence of varying the different parameters. Final goal is to use the model to design ductile fiber reinforced materials. Main purpose is to use it for cement based materials but it can also be applied to other matrices.

REFERENCES

Arslan, A., Schlangen, E. & van Mier, J.G.M. 1995. Effect of Model Fracture Law and Porosity on Tensile Softening of Concrete, Proc. FraMCoS-2 (ed. F.H. Wittmann), pp. 45–54.

Bolander, J.E. 2004. Numerical modeling of fiber reinforced cement composites: linking material scales, in Proc. Rilem Symposium BEFIB 2004, Varenna Italy, pp. 45–60.

Kabele, P. 2007. Multiscale framework for modeling of fracture in high performance fiber reinforced cementitious composites, Eng. Frac. Mech., 74, 194–209.

Li, V.C. 1998. Engineered Cementitious Composites (ECC)—Tailored Composites through micromechanical modeling, in Fiber Reinforced Concrete: Present and the Future edited by N. Banthia, A. Bentur, A. and A. Mufti, Canadian Society for Civil Engineering, Montreal, pp. 64–97.

Li, V.C. 2003. On Engineered Cementitious Composites (ECC)—A Review of the Material and its Applications. J. Advanced Concrete Technology. 1(3): pp. 215–230.

Li, Z., Perez Lara, M.A. & Bolander, J.E. 2006. Restraining effects of fibers during non-uniform drying of cement composites, in Cem. & Concr. Res. 36, 1643–1652.

Naaman, A.E. 2008. Development and evaluation of tensile strain hardening FRC composites, in Proc. Rilem symposium BEFIB 2008, Chennai, India (ed. R. Gettu).

Radtke, F.K.F. Simone, A. Stroeven, M. & Sluys, L.J. 2008. Multiscale framework to model fibre reinforced cementitious composite and study its microstructure, in proc. Rilem Symposium, CONMOD'08, (E. Schlangen & G. de Schutter eds.) Delft, The Netherlands, pp. 551–558.

Redon, C., Li, V.C., Wu, C., Hoshiro, H., Saito, T. & Ogawa, A. 2001. Measuring and Modifying Interface Properties of PVA Fibers in ECC Matrix. In J. of Mat. in Civil Eng., pp. 399–406.

Qian, S., Zhou, J., de Rooij, M.R., Schlangen, E., Ye, G. & van Breugel, K. 2009. Self-Healing Behavior of Strain Hardening Cementitious Composites Incorporating Local Waste Materials, Cem. Concr. Comp, Vol 31, 9, pp. 613–621.

Qian, Z. 2008. 3D Lattice Analysis of Cement Paste. MSc-thesis, Delft University of Technology, The Netherlands.

Schlangen, E. 2008. Crack Development in Concrete, Part 2: Modelling of Fracture Process, in Key Eng. Mat. Vols. 385–387, pp. 73–76.

Schlangen, E. & van Mier, J.G.M. 1992. Experimental and numerical analysis of micromechanisms of fracture of cement-based composites, Cem. Conc. Composites, 14: 105–118.

Schlangen, E. & van Mier, J.G.M. 1994. Fracture simulations in concrete and rock using a random lattice. In Siriwardane, H.J. and Zaman, M.M. (eds.), Computer Methods and Advances in Geomechanics, Balkema, Rotterdam, pp. 1641–1646.

Schlangen, E. & Garboczi, E.J. 1997. Fracture simulations of concrete using lattice models: computational aspects, Engineering Fracture Mechanics, 57(2/3): 319–332.

Schlangen, E., Qian, Z. Sierra-Beltran, M.G. & Zhou, J. 2009. Simulation of fracture in fibre cement based materials with a micro-mechanical lattice model. In Proceedings 12th International Conference on Fracture, July 12–17, Ottawa, Canada.

Sierra Beltran, M.G. & Schlangen, E. 2008. Wood fibre reinforced cement matrix: a micromechanical based approach, in Key Eng. Materials, Vols. 385–387, pp. 445–448.

Advances in Cement-Based Materials – van Zijl & Boshoff (eds)
© 2010 Taylor & Francis Group, London, ISBN 978-0-415-87637-7

Author index